T0180675

Quantum Mechanics

Quantum Mechanics

An Introduction for Device Physicists and Electrical Engineers

Third Edition

David Ferry

CRC Press is an imprint of the
Taylor & Francis Group, an informa business

Third edition published [2021]
by CRC Press
6000 Broken Sound Parkway NW, Suite 300,
Boca Raton, FL 33487-2742

and by CRC Press
2 Park Square, Milton Park, Abingdon, Oxon OX14 4RN

[Second edition published by CRC Press 2001]

CRC Press is an imprint of Taylor & Francis Group, LLC

Library of Congress Cataloging-in-Publication Data
A catalog record for this title has been requested

ISBN: 978-0-367-46915-3 (hbk)
ISBN: 978-0-367-46727-2 (pbk)
ISBN: 978-1-00-303194-9 (ebk)

Contents

Preface to the first edition

Most treatments of quantum mechanics have begun from the historical basis of the application to nuclear and atomic physics. This generally leaves the important topics of quantum wells, tunneling, and periodic potentials until late in the course. This puts the person interested in solid-state electronics and solid-state physics at a disadvantage, relative to their counterparts in more traditional fields of physics and chemistry. While there are a few books that have departed from this approach, it was felt that there is a need for one that concentrates primarily upon examples taken from the new realm of artificially structured materials in solid-state electronics. Quite frankly, we have found that students are often just not prepared adequately with experience in those aspects of quantum mechanics necessary to begin to work in small structures (what is now called mesoscopic physics) and nanoelectronics, and that it requires several years to gain the material in these traditional approaches. Students need to receive the material in an order that concentrates on the important aspects of solid-state electronics, and the modern aspects of quantum mechanics that are becoming more and more used in everyday practice in this area. That has been the aim of this text. The topics and the examples used to illustrate the topics have been chosen from recent experimental studies using modem microelectronics, heteroepitaxial growth, and quantum well and superlattice structures, which are important in today's rush to nanoelectronics.

At the same time, the material has been structured around a senior-level course that we offer at Arizona State University. Certainly, some of the material is beyond this (particularly Chapters 10 and 11), but the book could as easily be suited to a first-year graduate course with this additional material. On the other hand, students taking a senior course will have already been introduced to the ideas of wave mechanics with the Schrödinger equation, quantum wells, and the Krönig–Penney model in a junior-level course in semiconductor materials. This earlier treatment is quite simplified but provides an introduction to the concepts that are developed further here. The general level of expectation on students using this material is this prior experience plus the linear vector spaces and electromagnetic field theory to which electrical engineers have been exposed.

I would like to express thanks to my students who have gone through the course, and to Professors Joe Spector and David Allee, who have read the manuscript completely and suggested a great many improvements and changes.

David K. Ferry

Preface to the second edition

Many of my friends have used the first edition of this book, and have suggested a number of changes and additions, not to mention the many errata necessary. In the second edition, I have tried to incorporate as many additions and changes as possible without making the text overlong. As before, there remains far more material than can be covered in a single one-semester course, but the additions provide further discussion on many topics and important new additions, such as numerical solutions to the Schrödinger equation. We continue to use this book in such a one-semester course, which is designed for fourth-year electrical engineering students, although more than half of those enrolled are first-year graduate students taking their first quantum mechanics course.

I would like to express my thanks in particular to Dragica Vasileska, who has taught the course several times and has been particularly helpful in pointing out the need for additional material that has been included. Her insight into the interpretations has been particularly useful.

David K. Ferry

Preface to the third edition

At the turn of the new century in 1900, the world of (what we now call classical) mechanics was comfortable. Newton's laws worked very well, and scientists and engineers knew how to apply them. This was true for small objects such as screws and wires as well as astronomical objects subject to gravity. The latter, as well as Coulomb interactions, was apparently well understood in terms of the instantaneous force at a distance. Yet, within a scant five years, this comfortable understanding would be completely overturned in a manner that would take years to sort out. This would be true for very small objects, such as atoms, and for very large objects, such as the solar system, although the change would come from different theories and people.

At the small scale, the change would come before the end of the year and begin with Max Planck and the quantization of the photon required for his new theory of black body radiation. Planck was not working in isolation, as he had colleagues who had carried out new careful measurements on black body radiation, measurements which clearly showed that Wien's earlier theory just was not correct. Planck was able to fit the new data exceedingly accurately by recognizing two new fundamental constants—one he termed the Boltzmann's constant and the other came to be known as Planck's constant. From Planck's new theory would eventually come a first approximate new theory for atomic structure and then a full new theory that worked for atomic structure but also even for quantum gravity—the new quantum mechanics would appear only slightly less than three decades later. But the new quantum mechanics would be argued for longer, and even in the last third of the new century, Richard Feynman would declare that "… nobody understands quantum mechanics." How can this be? Understanding of a subject requires that a philosophy has arisen which describes how the new theory fits with the reality of the existing world. Perhaps the problem with the new theory was best explained by Murray Gell-Mann, who said, "The fact that an adequate philosophical interpretation has been so long delayed is no doubt caused by the fact that Niels Bohr brain-washed a whole generation of theorists into thinking that the job was done 50 years ago." Bohr had developed the first approximate theory for atomic structure and had then propagated strongly the view that the new theory was magic, and we could only find out what was true by measuring it. We will examine this somewhat further in the first chapter when describing the various understandings that led to the well-known Bohr-Einstein debates.

At the large scale, or perhaps I should say the high velocity scale, Newton's laws were forever changed by the introduction by Albert Einstein of his relativity theory. Now, there was a maximum velocity that matter could achieve, and this dramatically changed space and time forever. What was left was the understanding that Newton's laws worked fine, provided they were used within a well-defined set of parameters that were neither too fast nor too small. Nevertheless, it is not at all clear that the two theories work together. Relativity theory rolls mechanics, electromagnetics, and gravity all together. From this, we have learned to consider four forces in nature—the electromagnetic, strong, weak, and gravity forces. A form of quantum field theory has developed to treat the first three, but inclusion of gravity has been quite difficult, and many of the assumptions in quantum field theory are not applicable in many everyday problems.

Leaving all that aside, one can find a great deal of satisfaction in working with a form of nonrelativistic quantum mechanics that provides satisfactory applications in many real-world problems. In fact, most of the population employs tools, particularly electronics tools, that depend upon the quantum properties of matter, even if they do not realize it. The information revolution that began around the middle of the last century depends entirely upon the development of the transistor and integrated circuits, and the continued development and expansion of the latter. Yet, the transistor and integrated circuit themselves explicitly depend upon the quantum properties of the materials from which they are made. Thus, an understanding of quantum mechanics is essential to the engineers and physicists who design these devices. And, the purpose of this book has been, and continues to be, to provide an adequate understanding of quantum mechanics. And, despite the expressed doubts of Feynman, it is my belief that an adequate understanding of quantum mechanics exists for this purpose. The success of the first two editions of this book, and of many other fine books on the topic, supports this view. The difference in this book is the focus on device physics and electrical engineering applications to which the theory is applied. That is, what is very important to these people may be quite esoteric to physicists in other fields, and vice versa.

As an example of the last comment is quantum tunneling. A physics colleague, who has written a fine textbook on quantum mechanics, was told by his colleagues to leave tunneling out of his new edition, as it was too esoteric and unimportant. But the topic is crucial to considering leakage in modern transistors, and it is of fundamental importance to Josephson junctions, that may well be the basis for future quantum computing. Hence, it is a topic of fundamental importance to device engineers and physicists. In fact, many of the changes in this third edition have been made to allow us to describe the new applications that are required for quantum computing, including a discussion of qubits (as well as Josephson junctions).

I would like to thank Prof. Dragica Vasileska for many comments arising from her reading of various sections of this manuscript.

David K. Ferry

Author

David K. Ferry is Regents' Professor Emeritus in the School of Electrical, Computer, and Energy Engineering at Arizona State University. He was also graduate faculty in the Department of Physics and the Materials Science and Engineering program at ASU, as well as Visiting Professor at Chiba University in Japan. He came to ASU in 1983 following shorter stints at Texas Tech University, the Office of Naval Research, and Colorado State University. In the distant past, he received his doctorate from the University of Texas, Austin, and spent a postdoctoral period at the University of Vienna, Austria. He enjoyed teaching (which he refers to as 'warping young minds') and continues active research. The latter is focused on semiconductors, particularly as they apply to nanotechnology and integrated circuits, as well as quantum effects in devices. In 1999, he received the Cledo Brunetti Award from the Institute of Electrical and Electronics Engineers and is a Fellow of this group as well as the American Physical Society and the Institute of Physics (UK). He has been a Tennessee Squire since 1971 and an Admiral in the Texas Navy since 1973. He is the author, coauthor, or editor of some 40 books and about 900 refereed scientific contributions. More about him can be found on his home pages http://ferry.faculty.asu.edu/ and http://dferry.net/.

1 Waves and Particles

1.1 INTRODUCTION

Science has developed through a variety of investigations more or less over the time scale of human existence. On this scale, quantum mechanics is an extremely young field, existing essentially only about a century. Even our understanding of classical mechanics has existed for a comparatively long period—roughly having been formalized with Newton's equations published in his *Principia Mathematica*, in April 1686. Up until the very end of the nineteenth century, mechanics in the form of Newton's laws were comfortable, being understood and accepted by all who dealt with the motion of reasonable bodies. With the turn of the new century, however, mechanics would be shaken to its roots. The change began already on December 14, 1900, when Planck (1900) presented his new theory of black-body radiation to the German Physical Society, and this change would affect mechanics in physically small structures (and lead to quantum mechanics). Mechanics of large bodies would similarly be changed in 1905 with the arrival of Albert Einstein's (1905a) relativity theory. In this book, we will deal with only the first of these two great breakthroughs.

Black-body radiation had been studied for many years prior to Planck's discovery, but it lacked an adequate theory. One of the early theories was due to Wien, but it was approximate, and it was discovered that it really didn't fit to the observations. The fact that the theory did not fit led to new, very careful experiments at one of Berlin's national laboratories, and fortunately, Planck was able to interact with the latter group. This provided the data that allowed Planck to finalize his theory in what is now known as the Planck black-body radiation law:

$$I(f)df \sim \frac{f^3 df}{exp\left(\frac{hf}{k_B T}\right) - 1}, \tag{1.1}$$

where f is the frequency of the radiation, T is the temperature, and I is the intensity of the radiation. In addition to these parameters, Planck defined two new constants which came from comparing the data to the theory. One of these, he named as the Boltzmann constant, $k_B = 1.380649 \times 10^{-23}$ Joules/Kelvin. The second, he did not name, but it has since come to be known as Planck's constant $h \sim 6.62607015 \times 10^{-34}$ Joule-second. (While these constants historically were more uncertain, the redefinitions that occurred after 2019 in the SI base units have given us these values.) In coming to his radiation law, Planck had to make one more connection and that is a relation between the frequency and the energy, as

$$E = hf, \tag{1.2}$$

where E is the energy of light at the frequency f. While Planck had given us the idea of quanta of energy, he was not comfortable with this idea. But it took only a decade for Einstein's theory of the photoelectric effect (discussed later) to confirm that radiation indeed was composed of quantum particles of energy given by (1.2).

While we commented that (1.2) leads us to conclude that Planck had quantized the radiation with which he was dealing, this is not evident from (1.2) alone. We also have to consider the amount of power that is provided by the radiation. For example, let us assume that the light from the sun (which is a black-body radiator) arriving at the surface of the earth is, on average, about 0.1 W/cm^2. If the average photon has an energy of 2.5 eV, then the number of photons arriving at the surface of the earth is about 2.5×10^{17} per second per square centimeter. It is this combination of quantities, power and energy, that together requires the light to be quantized. That is, about 2.5×10^{17} photons/cm^2 arrive every second from the sun, and this requires that the power falling on the earth's surface must jump with the arrival of each photon.

But, to be truthful, this is not such a revolution, as the ancient Greeks always believed that light was composed of small corpuscular units. Even the great physicist Newton tended to believe that light was corpuscular in nature. It was only the work of Young (1809) in the late eighteenth century that showed that light would give rise to an interference effect, a property of waves. Let us examine this important aspect of light as a wave. If we create a source of coherent light (a single frequency), and pass this through two slits, the wavelike property of the light will create an interference pattern, as shown in Figure 1.1. Now, if we block one of the slits, so that light passes through just a single slit, this pattern disappears, and we see just the normal passage of the light waves. It is this interference between the light, passing through two different paths so as to create two different phases of the light wave, that is an essential property of the single wave. When we can see such an interference pattern, it is said that we are seeing the wavelike properties of light. Of course, the wavelike view was reinforced in the middle of the nineteenth century with the arrival of Maxwell's equations for electromagnetic waves.

1.2 LIGHT AS PARTICLES—THE PHOTOELECTRIC EFFECT

Proof of the fact that light was a particle came quick enough, and it came as an explanation of the photoelectric effect. It was known that when light was shone upon the surface of a metal, or some other conducting medium, electrons could be emitted from the surface provided that the frequency of the incident light was sufficiently high. The curious effect is that the velocity of the emitted electrons depends only

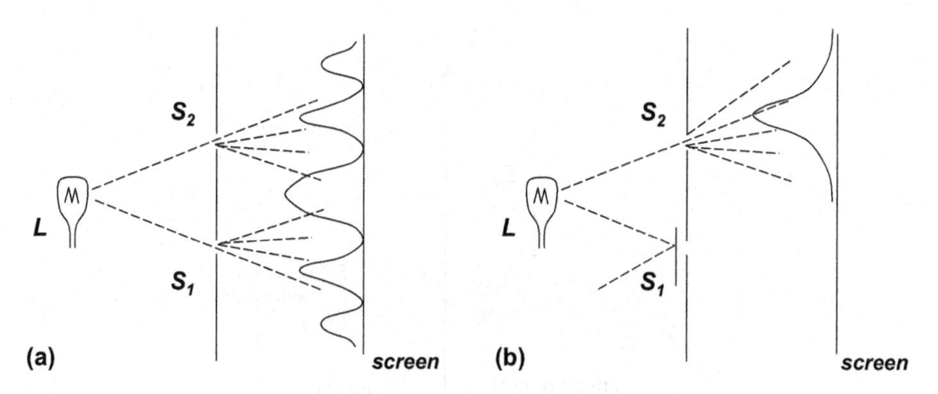

FIGURE 1.1 In panel (a), we illustrate how light coming from the source L and passing through the two slits S_1 and S_2 interferes to cause the pattern indicated on the 'screen' on the right. If we block one of the slits, say S_1, then we obtain only the light intensity passing through S_2 on the 'screen' as shown in panel (b).

upon the wavelength of the incident light, and *not upon the intensity of the radiation.* In fact, the energy of the emitted particles varies inversely with the wavelength of the light waves. On the other hand, the *number* of emitted electrons does depend upon the intensity of the radiation, and not upon its wavelength. Today, of course, we do not consider this surprising at all, but this is after it was explained in the Nobel Prize–winning work of Einstein. What Einstein concluded was that the explanation of this phenomenon required a treatment of light in terms of its 'corpuscular' nature; that is, we need to treat the light wave as a beam of particles impinging upon the surface of the metal. In fact, it is important to describe the energy of the individual light particles, which we call *photons*, using the relation (1.2) (Einstein 1905b), which can be rewritten as

$$E = hf = \hbar\omega \tag{1.2}$$

where $\hbar = h/2\pi$. The photoelectric effect can be understood through consideration of Figure 1.2. However, it is essential to understand that we are talking about the flow of 'particles' as directly corresponding to the wave intensity of the light wave. Where the intensity is 'high,' there is a high density of photons. Conversely, where the wave amplitude is weak, there is a low density of photons.

A metal is characterized by a work function E_W, which is the energy required to raise an electron from the Fermi energy to the vacuum level, from which it can be emitted from the surface. Thus, in order to observe the photoelectric effect, or photoemission as it is now called, it is necessary to have the energy of the photons greater than the work function, or $E > E_W$. The excess energy, that is the energy difference between that of the photon and the work function, becomes the kinetic energy of the emitted particle. Since the frequency of the photon is inversely proportional to the wavelength, the kinetic energy of the emitted particle varies inversely as the wavelength of the light. As the intensity of the light wave is increased, the number of

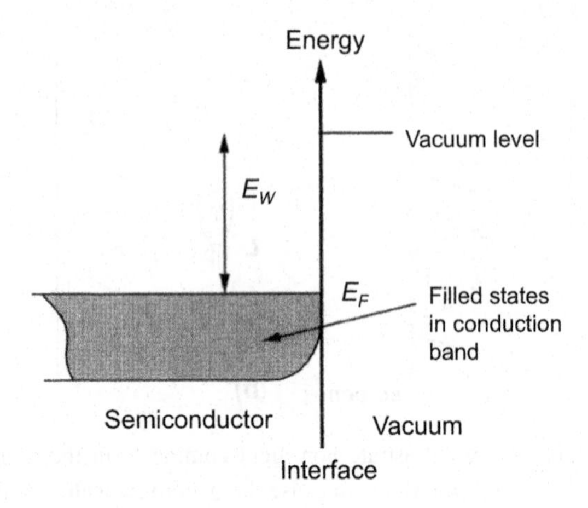

FIGURE 1.2 The energy bands for the surface of a metal. An incident photon with an energy greater than the work function, E_W, can cause an electron to be raised from the Fermi energy, E_F, to above the vacuum level, whereby it can be photo-emitted.

incident photons increases, and therefore the number of emitted electrons increases. However, the momentum of each emitted electron depends upon the properties of a single photon, and therefore is independent of the intensity of the light wave.

A corollary of the acceptance of light as particles is that there is a momentum associated with each of the particles. It is well known in field theory that there is a momentum associated with the (massless) wave, which is given by $p = hf/c$, which leads immediately to the relationship

$$p = \frac{hf}{c} = \frac{h}{\lambda} = \hbar k, \tag{1.3}$$

where in the last form we have used the well-known relation $k = 2\pi/\lambda$.

It is finally clear from the interpretation of light waves as particles that there exists a relationship between the 'particle' energy and the frequency of the wave, and a connection between the momentum of the 'particle' and the wavelength of the wave. Equations (1.2′) and (1.3) give these relationships. The form of (1.3) has usually been associated with de Broglie, and the wavelength corresponding to the particle momentum is usually described as the *de Broglie wavelength.* However, it is worth noting that de Broglie (1937) referred to the set of Equations (1.2) and (1.3) as the Einstein relations! In fact, de Broglie's great contribution was the recognition that atoms localized in orbits about a nucleus must possess these same wavelike properties. Hence, the electron orbit must be able to incorporate an exact integer number of wavelengths, given by (1.3) in terms of the momentum. This then leads to quantization of the energy levels. We shall return to de Broglie later.

Now, at this point, we have a conundrum. Is light a wave, as suggested by the interference phenomena of Young, or is it a particle, as suggested by Planck and the photoelectric experiments? Needless to say, this would be investigated. Taylor (1909), following an idea of J. J. Thomson (1907), initiated experiments using very weak light. Thomson was studying the ionization of gases under ultra-violet and X-ray illumination and noticed that only a small fraction of the gas molecules were being ionized. This led him to conjecture that the energy of the light was localized in Planck's packets, which also were *physically small in dimension*, so that these corpuscles of light missed the majority of the molecules. This led him to suggest that light striking a metal surface, or a photographic plate, would actually cause a "... series of bright specks on a dark ground." He supposed that if the plate was moved further from the source, "... we shall diminish the number of these [specks] ... but not the energy in the individual units." This went beyond Planck's quantization by also suggesting that the photons were physically quite small. By attenuating the light in his measurements, Taylor could get a situation in which the photons came through one at a time and were separated from one another by a large distance. He used a flame passing through a single slit as his source, and then diffracted the light around a needle. He attenuated the light with various smoked glass windows, which he calibrated first to determine the attenuation of each window. With the weakest light, the photographic plate had to be exposed for 2,000 hours. Taylor states that he saw no diminution in the diffraction pattern for any of his exposures, thus confirming Thomson's ideas. Even though light arrived at each photographic plate as a single photon at a time, and exposed a single spot on the plate, the net pattern formed over time matched the expected wave interference patterns. Consequently, the light in this experiment was acting as both particle (upon striking the plate) and wave (when passing the needle). From this, and from later experiments on two slits, we know that what is observed is each photon creates a single point on the screen behind the two slits. That is, a single photon impacting the screen does not create the interference pattern. But, after thousands of photons impact the screen, each one going through the two slits as an individual photon, the interference pattern emerges. Thus, it was clear that light had both wave *and* particle aspects and could exhibit both in a single experiment!

1.3 ELECTRONS AS WAVES

In the previous section, we discussed how light, or electromagnetic waves, can behave as either a wave or a collection of particles. It is fair to then ask whether particles, such as electrons, behave as waves? This question remained unanswered for several years. In 1921, Clinton Davisson and Charles Kunsman (1921, 1923), working at the AT&T research laboratories in America, had noticed that the scattering of electrons from Ni and other metals showed unexpected peaks in the angular variation of the scattering. In sending a beam of electrons into a crystal, such as Ni, they found these unexpected results. If the energy of the entering electrons is sufficiently large, they can knock other electrons out of their atomic orbits and these electrons can be emitted in complete analogy to the photoelectric effect, with the incident electron replacing the photon. But Davisson and Kunsman also observed electrons with almost the same energy as the incident beam being reflected from the crystal over a large range of angles. Studies of the angular distribution of the emitted/scattered electrons showed the appearance of peaks much like diffraction peaks, apparently due to the periodic nature of the atomic arrangement. In these studies, the scattering cross-section had an oscillatory behavior with peaks and valleys as the energy was varied as in the case with Ni. Walter Elsasser (1925) proposed that these effects were the result of the diffraction of electron waves, and his paper was published only after the editor consulted with Einstein. Just after this, George Thomson (1927), the son of J. J. Thomson, obtained clear diffraction patterns of electrons striking thin metal films. Sending the electrons through the thin films, he observed the diffraction rings of the exciting electrons. Thus, it was finally clear that the electrons could show wavelike properties.

From these experiments, we have a similarity between light and particles. Consequently, there are times when it is clearly advantageous to describe particles, such as electrons, as waves. In the correspondence between these two viewpoints, it is important to note that the varying intensity of the wave reflects the presence of a varying number of particles; the particle density at a point x, at time t, reflects the varying intensity of the wave at this point and time. For this to be the case, it is important that quantum mechanics describe both the wave and particle pictures through the principle of superposition. That is, the amplitude of the composite wave is related to the sum of the amplitudes of the individual waves corresponding to each of the particles present. Note that it is the amplitudes, and not the intensities, that are summed, so there arises the real possibility for *interference* between the waves of individual particles. Thus, for the presence of two (noninteracting) particles at a point x, at time t, we may write the composite wave function as

$$\Psi(x,t) = \Psi_1(x,t) + \Psi_2(x,t). \tag{1.4}$$

This composite wave may be described as a *probability wave*, in that the square of the magnitude describes the probability of finding an electron at a point (we will go further into this point below).

It may be noted from (1.3) that the momentum of the particles goes immediately into the so-called *wave vector k* of the wave. A special form of (1.4) is

$$\Psi(x,t) = Ae^{i(k_1x-\omega t)} + Be^{i(k_2x-\omega t)}, \tag{1.5}$$

where it has been assumed that the two components may have different momenta (but we have taken the energies to be equal). For the moment, only the time-independent steady state will be considered, so the time-varying parts of (1.5) will be suppressed. It is known, for example, that a time-varying magnetic field that is enclosed by a conducting loop will induce an electric field (and voltage) in the loop through Faraday's law. Can this happen for a time-independent magnetic field? We know that a d.c. magnetic field can affect the motion of electrons through the Lorentz force. The question is how this arises for a wavelike electron.

A particularly remarkable illustration of the importance of the phase is the magnetic Aharonov-Bohm (1959) (AB) effect. The effect was predicted in the mid-twentieth century and was verified by a number of experiments almost immediately. Here, we illustrate the effect with an experiment conducted on an appropriate nanostructure. The basic structure of the experiment is illustrated in Figure 1.3. The basic structure is formed on an AlGaAs/GaAs semiconductor heterostructure. A quasi-one-dimensional (Q1D) conducting channel is fabricated on the surface of a semiconductor by using electron-beam lithography to deposit a NiCr pattern by liftoff, and then using this pattern as a mask for reactive-ion etching away parts of the heterostructure to leave electrons in the ring structure [Mankiewich et al., 1988]. The waveguide is sufficiently small so that only one or a few electron modes are possible. The incident electrons, from the left of the ring in Figure 1.3(a), have their wave split at the entrance to the ring. The waves propagate around the two halves of the ring to recombine (and interfere) at the exit port. The overall transmission through the structure, from the left electrodes to the right electrodes, depends upon the relative size of the ring circumference in comparison to the electron wavelength. If the size of the ring is small compared to the inelastic mean free path, the transmission depends on the phase of the two fractional paths. In this effect, a magnetic field is passed through the annulus of the ring, and this magnetic field will modulate the phase interference at the exit port. There are two types of resistance measurements, which are labeled R_{xx} and R_{xy} in Figure 1.3(b). The R_{xx} measurement is made directly across the ring and is a direct measure of the resistance of the ring as measured by the voltage drop across it. The other measurement is a transverse measurement, which in some sense is a nonlocal one as it does not measure the voltage drop across the ring but the effect the ring voltage has on the rest of the circuit.

We understand the measured behavior from the assumption that the magnetic field passes vertically through the ring. The vector potential for a magnetic field passing through the annulus of the ring is azimuthal, so that electrons passing through either side of the ring will travel either parallel or antiparallel to the vector potential, and this difference produces the phase modulation, as indicated in Figure 1.3(b). The vector potential will be considered to be directed counterclockwise around the ring. (We adopt cylindrical coordinates, with the magnetic field directed in the z-direction and the vector potential in the θ-direction.) To understand the phase interference, we note that the wave momentum is affected by any electric field via (this follows classical physics, discussed below in Section 1.5, where the time derivative of the momentum is given by the force)

$$\hbar\frac{dk}{dt} = F = -eE \quad \rightarrow \quad \hbar k = \hbar k_0 - e\int \mathbf{E}dt, \tag{1.6}$$

where, here, **E** is the electric field (not to be confused with the energy). Now, in electromagnetics, we know that in the proper gauge, the electric field is related to the vector potential as

$$\mathbf{E} = -\frac{\partial \mathbf{A}}{\partial t}. \tag{1.7}$$

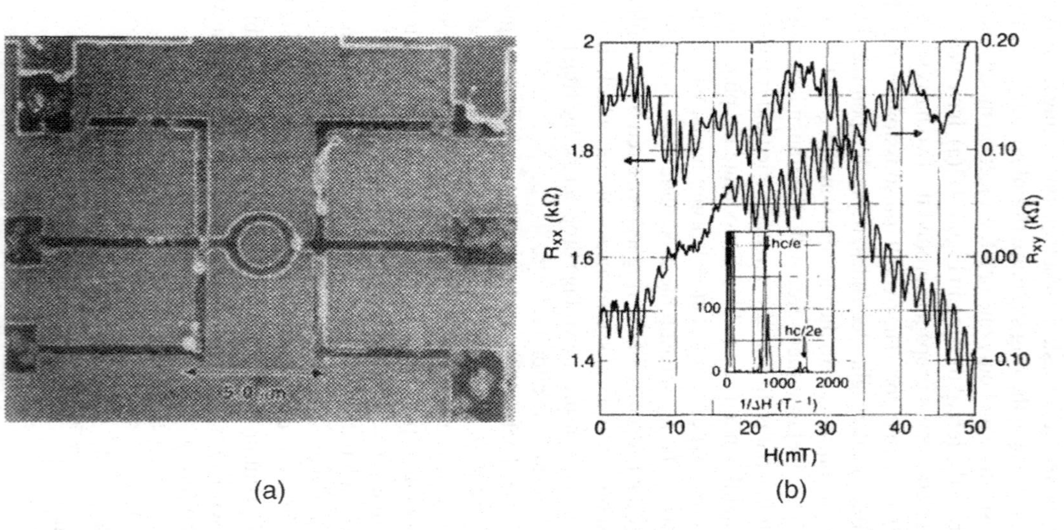

FIGURE 1.3 (a) Micrograph of a conducting semiconductor ring etched into an AlGaAs/GaAs heterostructure. (b) The magnetoresistance and Hall resistance measured for the ring of panel (a). The inset shows the Fourier transform of the resistance. Reprinted with permission from (Mankiewich et al., 1988). Copyright 1988 American Vacuum Society.

Equations (1.6) and (1.7) can now be combined to show that the proper momentum arises from the Peierts' substitution in which the normal momentum vector is replaced by $\mathbf{p} - e\mathbf{A}/\hbar$. This means that the proper phase for the two parts of the wave function are determined from

$$\phi = \phi_0 + \frac{1}{\hbar}(\mathbf{p} - e\mathbf{A}) \cdot \mathbf{r}, \tag{1.8}$$

so that the exit phases for the upper and lower arms of the ring can be expressed as

$$\begin{aligned} \phi &= \phi_0 + \int_{\pi}^{0} \left(\mathbf{k} + \frac{e}{\hbar}\mathbf{A} \right) \cdot \mathbf{a}_{\vartheta} r d\varphi \\ \phi &= \phi_0 - \int_{\pi}^{0} \left(\mathbf{k} - \frac{e}{\hbar}\mathbf{A} \right) \cdot \mathbf{a}_{\vartheta} r d\varphi \end{aligned} \tag{1.9}$$

The two waves from (1.9) will now have a phase difference

$$\delta\phi = \frac{e}{\hbar} \int_0^{2\pi} \mathbf{A} \cdot \mathbf{a}_{\vartheta} r d\varphi = \frac{e}{\hbar} \int \boldsymbol{B} \cdot \mathbf{a}_z dS = 2\pi \frac{\Phi}{\Phi_0}, \tag{1.10}$$

where $\Phi_0 = h/e$ is the quantum unit of flux and Φ is the flux enclosed in the ring. The phase interference term in Equation (1.10) goes through a complete oscillation each time the magnetic field is increased by one flux quantum unit. This produces a modulation in the conductance (resistance) that is periodic in the magnetic field, with a period h/eS, where S is the area of the ring. This periodic oscillation is the AB effect, and in Figure 1.3(b) results are shown for such a semiconductor structure. One can see that exactly this 'frequency' appears in the Fourier transform shown as the inset to this figure. More interestingly, the transverse resistance, which is not measured across the ring also shows such an oscillation. This appears to be a nonlocal effect but is a reflection of the fact that the voltage oscillations across the ring directly affect the rest of the circuit.

The preceding paragraphs describe how we can 'measure' the phase interference between the electron *waves* passing through two separate arms of the system. In this regard, these two arms serve as the two *slits* for the optical waves of Figure 1.1. Observation of the interference phenomena shows us that the electrons must be considered as waves, and not as particles, for this experiment. Once more, we have a confirmation of the *correspondence* between waves and particles as two views of a coherent whole.

1.4 REALITY AND CAUSALITY

The anomaly of light and particles having both wavelike and particle-like properties was confusing to the world of the early twentieth century and, not in the least, to the scientists of the day. It challenged the understanding of our views of mechanics and electromagnetics. In the first, particles were definitely only particles, while in the second, optical and radio waves were definitely waves. It would only get more complicated as the spectra of light absorption and emission from atoms was considered.

By this time, it was well known that the absorption of light by atoms, and the subsequent emission of light from these atoms, was definitely quantized into a set spectrum of discrete frequencies or wavelengths. The reason for this discreteness was not yet known. One of the early pioneers was the Swedish physicist Anders Jonas Ångström. In the middle of the nineteenth century, he developed tools with which he studied the emission spectrum of the sun, and determined that hydrogen, as well as other atoms, were in this spectrum (Ångström 1862). In 1868, he published what is known as the great map of the solar spectrum (Ångström 1868). Working with the data from these publications, Johann Balmer (1885) devised a formula to empirically fit this data. Balmer was a Swiss mathematician, and this seems to be his greatest contribution, done when he was already in his 60s. He wrote the formula in terms of the wavelength of light λ as

$$\lambda = \frac{Bn^2}{n^2 - 2}, \tag{1.11}$$

where B is a constant and $n = 3,4,5, \ldots$. Of course, he was not clear why the second number in the denominator was 2, but it fit all the data in the visible spectrum to which he had access.

Working almost independently, Johannes Rydberg (1889), a Swedish physicist, developed a somewhat more general formula, as he was trying to fit to a wide range of materials. He recognized that Balmer had found a special case of a more general formula that he had created. Rydberg's formula read

$$\frac{1}{\lambda} = RyZ^2\left(\frac{1}{m^2} - \frac{1}{n^2}\right), \tag{1.12a}$$

which reproduced Balmer's formula when $m = 2$. Here, Ry is known as the Rydberg constant

$$Ry = \frac{e^4}{8\epsilon_0^2 h^3 c} \frac{m_e m_p}{m_e + m_p} \sim 13.605eV, \tag{1.12b}$$

which he thought was the same for all materials (as it turned out to be), and Z is the atomic number (the number of protons in the atom). As Balmer's series was mainly in the visible, it is natural to understand that it was this part of the hydrogen spectrum that was the first one found and remained the only known set of emission lines for a great many years. Other series would be found as time passed onward.

To try to explain these observations, Nagaoka (1904) modeled a planar atom in which the electrons all resided on a circle around a central positive nucleus. The latter had shown that this was stable provided the attractive force between the electrons and the nucleus was large enough, and the electrons all had about the same velocity. Nagaoka also had considered already that the electrons may lie in many rings, all of which may or may not lie in the same plane, so it could be a planetary-like model or a more spherical model. He noted that the many rings would align better with studies of the optical absorption measurement in atoms, which showed a series of very sharp lines in the spectra, which he connected to the rings of his atomic model. Subsequently, Niels Bohr would work on the problem. He had gone to Manchester to work with Rutherford as a postdoctoral fellow. There he studied Rutherford's planetary model of the atom but was disturbed by the question of why the electrons didn't simply collapse into the nucleus as they radiated their energy. Certainly, he knew of Nagaoka's work with the electrons outside the nucleus. This led him to ask whether the electrons would be restricted into very well-defined energy levels using the same idea of quantization that Planck had used. He was still working on the problem when his fellowship expired, and he returned to Denmark. He continued to work on his atomic model and the first paper was finally published (Bohr 1913). Surprisingly, Bohr had not learned about the Balmer and Rydberg formulae until about this time but learned just in time to use the experimental results in his paper. This would be a fortunate addition, because it clarified his thoughts and provided the final justification for the paper. The result of using Planck's quantization of the photon energy led also to the result that the angular momentum of the electrons in their motion around the nucleus was also quantized, but for reasons which he did not understand.

Here, though, we are facing the fact that this was not a solid theory. Rather, it was a hand-waving exercise which fortuitously gave an understanding of the source of the Rydberg formula. It explained the series of lines and fit the experiments. But, if you believed that reality existed without the need for observers, then Bohr's work was unsatisfying. So, now the dilemma arises. For Bohr adopted the view that it was the experiment *and the apparatus with which it was performed* that provided the reality of the result. It was the reality of the measured atomic spectra that defined the reality of his work. Reality arrived with the measurement and different measurements could have different realities. If the measured atomic spectra defined reality, then his theory was good to go. He no longer needed to explain why the momentum was quantized, or why the electrons did not collapse into the nucleus. There was no need to have any underlying theoretical derivation—his approach fit the experimental results and thus was as real as it gets. Not only did Bohr redefine what reality meant, he also sent the idea of causality to the dustbin. Now, his version of the state of quantum effects was that one did not have to comprehend the nature of the atom, we only had to measure it. Quantum phenomena now required an observer to provide reality, and this observer provided the needed evidence to interpret the experiment. And, it would be another decade before the proper quantum theories arrived; by then, Bohr had cemented his philosophical view onto the quantum community, mostly by a process a later physicist would refer to as "brain washing" (Gell-Mann 1979).

As students and postdocs passed through Copenhagen, they were indoctrinated into Bohr's philosophical views.

In the early 1920s, Louis de Broglie (1923) developed an alternative to the wave *or* particle picture. His inspiration was the experiments of Taylor mentioned above. This led him to propose that the corpuscles of light were coupled to a non-material light wave of the appropriate frequency. He viewed the wave as guiding the particle in the mode of a "pilot" wave. This was a wave *and* particle viewpoint, and would be a major sticking point between the coming two groups working on the new quantum theory. One camp would be led by Bohr, while the other would involve Einstein and be based upon what would be called the wave theory. In 1925, we would get the first version of the new quantum mechanics from the work of Heisenberg (1925). At the time, Heisenberg was a postdoc in Copenhagen working with Bohr, but also with a position in Göttingen with Born. His theory was one of kinematics, the mechanics of moving bodies. In principle, he was dealing with measurable quantities, but his work became known as matrix mechanics. Importantly, he developed the quantization of the angular momentum through

$$\oint p dq = J = nh. \tag{1.13}$$

He then pointed out that the derivative of this equation, with respect to the integer n, would yield a value of the angular momentum defined by Planck's constant h. Importantly, as a disciple of Bohr, he wanted nothing to do with the wave picture put forward by de Broglie, and firmly supported Bohr's observer requirement for reality. His colleague von Neumann (1932) would publish a textbook describing the Heisenberg approach in terms of sets in a linear vector space and matrix mechanics, which became highly read and cited. But Heisenberg's kinematics remained somewhat obtuse and confusing to most scientists.

Following his initial work, de Broglie realized that his wave and particle approach could also be applied to particles. And this meant that the particle wave had to have a momentum associated with the accompanying wave. Einstein had already shown that the photon had a momentum given in (1.3), which meant that the particle would have a corresponding wavelength given by its momentum as

$$\lambda = \frac{h}{p}, \tag{1.14}$$

which arrives directly from (1.3) (de Broglie 1924). This important result provided precisely that reason for Bohr's angular momentum to be quantized. If the electrons in the atom acted as waves, then it was important that the motion around the circumference of the orbit was exactly a multiple of the wavelength, in order to fit the wave function into the orbit. Now, there was a fundamental reason for Bohr's results.

Only a short time later, it would be Erwin Schrödinger (1926) who would give us the equation for the waves:

$$i\hbar\frac{d\psi(x,t)}{dt}=-\frac{\hbar^2}{2m}\nabla^2\psi(x,t)+V(x,t)\psi(x,t). \tag{1.15}$$

It is clear that solution of this equation depends upon an initial condition, which means that the wave function ψ evolves from this initial condition. In this, and in de Broglie's work, this evolution from the initial condition restores causality to the mechanics. Hence, wave mechanics was unacceptable to the Bohr acolytes. With causality and determinism, one does not depend upon the observer to establish reality; reality exists whether the results are measured or not. Moreover, a great many scientists were already familiar with waves, so that the Schrödinger approach would be easy for them to understand and to use.

1.5 CONNECTION TO THE CLASSICAL WORLD

How does the new quantum mechanics connect to the classical world, which is our world of Newtonian mechanics? If we are to believe the philosophy of Bohr, there is no such connection. And, while we may think that the wave view of de Broglie and Schrödinger has no such connection either, this would be wrong. In fact, in both Heisenberg's matrix approach and Schrödinger's wave approach, there is a strong connection to classical mechanics that runs throughout the theories. In this section, we want to develop this connection further. Our approach is best developed with the Hamiltonian mechanics, in which we can define the total energy as (we work in one dimension for clarity, and consider the normal case where the energy is conserved and therefore the total energy is not a function of time)

$$E=T+V=\frac{p^2}{2m}+V(x). \tag{1.16}$$

That is, the total energy of a particle, or a system of particles if we sum over all the particles, is given by the sum of the kinetic energy and the potential energy. The potential energy V is the same as that appearing in the Schrödinger equation (1.15). In this approach, the time rate of change of the position and momentum give rise to Newton's equations as

$$\begin{aligned}\frac{dx}{dt}&=v=\frac{\partial H}{\partial p}=\frac{p}{m}\\ \frac{dp}{dt}&=-\frac{\partial H}{\partial x}=-\frac{\partial V}{\partial x}\end{aligned}. \tag{1.17}$$

The two variables x and p are known as a conjugate pair. These are our common variables and will carry over to quantum mechanics, although our understanding of them will change.

1.5.1 Position and Momentum

We begin by looking at how position and momentum are characterized in wave mechanics. The wave function $\psi(x,t)$ is a complex quantity, while the classical position and momentum of the particle are real quantities. Moreover, the wave is a distributed quantity, while we expect the particle to be relatively localized in space. This suggests that we relate the *probability* of finding the electron at a position x to the square of the magnitude of the wave. Probability was relatively well known in classical physics although it was also new. It appears in statistical physics, which was developed in the last half of the nineteenth century, most famously by Boltzmann. In statistical physics, one dealt with probabilities; e.g., the probability of a particle's position when there was noise and uncertainty in a large ensemble of such particles. In this world, Newton's law was extended to an ensemble, for which we use Langevin's equation (1908)

$$\frac{dp}{dt} = \frac{d(mv)}{dt} = F - \frac{mv}{\tau} + R(t). \tag{1.18}$$

Here, the acceleration of a given particle is given by the applied force F (determined from the potential) plus a random force R, and it was decelerated by collisions with a mean time between collisions of τ. The random force has an average value of zero, but it imparted a random motion characterized by the temperature, with the random thermal velocity being given by

$$v_T = \sqrt{\frac{3k_B T}{m}} \tag{1.19}$$

in three dimensions. To find the average acceleration, we summed over all the various particles, say N of them, and

$$\frac{d\langle v \rangle}{dt} = \frac{1}{m}F - \frac{\langle v \rangle}{\tau}. \tag{1.20}$$

The major point of statistical physics is that the N particles could be characterized by a distribution function, such as the Maxwell-Boltzmann distribution, itself characterized by a temperature T. This distribution would tell us the probability that a particle existed at a given energy would vary as

$$exp\left(-\frac{E}{k_B T}\right). \tag{1.21}$$

Thus, the concepts of probability and distribution functions and taking ensemble averages were not new in classical physics. What was new in wave mechanics was to say that the probability of finding the particle at position x was given by $|\psi(x,t)|^2$. As a proper probability distribution function, it is required to normalize the probability with

$$\int_{-\infty}^{\infty} |\psi(x,t)|^2 dx = 1. \tag{1.22}$$

With the normalization that we have now introduced, it is clear that we are equating the square of the magnitude of the wave function with a probability density function. This allows us to compute immediately the expectation value, or average value, of the position of the particle with the normal definitions introduced in probability theory. That is, the average value of the position is given by (we denote the *expectation* value with the sharp brackets)

$$x = \int_{-\infty}^{\infty} x|\psi(x,t)|^2 dx = \int_{-\infty}^{\infty} \psi^\dagger(x,t) x \psi(x,t) dx, \tag{1.23}$$

where the small "cross" on the first wave function denotes the adjoint (usually complex conjugate of the wave function). The second new point is to recognize that the position itself is an *operator*. That is, the x inside the integral is an *operator*, and that is why it is placed between the two wave functions in the last integral of (1.23). This is going to be true for the momentum as well, but we have to ascertain the form of that operator. But it is important to note that the expectation value of an operator is evaluated with the operator *between* the two wave functions.

The wave function described in (1.15) contains only variations in space and time and is not a uniform quantity. In fact, if it is to describe a localized particle, it must vary quite rapidly in space. But there is no discussion of the momentum in this equation. It is possible to Fourier transform this wave function in order to get a representation that describes the spatial frequencies that are involved, and this Fourier transform will then describe a wave function in momentum space, as the Fourier transform of one of a pair of conjugate operators will produce a description in the space of the other operator. Then, the wave function can be written in terms of the spatial frequencies as an inverse transform (again, in one spatial dimension for clarity)

$$\psi(x,t) = \frac{1}{\sqrt{2\pi}} \int_{-\infty}^{\infty} \varphi(k,t) e^{ikx} dk. \tag{1.24}$$

We can now invert this Fourier transform, by standard procedures, to give

$$\varphi(k,t) = \frac{1}{\sqrt{2\pi}} \int_{-\infty}^{\infty} \psi(x,t) e^{-ikx} dx. \tag{1.25}$$

Suppose now that we are using the position representation wave functions. How then are we to interpret the expectation value of the momentum? The wave functions in this representation are functions only of x and t. To evaluate the expectation value of the momentum operator, it is necessary to develop the operator corresponding to the momentum in the position representation. To do this, we use the analog of (1.23) and then introduce the Fourier transforms corresponding to the functions φ. Then, we may write the expectation of the momentum as

$$\begin{aligned}\langle p\rangle &= \int_{-\infty}^{\infty} \varphi^{\dagger}(k,t)\,p\varphi(k,t)\,dk = \int_{-\infty}^{\infty} \varphi^{\dagger}(k,t)\,\hbar k\varphi(k,t)\,dk \\ &= \frac{\hbar}{2\pi}\int_{-\infty}^{\infty}\int_{-\infty}^{\infty} \psi^{\dagger}(x',t)e^{ikx'}dx'k\int_{-\infty}^{\infty} \psi(x,t)e^{-ikx}dxdk \\ &= \frac{\hbar}{2\pi}\int_{-\infty}^{\infty}\int_{-\infty}^{\infty} \psi^{\dagger}(x',t)e^{ikx'}dx'\int_{-\infty}^{\infty} \psi(x,t)i\frac{\partial}{\partial x}e^{-ikx}dxdk \\ &= -i\hbar\int_{-\infty}^{\infty} \psi^{\dagger}(x,t)\frac{\partial}{\partial x}\psi(x,t)\,dx\end{aligned} \tag{1.26}$$

In arriving at the final form of (1.26), an integration by parts has been done (the evaluation at the limits is assumed to vanish) after replacing k by the partial derivative. The last line is achieved by recognizing the delta function:

$$\delta(x-x') = \frac{1}{2\pi}\int_{-\infty}^{\infty} e^{-ik(x-x')}dk. \tag{1.27}$$

Now, by comparing (1.26) with (1.23), we determine that the momentum, in the position representation, is given by the differential operator

$$p = -i\hbar\frac{\partial}{\partial x}. \tag{1.28}$$

1.5.2 Noncommuting Operators

The description of the momentum operator in the position representation is that of a differential operator. This means that the operators corresponding to the position and to the momentum will not commute, just as suggested by Heisenberg. But this is a result of the Schrödinger equation as well, since that is how we determined that the momentum was a differential operator. To see this, we define the commutator relation for position and momentum as

$$[x, p] = xp - px. \tag{1.29}$$

The left-hand side of (1.29) defines a quantity that is called the *commutator bracket*. However, by itself it only has no implied meaning. The terms contained within the brackets are operators and must actually operate on some wave function. Thus, the role of the commutator can be explained by considering the inner product, or expectation value. This gives

$$\begin{aligned}\langle[x,p]\rangle &= -i\hbar\int_{-\infty}^{\infty}\psi^{\dagger}(x,t)\left(x\frac{\partial}{\partial x}-\frac{\partial}{\partial x}x\right)\psi(x,t)dx \\ &= -i\hbar\int_{-\infty}^{\infty}\psi^{\dagger}(x,t)\left[\left(x\frac{\partial}{\partial x}-x\frac{\partial}{\partial x}\right)\psi(x,t)-\psi(x,t)\frac{\partial}{\partial x}x\right]\psi(x,t)dx = i\hbar\end{aligned} \tag{1.30}$$

If conjugate variables, or operators, do not commute, there is an implication that these quantities cannot be measured simultaneously. Here again, there is another and deeper meaning. In the previous section, we noted that the operation of the position operator x on the wave function in the position representation produced an eigenvalue $\underline{x}$, which is actually the expectation value of the position. The momentum operator does not produce this simple result with the wave function of the position representation. Rather, the differential operator produces a more complex result. For example, if the differential operator were to produce a simple eigenvalue, then the wave function would be constrained to be of the form $exp(ipx/\hbar)$ (which can be shown by assuming a simple eigenvalue form with the differential operator and solving the resulting equation). This form is not integrable (it does not fit our requirements on normalization), and thus the same wave function cannot simultaneously yield eigenvalues for both position and momentum. Since the eigenvalue relates to the expectation value, which corresponds to the most likely result of an experiment, these two quantities cannot be simultaneously measured. We will return to this in Chapter 5.

There is a further level of information that can be obtained from the Fourier transform pair of position and momentum wave functions. If the position is known, for example if we choose the delta function of (1.27), then the Fourier transform has unit amplitude everywhere; that is, the momentum has equal probability of taking on any value. Another way of looking at this is to say that since the position of the particle is completely determined, it is impossible to say anything about the momentum, as any value of the momentum is equally likely. Similarly, if a delta function is used to describe the momentum wave function, which implies that we know the value of the momentum exactly, then the position wave function has equal amplitude everywhere. This means that if the momentum is known, then it is impossible to say anything about the position, as all values of the latter are equally likely. As a consequence, if we want to describe both of these properties of the particle, the position wave function and its Fourier transform must be selected carefully to allow this to occur. Then there will be an uncertainty Δx in position and Δp in momentum. We will return to this topic in the next chapter.

1.5.3 Another View

In the discussion above, it was suggested that the introductory basics of quantum mechanics arise from the changes from classical mechanics that are brought to an observable level by the smallness of some parameter, such as the size scale. The most important effect is the appearance of operators for dynamical variables, and the noncommuting nature of these operators. We also discussed a wave function, either in the position or momentum representation, introduced by Erwin Schrödinger (1.15). The squared magnitude is thought to be related to the probability of finding the equivalent particle at a particular point in space. Shortly after this, Erwin Madelung (1926) and Earle Kennard (1928) separately decided to seek further understanding of Schrödinger's equation. This new wave equation is a complex equation, in that it the wave function involves both real and imaginary parts. They decided to look at these two parts separately to investigate further how these parts represented the understanding of quantum physics in relation to classical physics. Some two decades later, David Bohm (1952) would rediscover this approach and draw significantly more attention to this understanding.

In order to separate the real and imaginary parts, we need to define the wave function in a manner that allows this separation. For this, we chose to let

$$\psi(x,t) = A(x,t)exp\left(iS(x,t)/\hbar\right). \tag{1.31}$$

The quantity S is known as the *action* in classical mechanics (but familiarity with this will not be required). Let us put this form for the wave function into (1.15), which gives (the exponential factor is omitted as it cancels equally from all terms)

$$-A\frac{\partial S}{\partial t} + i\hbar\frac{\partial A}{\partial t} = \frac{A}{2m}\left(\frac{\partial S}{\partial x}\right)^2 - \frac{i\hbar A}{2m}\frac{\partial^2 S}{\partial x^2} - \frac{i\hbar}{m}\frac{\partial S}{\partial x}\frac{\partial A}{\partial x} - \frac{\hbar^2}{2m}\frac{\partial^2 A}{\partial x^2} + VA. \tag{1.32}$$

For this equation to be valid, it is necessary that the real parts and the imaginary parts balance separately, which leads to

$$\frac{\partial S}{\partial t} + \frac{1}{2m}\left(\frac{\partial S}{\partial x}\right)^2 + V - \frac{\hbar^2}{2mA}\frac{\partial^2 A}{\partial x^2} = 0 \tag{1.33}$$

and

$$\frac{\partial A}{\partial t} + \frac{A}{2m}\frac{\partial^2 S}{\partial x^2} + \frac{1}{m}\frac{\partial S}{\partial x}\frac{\partial A}{\partial x} = 0. \tag{1.34}$$

The second equation, (1.34), can be rearranged by multiplying by A, as

$$2A\frac{\partial A}{\partial t} + \frac{A^2}{m}\frac{\partial^2 S}{\partial x^2} + \frac{2A}{m}\frac{\partial S}{\partial x}\frac{\partial A}{\partial x} = 0, \tag{1.34'}$$

and then

$$\frac{\partial A^2}{\partial t}+\frac{\partial}{\partial x}\left(\frac{A^2}{m}\frac{\partial S}{\partial x}\right)=0. \tag{1.35}$$

If take the partial derivative with respect to x into the three-dimensional vector form, this can be rewritten as

$$\frac{\partial A^2}{\partial t}+\nabla\bullet\left(\frac{A^2}{m}\nabla S\right)=0. \tag{1.35'}$$

The form of this equation is known as a *continuity* equation, and may be recognized as the Kirchoff current equation (if we multiply by the charge) where the sum of the currents flowing into, or out of, a node yields the time rate of the charge on the node. It is also recognizable from a normal result for conservation of charge in Maxwell's equations. Hence, we are led to conclude that, if $A^2=|\psi|^2$, then this equation tells us that the increase in probability at a spatial point is given by the *flow* of probability into that point (inward flow adds a minus sign to the second term). Connecting with the normal continuity equation, leads us to also conclude that we can define a probability flow velocity as

$$v=\frac{1}{m}\nabla S. \tag{1.36}$$

Obviously, then ∇S represents the momentum associated with the probability flow. In essence, this is a treatment of the quantum wave function as a form of hydrodynamics.

To gain an understanding of the first of the equations, (1.33), let us return to (1.15) and use a separation of variables. For this, we write the wave function as

$$\psi(x,t)=\varphi(x)\chi(t), \tag{1.37}$$

for which (1.15) can be rewritten as

$$\frac{i\hbar}{\chi(t)}\frac{d\chi(t)}{dt}=\frac{1}{\varphi(x)}\left[-\frac{\hbar^2}{2m}\nabla^2\varphi(x)+V(x)\varphi(x)\right]=E. \tag{1.38}$$

Using just the time dependent term on the left and the constant (for energy), we obtain

$$\chi(t)\sim exp\left(-\frac{iEt}{\hbar}\right). \tag{1.39}$$

Upon comparing this with our assumed form for the wave function (1.31), we recognize that the action S is just the negative of the time integral of the energy. Hence, we can rewrite (1.33) as

$$E = \frac{1}{2m}\left(\frac{\partial S}{\partial x}\right)^2 + V - \frac{\hbar^2}{2mA}\frac{\partial^2 A}{\partial x^2}. \tag{1.40}$$
$$= T + V + Q$$

Using the above determination that the spatial derivative of S is the momentum, we recognize the first term on the right to be the kinetic energy. The potential energy remains the potential energy of course. But, when we compare to the classical form (1.16), we have a new term

$$Q = -\frac{\hbar^2}{2mA}\frac{\partial^2 A}{\partial x^2}. \tag{1.41}$$

Kennard (1928) expressed this as a quantum potential. It gives rise to the fact that each element of the probability would move as a particle would move according to Newton's laws under the classical forces, those arising from the potentials in (1.16), plus any *quantum force*. Q is one form of that quantum potential that gives rise to the quantum force. We note that, because of the amplitude in the denominator of (1.41), the quantum force depends upon the shape of the wave function and not on the amplitude of it. Different forms of a quantum potential have been discussed by Wigner (1932), Feyman (1965), and Ferry (2000).

But there is more to say about S. Examination of (1.39) and (1.31) tells us that S is defined modulo h (Planck's constant). Since the momentum is normally described by ∇S, the periodic nature of S leads to a quantization condition

$$\oint \boldsymbol{p}\cdot d\boldsymbol{r} = nh, \tag{1.42}$$

where n is an integer. This is precisely Heisenberg's quantization condition (1.13). Thus, S is not determined entirely by the local dynamics, as would be the classical case, but must also satisfy a topological condition which follows from its origin as the phase of an independent complex field—the wave function (Ferry 1998). Another way of expressing this is that, if there are any vortices that form in the quantum flow, their angular momentum must be quantized as well.

Equation (1.42) was not original to Heisenberg. It was originally proposed by Einstein (1917), and further developed by Brillouin (1926) and Keller (1958). It is usually referred to as EBK quantization due to these authors. But it is important to understand how this simple quantization relates to the topological nature of the wave function. This importance was highlighted in discussions of the geometrical phase associated with the wave function by Berry (1989). We consider motion around a closed path according to (1.42), which has a period of T around this path such that $\mathbf{r}(T) = \mathbf{r}(0)$. The state of the system then evolves according to (1.15). While the energy may be slowly varying in space and time (despite the constant energy assumption used previously), quite generally there is no firm relation between the phases at different points in space. If the system evolves adiabatically, then the states will evolve in a manner in which the wave function can be written as

$$\psi(t) \sim e^{-\frac{i}{\hbar}\int_0^t E dt'} e^{i\gamma(t)} \psi(\boldsymbol{r}(t)), \tag{1.43}$$

which differs from (1.37) and (1.39) only by the second exponential. Berry points out that this phase factor, $\gamma(t)$, is nonintegrable and cannot be written as a function of the position variable, as it is usually not single-valued as one moves around the circuit. This phase factor is determined by the requirement that Equation (1.43) satisfies (1.15), and direct substitution leads to the relationship

$$\frac{\partial \gamma(t)}{\partial t} = i\langle \psi(\boldsymbol{r}(t)) | \nabla \psi(\boldsymbol{r}(t)) \rangle \cdot \frac{\partial \boldsymbol{r}(t)}{\partial t}. \tag{1.44}$$

Integrating this around the closed contour discussed above gives

$$\gamma(C) = i\oint_C \langle \psi(\boldsymbol{r}(t)) | \nabla \psi(\boldsymbol{r}(t)) \rangle \cdot d\boldsymbol{r}. \tag{1.45}$$

If we now use the momentum relation (1.28), this becomes

$$\gamma(C) = -\frac{1}{\hbar}\oint \langle \boldsymbol{p} \rangle \cdot d\boldsymbol{r}. \tag{1.46}$$

This phase is often referred to as the Berry phase, with the previous expectations denoted as the Berry connection. Comparing this to (1.42) tells us that the left-hand side of (1.46) is usually an integer.

We can go a little further and relate this to the AB effect discussed above. Let us replace the momentum in (1.42) by the adjusted momentum due to the Peierls' substitution. Then, (1.42) becomes

$$\oint \boldsymbol{p} \cdot d\boldsymbol{r} - e\oint \boldsymbol{A} \cdot d\boldsymbol{r} = nh. \tag{1.47}$$

Let us now take the second term on the left and rewrite it using Stoke's theorem as

$$e\oint \boldsymbol{A} \cdot d\boldsymbol{r} = e\int (\nabla \times \boldsymbol{A}) \cdot \boldsymbol{n} d\Omega = eB\Omega = \frac{h\Phi}{\Phi_0}, \tag{1.48}$$

where Ω is the area enclosed by the contour discussed above (the ring in the AB effect), and the other symbols have their normal meanings. Now, we see that the AB effect is an example of the Berry phase and the topology of the ring. As the flux enclosed by the ring (Φ) changes, this requires the momentum in the ring to also change in order to maintain the constant right-hand side of (1.47). This change in momentum is sensed outside the ring as a change in conductance that gives rise to the oscillations of Figure 1.3(b).

1.5.4 Wave Packets

While we have assumed that the momentum wave function obtained in (1.25) is centered at zero momentum, this is not the general case. If we look at (1.15) and (1.36), the wave function may in fact be moving, especially if it corresponds to a moving particle. Now, the question is just how we relate this wave function to the localized idea of a particle. The answer to this is the wave packet. Such a localized wave packet is shown in Figure 1.4. Suppose that we assume that the momentum wave function is centered at a displaced value of k, given by k_0. Then, the entire position representation wave function moves with this average momentum and shows an average velocity $v_0 = \hbar k_0 / m$. We can expect that the peak of the position wave function, x_{peak}, moves, but does it move with this velocity? The position wave function is made up of a sum of a great many Fourier components, each of which arises from a different momentum. Does this affect the uncertainty in position that characterizes the half-width of the position wave function? The answer to both of these questions is yes, but we will try to demonstrate that these are the correct answers in this section.

Our approach is based upon the definition of the Fourier inverse transform (1.24). This latter equation expresses the position wave function $\psi(x)$ as a summation of individual Fourier components, each of whose amplitudes is given by the value of $\varphi(k)$ at that particular k. From the earlier work, we can extend each of the Fourier terms into a plane wave corresponding to that value of k, by introducing the frequency term via

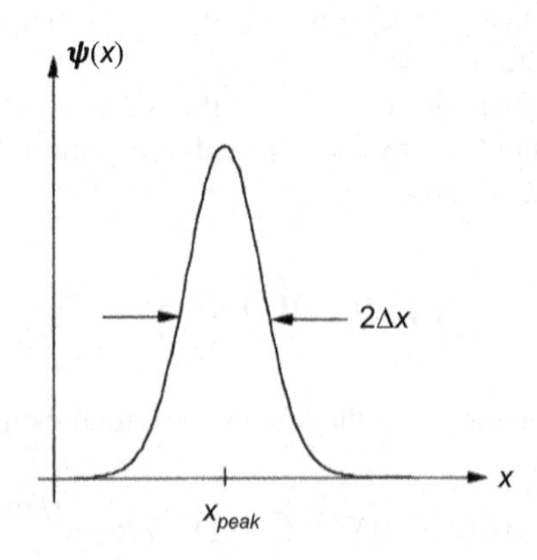

FIGURE 1.4 The positional variation of a typical wave packet.

$$\psi(x,t) = \frac{1}{\sqrt{2\pi}} \int_{-\infty}^{\infty} \varphi(k,t) e^{i(kx-\omega t)} dk. \tag{1.49}$$

While the frequency term has not been shown with a variation with k, it must be recalled that each of the Fourier components may actually possess a slightly different frequency. If the main frequency corresponds to the peak of the momentum wave function, then the frequency can be expanded as

$$\omega(k) = \omega(k_0) + (k - k_0) \left. \frac{\partial \omega}{\partial k} \right|_{k=k_0} + \ldots. \tag{1.50}$$

The interpretation of the position wave function is now that it is composed of a group of closely related waves, all propagating in the same direction (we assume that $\varphi(k) = 0$ for $k < 0$, but this is merely for convenience and is not critical to the overall discussion). Thus, $\psi(x, t)$ is now defined as a *wave packet*. Equation (1.50) defines the *dispersion* across this wave packet, as it gives the gradual change in frequency for different components of the wave packet.

To understand how the dispersion affects the propagation of the wave functions, we insert (1.50) into (1.49), and define the difference variable $u = k - k_0$. Then, (1.49) becomes

$$\psi(x,t) = \frac{1}{\sqrt{2\pi}} e^{i(k_0 x - \omega_0 t)} \int_{-\infty}^{\infty} \varphi(u + k_0, t) e^{i(ux - \omega' ut)} du, \tag{1.51}$$

where ω_0 is the leading term in (1.50) and ω' is the partial derivative in the second term of (1.50). The higher-order terms of (1.50) are neglected, as the first two terms are the most significant. If u is factored out of the argument of the exponential within the integral, it is seen that the position variable varies as $x - \omega' t$. This is our guide as to how to proceed. We will reintroduce k_0 within the exponential, but multiplied by this factor, so that

$$\begin{aligned}
\psi(x,t) &= \frac{1}{\sqrt{2\pi}} e^{-i(x-\omega' t)} e^{i(k_0 x - \omega_0 t)} \int_{-\infty}^{\infty} \varphi(u + k_0, t) e^{i(x-\omega' t)} e^{i(ux - \omega' ut)} du \\
&= \frac{1}{\sqrt{2\pi}} e^{i(\omega_0 - \omega' k_0)t} \int_{-\infty}^{\infty} \varphi(u + k_0, t) e^{i(u+k_0)(x-\omega' t)} du \\
&= e^{i(\omega_0 - \omega' k_0)t} \psi(x - \omega' t, t).
\end{aligned} \tag{1.52}$$

The leading exponential provides a phase shift in the position wave function. This phase shift has no effect on the square of the magnitude, which represents the expectation value calculations. On the other hand, the entire wave function moves with a velocity given by ω'. This is not surprising. The quantity ω' is the partial derivative of the frequency with respect to the momentum wave vector, and hence describes the group velocity of the wave packet. Thus, the average velocity of the wave packet in position space is given by the group velocity

$$v_g = \omega' = \left.\frac{\partial \omega}{\partial k}\right|_{k=k_0}. \tag{1.53}$$

This answers the first question: the peak of the position wave function remains the peak and moves with an average velocity defined as the *group velocity* of the wave packet. Note that this group velocity is defined by the frequency variation with respect to the wave vector. Is this related to the average momentum given by k_0? The answer again is affirmative, as we cannot let k_0 take on any arbitrary value. Rather, the peak in the momentum distribution must relate to the average motion of the wave packet in position space. Thus, we must impose a value on k_0 so that it satisfies the condition of actually being the average momentum of the wave packet:

$$v_g = \frac{\hbar k_0}{m} = \frac{\partial \omega}{\partial k}. \tag{1.54}$$

If we integrate the last two terms of (1.47) with respect to the wave vector, we recover the other condition that ensures that our wave packet is actually describing the dynamic motion of the particles:

$$E = \hbar \omega = \frac{\hbar^2 k^2}{2m} = \frac{p^2}{2m}. \tag{1.55}$$

It is clear that it is the group velocity of the wave packet that describes the average momentum of the momentum wave function and also relates the velocity (and momentum) to the energy of the particle.

1.6 SUMMARY

Quantum mechanics furnishes a methodology for treating the wave-particle duality. The main importance of this treatment is for structures and times, both usually small, for which the *interference* of the waves can become important. The effect can be either the interference between two wave packets, or the interference of a wave packet with itself, such as in boundary value problems. In quantum mechanics, the boundary value problems deal with the equation that we will develop in the next chapter for the Schrödinger equation.

The result of dealing with the wave nature of particles is that dynamical variables have become operators which in turn operate upon the wave functions. As operators, these variables often no longer commute, and there is a basic uncertainty relation between noncommuting operators. The noncommuting nature arises from it being no longer possible to generate a wave function that yields eigenvalues for *both* of the operators, representing the fact that they cannot be simultaneously measured. It is this that introduces an uncertainty relationship that will be dealt with in the next chapter.

We can draw another set of conclusions from this behavior that will be important for the differential equation that can be used to find the actual wave functions in different situations. The entire time variation has been found to derive from a single initial condition, which implies that the differential equation must be only first order in the time derivatives. Second, the motion has diffusive components, which suggests that the differential equation should bear a strong resemblance to a diffusion equation (which itself is only first order in the time derivative). These points will be expanded upon in the next chapter.

REFERENCES

Ångström, A. J. 1862. On the Fraunhofer-lines visible in the solar spectrum. *Phil. Mag., Ser.* 4, 24: 1–11

Ångström, A. J. 1868. *Recherches sur le spectre solaire*. Uppsala: W. Schultz

Balmer, J. 1885. Notiz über die Spektrallinien des Wasserstoffs. *Verh. Naturfors. Basel.* 7: 548–560

Berry, M. V. 1989. Quantum scars of classical closed orbits in phase space. *Proc. Roy. Soc. London.* 423: 219–231

Bohm, D. 1952. A suggested interpretation of the quantum theory in terms of "hidden" variables. *I. Phys. Rev.* 85: 166–179

Bohr, N. 1913. On the constitution of atoms and molecules. *Phil. Mag., Ser.* 6, 26: 1–25

Brillouin, L. 1926. Remarques sur la mécanique ondulatoire. *J. Phys. Radium.* 7: 353–367

Davisson, C., and Kunsman, C. H. 1921. The scattering of electrons by nickel. *Science.* 54: 522–524

Davisson, C., and Kunsman, C. H. 1923. The scattering of low speed electrons by platinum and magnesium. *Phys. Rev.* 22: 242–258

de Broglie, L. 1923. Waves and quanta. *Nature* 112: 540-540

de Broglie, L. 1924. A tentative theory of light quanta. *Phil. Mag., Ser.* 6, 47: 446–458

de Broglie, L. 1937. *Matière et Lumière*. Paris: Albin Michel

Einstein, A. 1905a. Zur Elektrodynamik bewegter Körper. *Ann. Phys.* 17: 891–921

Einstein, A. 1905b. Über einen die Erzeugung und Verwandlung des Lichtes betreffenden heuristischen Gesichtspunkt. *Ann. Phys.* 17:132–148

Einstein, A. 1917. Zum Quantensatz von Sommerfeld and Epstein. *Verh. Deutsche Phys. Gesell.* 19: 82–92

Elsasser, W. 1925. Bemerken zur Quantenmechanik freier Elektronen. *Naturwiss.* 13: 711–711

Ferry, D. K. 1998. Open problems in quantum simulation in ultra-submicron devices. *VLSI Design.* 8: 165–172

Ferry, D. K. 2000. Effective potentials and the onset of quantization in ultrasmall MOSFETs. *Superlatt. Microstruc.* 28: 419–423

Feyman, R. P. and Hibbs, A. R. 1965. *Quantum Mechanics and Path Integrals.* New York: McGraw-Hill. pp. 279–285

Gell-Mann, M. 1979. *The Nature of the Physical Universe: 1976 Nobel Conference.* Ed. by D. Huff and O. Prewett. New York: Wiley-Interscience. p. 29

Heisenberg, W. 1925. Über quantentheoretische Umdeutung kinematischer und mechanischer Beziehunger. *Z. Phys.* 33: 879–893; Tr. in B. L. van der Warden. 1967. *Sources of Quantum Mechanics*. Amsterdam: North-Holland, 261–276

Keller, J. B. 1958. Corrected Bohr-Sommerfeld quantum conditions for nonseparable systems. *Ann. Phys.* 4: 180–188

Kennard, E. H. 1928. On the quantum mechanics of a system of particles. *Phys. Rev.* 31: 876–890

Langevin, P. 1908. Sur la théorie du movement brownien. *C. R. Acad. Sci. Paris.* 148: 530–533

Madelung, E. 1926. Quantentheorie in hydrodynamischer Form. *Z. Phys.* 40: 322–326

Mankiewich, P. M., Behringer, R. E., Howard, R. E. *et al.* 1988. Observation of Aharonov-Bohm effect in quasi-one dimensional GaAs/AlGaAs rings. *J. Vac. Sci. Technol. B.* 6: 131–134

Nagaoka, H. 1904. Kinetic of a system of particles illustrating the line and the band spectrum and the phenomena of radioactivity. *Phil. Mag., Ser.* 6, 7: 445–455

Planck, M. 1900. Zur Theorie des gesetses der Energievertheilung im Normalspektrum. *Verhanlungen der Deutschen Physikalischen Gesellschaft* 2: 237–245

Rydberg, J. R. 1889. La constitutions des spectres d'emission des elements chimiques. *Kongl. Swedish Vet.-Akad. Handl.* 23(11): 1–177

Taylor, G. I. 1909. Interference fringes with feeble light. *Proc. Cambridge Phil. Soc.* 15: 114–115

Thompson, G. P. 1927. Diffraction of cathode rays by a thin film. *Nature.* 119: 890–890

Thompson, J. J. 1907. On the ionization of gases by ultra-violet light and on the evidence as to the structure of light afforded by its electrical effects. *Proc. Cambridge Phil. Soc.* 14: 417–424

von Neumann, J. 1932. *Mathematische Grundlagen der Quantenmechanik*. Berlin: Springer

Wigner, E. 1932. On the quantum correction for thermodynamic equilibrium. *Phys. Rev.* 40: 749–759

Young, T. 1809. *Course of Lectures on Natural Philosophy and Mechanical Arts*. London: J. Johnson. Lecture XXXIX.

PROBLEMS

1 Calculate the energy density for the plane electromagnetic wave described by the complex field strength:

$$E_c = E_0 \mathrm{e}^{i(\omega t - kx)}$$

and show that its average over a temporal period T is $\hbar\omega = (E/2)|E_c|^2$.

2 What are the de Broglie frequencies and wavelengths of an electron and a proton accelerated to 100 eV? What are the corresponding group and phase velocities?

3 Show that the position operator x is represented by the differential operator:

$$i\hbar \frac{\partial}{\partial \mathrm{p}}$$

in momentum space, when dealing with momentum wave functions.

4 Express the expectation value of the kinetic energy of a Gaussian wave packet in terms of the expectation value and the uncertainty of the momentum wave function. You may express the uncertainty in position from the width of the Gaussian in position (width at half maximum value), as per Figure 1.4. The uncertainty of the momentum wave function comes from the width at half maximum of the resulting Gaussian in momentum space.

5 A particle is represented by a wave packet propagating in a dispersive medium, described by:

$$\omega = \frac{A}{\hbar}\left\{\sqrt{1+\frac{\hbar^2 k^2}{mA}}-1\right\}.$$

What is the group velocity as a function of $\hbar k$.

6 The longest wavelength that can cause the emission of electrons from silicon is 296 nm. *(a)* What is the work function of silicon? *(b)* If silicon is irradiated with light of 250 nm wavelength, what is the energy and momentum of the emitted electrons? What is their wavelength? (c) If the incident photon flux is 5 mWcm–2, what is the photoemission current density?

7 For particles which have a thermal velocity, what is the wavelength at 300 K of electrons, helium atoms, and the *a*-particle (which is ionized ^{4}He)?

8 A wave function has been determined to be given by the spatial variation:

$$\Psi(x) = \begin{cases} 2A & -a < x < 0 \\ 2A(a-x) & 0 < x < a \\ 0 & \text{elsewhere.} \end{cases}$$

Determine the value of A, the expectation value of x, x^2, p, and p^2. What is the value of the uncertainty in position and momentum?

9 A wave function has been determined to be given by the spatial variation:

$$\Psi(x) = \begin{cases} 2A\sin\left(\frac{\pi}{a}\right) & -a < x < \mathrm{a} \\ 0 & \text{elsewhere.} \end{cases}$$

Determine the value of A, the expectation value of x, x^2, p, and p^2.

2 The Schrödinger Equation

In the first chapter, it was explained that the introductory basics of quantum mechanics arise from the changes from classical mechanics that are brought to an observable level by the smallness of some parameter, such as the size scale. The most important effect is the appearance of operators for dynamical variables, and the noncommuting nature of these operators. We also found a wave function, either in the position or momentum representation, whose squared magnitude is related to the probability of finding the position or momentum of the equivalent particle. The properties of the wave could be expressed as basically arising from a linear differential equation of a diffusive nature. In particular, because any subsequent form for the wave function evolved from a single initial state, the equation can only be of first order in the time derivative (and, hence, diffusive in nature).

It must be noted that the choice of a wave-function-based approach to quantum mechanics is not the only option. As we remarked in the first chapter, there were two separate formulations of the new quantum mechanics appeared almost simultaneously. One was developed by Werner Heisenberg (1925), at that time a lecturer in Göttingen (Germany) and a postdoctoral fellow in Copenhagen (Denmark). In this approach, a calculus of noncommuting operators was developed. This approach was quite mathematical and required considerable experience to work through in any detail. It remained until later to discover that this calculus was actually representable by normal matrix calculus. The second formulation was worked out by Erwin Schrödinger (1926), at the time a Professor in Vienna. In Schrödinger's formulation, a wave equation was found to provide the basic understanding of quantum mechanics. Although not appreciated at the time, Schrödinger presented the connection between the two approaches in subsequent papers. In a somewhat political environment, Heisenberg received the 1932 Nobel Prize for 'discovering' quantum mechanics, while Schrödinger was forced to share the 1933 prize with Paul Dirac for advances in atomic physics. Nevertheless, it is Schrödinger's formulation which is almost universally used today, especially in textbooks. This is especially true for students with a background in electromagnetic fields, as the concept of a wave equation is not completely new to them.

In this chapter, we want now to deal with the wave equation—the Schrödinger equation, from which one version of quantum mechanics—wave mechanics—has evolved. While the equation was introduced in Chapter 1, we want here to justify how the equation connects to classical physics. Here we want to gain insight into the quantization process, and the effects it causes in normal systems. In the following section, we will give a justification for the wave equation, but no formal derivation is really possible (as in the case of Maxwell's equations); rather, the equation is found to explain experimental results in a correct fashion, and its validity lies in that fact. In subsequent sections, we will then apply the Schrödinger equation to a variety of problems to gain the desired insight. But let us be clear. The quantum aspect of treating e.g. an electron and its behavior arises from the decision to treat the particle as a wave. It is a wave equation that we are solving. In particular, when we ignore the time variation of the wave function, we arrive at a nearly classical boundary value problem that must be solved. We are doing no more than solving a version of the Sturm-Liouville problem, which differs little from solutions of other boundary value problems in mathematics and electromagnetics. Again, the quantum nature lies entirely in the decision to deal with the wave approach.

2.1 WAVES AND THE DIFFERENTIAL EQUATION

In essence, the energy is the eigenvalue of the time derivative in the separation of variables done in (1.37), although the time derivative is not a true operator, as time is not a dynamic variable. Thus, it may be thought that the energy represents a set of other operators that do represent dynamic variables. It is common to express the energy as a sum of kinetic and potential energy terms; for example, as done in (1.16) as

$$E = T + V = \frac{p^2}{2m} + V(x,t). \tag{2.1}$$

We also showed in Chapter 1 that the momentum becomes a differential operator in quantum mechanics for which

$$p = -i\hbar\nabla \tag{2.2}$$

in the position representation. From the fact that $E = \hbar\omega$ and the frequency can be related to a time derivative, we arrive at the Schrödinger equation

$$i\hbar\frac{d\psi(x,t)}{dt} = -\frac{\hbar^2}{2m}\nabla^2\psi(x,t) + V(x)\psi(x,t). \tag{2.3}$$

In this section, we want to deal principally with the time-independent version, where the left-hand side of (2.3) is just the energy and wave function as

$$-\frac{\hbar^2}{2m}\nabla^2\psi(x,t) + V(x)\psi(x,t) = E\psi(x,t). \tag{2.4}$$

It is this version of the equation that is principally a Sturm-Liouville equation, which is solved as a boundary value problem. In principle, any solution of (2.4) that satisfies the boundary conditions is a valid solution, because the equation itself is linear. So, we will proceed to treat some simple cases of this equation.

2.1.1 The Free Particle

We begin by first considering the situation in which the potential is zero. Then the time-independent equation becomes (in one dimension)

$$-\frac{\hbar^2}{2m}\frac{\partial^2}{\partial x^2}\psi + k^2\psi = 0, \tag{2.5}$$

where

$$\frac{\hbar^2 k^2}{2m} = E, \quad k = \sqrt{\frac{2mE}{\hbar^2}}. \tag{2.6}$$

The solution to (2.5) is clearly of the form of sines and cosines, but here we will take the exponential terms, and

$$\psi(x) = Ae^{ikx} + Be^{-ikx}. \tag{2.7}$$

These are just the plane-wave solutions with which we began our treatment of quantum mechanics. The plane-wave form becomes more obvious when the time variation $e^{-i\omega t}$ is reinserted into the total wave function.

Then in looking at the solutions of this equation, and with the time variation restored, so that

$$\psi(x) = Ae^{ikx}e^{-i\omega t} + Be^{-ikx}e^{-i\omega t}, \tag{2.8}$$

We see that the first wave is traveling to the right (toward positive x) while the second solution is traveling to the left (toward negative x). Thus, in free space, the solutions to the Schrödinger equation is just a pair of plane waves, one each traveling in the positive and negative directions. We note that each has a positive amplitude and a single value of the wave number k, so that its formulation in momentum space is a delta function at the proper momentum. We cannot determine its position in real space, because the amplitude is uniform everywhere.

2.1.2 A Potential Step

To begin to understand the role that the potential plays, let us investigate a simple potential step, in which the potential is defined as

$$V(x) = V_0\theta(x), \quad V_0 > 0, \tag{2.9}$$

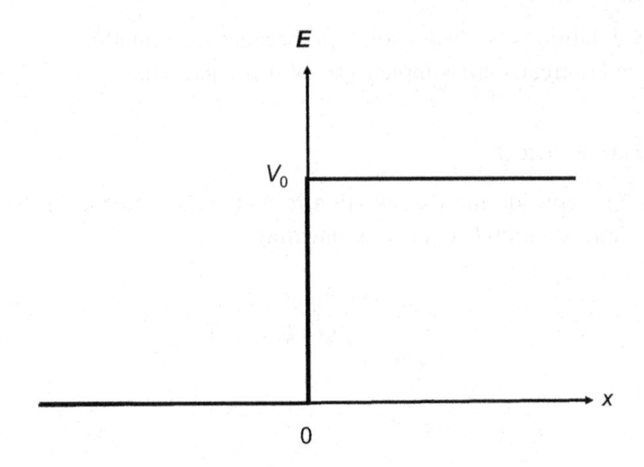

FIGURE 2.1 Schematic view of the potential of (2.9), which is nonzero (and constant) only in the positive half-space.

where $\theta(x)$ is the Heaviside step function in which $\theta = 1$ for $x \geq 0$, and $\theta = 0$ for $x < 0$. The potential is shown in Figure 2.1. The potential has a height of V_0 for positive x and is zero for the negative-x region. This potential creates a barrier to the wave function, and a wave incident from the left (the negative region) will have part (or all) of its amplitude reflected from the barrier. The results that are obtained depend upon the relative energy of the wave. If the energy is less than $V_{0,}$ the wave cannot propagate in the region of positive x. This is clearly seen from incorporating the potential into the definition of k via

$$\frac{\hbar^2 k^2}{2m} = E - V(x), \quad k = \sqrt{\frac{2m(E - V(x))}{\hbar^2}}, \tag{2.10}$$

and the wave vector is imaginary for $E < V_0$. Only one exponent can be retained, as we require that the wave function remain finite (but zero) as $x \to \infty$. So, we will examine two cases, one for which the energy is larger than the potential and one for which the energy is smaller than the potential step. These are illustrated in Figure 2.2.

Case I. $E < V_0$

Let us first consider the low-energy case, where the wave is a nonpropagating wave for $x >> 0$. In the negative half-space, we consider the wave function ψ_1 to be of the form of (2.7), composed of an incident wave (the positive-exponent term) and a reflected wave (the negative-exponent term). That is, we write the wave function for $x < 0$ exactly as (2.7), where the energy and the wave vector k are given by (2.6). For the positive half-space, where the potential exists, we write the wave function as

$$\psi_2(x) = Ce^{-\gamma x}, \tag{2.11}$$

where

$$\gamma = \sqrt{\frac{2m\left(V(x)-E\right)}{\hbar^2}}. \tag{2.12}$$

Here, we have defined a wave function in two separate regions, in which the potential is constant in each region. These two wave functions must be smoothly joined where the two regions meet.

While three constants are defined (A, B, C), one of these is defined by the resultant normalization of the wave function (we could e.g. let $A = 1$ without loss of generality). Two boundary conditions are required to evaluate the other two coefficients in terms of A. The boundary conditions can vary with the problem, but one must describe the continuity of the probability across the interface between the two regions. Thus, one boundary condition is that the wave function itself must be continuous at the interface, or

$$\psi_1(0) = \psi_2(0) \rightarrow A + B = C. \tag{2.13}$$

To obtain a second boundary condition, we shall require that the derivative of the wave function is also continuous (we can have only two boundary conditions, as we have only one boundary at $x = 0$, and the equation is only of second order in the derivative). In some situations, we cannot specify such a boundary condition, as there may not be a sufficient number of constants to evaluate (this will be the case in the next section). Equating the derivatives of the wave functions at the interface leads to

$$\left.\frac{d\psi_1}{dx}\right|_{x=0} = \left.\frac{d\psi_2}{dx}\right|_{x=0} \rightarrow ik\left(A-B\right) = -\gamma C. \tag{2.14}$$

This last equation can be rearranged by placing the momentum term in the denominator on the right-hand side. Then adding (2.13) and (2.14) leads to

$$\frac{C}{A} = \frac{2ik}{ik-\gamma}. \tag{2.15}$$

This result can now be used in (2.13) to yield

$$\frac{B}{A} = \frac{ik+\gamma}{ik-\gamma}. \tag{2.16}$$

The amplitude of the reflected wave is unity, so there is no probability amplitude transmitted across the interface. In fact, the only effect of the interface is to phase shift the reflected wave; that is, the wave function is ($x < 0$)

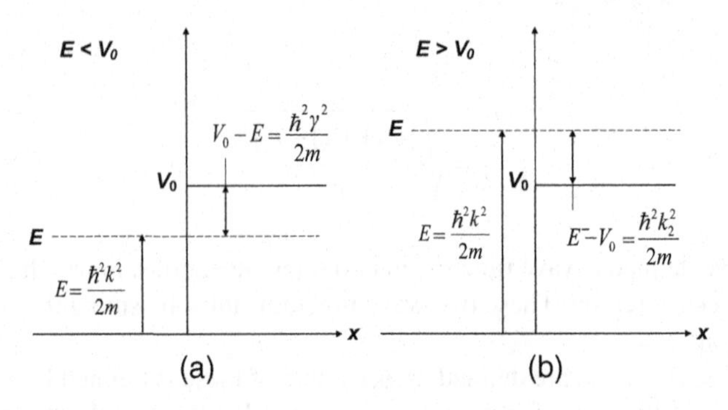

FIGURE 2.2 The various wave vectors are related to the energy of the wave: (*a*) the case for $E < V_0$; (*b*) the case for $E > V_0$.

$$\psi_1(x) = A\left(e^{ikx} + e^{-i(kx+\xi)}\right) \tag{2.17}$$

where

$$\xi = 2arctan\left(\frac{\gamma}{k}\right), \tag{2.18}$$

The probability amplitude is given by

$$|\psi_1(x)|^2 = 2A^2,\ x < 0. \tag{2.19}$$

As may have been expected, this is a *standing-wave* pattern, with the probability oscillating from 0 to twice the value of A^2. The first peak occurs at a distance $x = -\xi/2k$, that is, the distance to the first peak is dependent upon the phase shift at the interface. If the potential amplitude is increased without limit, $V_0 \rightarrow \infty$, the damping coefficient $\gamma \rightarrow \infty$, and the phase shift approaches π. However, the first peak occurs at a value of $kx = \pi/2$, which also leads to the result that *the wave function becomes zero* at $x = 0$. We cannot examine the other limit ($V_0 \rightarrow 0$), as we do not have the proper transmitted wave, but this limit can be probed when the transmission mode is examined. It may also be noted that a calculation of the probability current for $x > 0$ leads immediately to zero as the wave function is real. Thus, no probability current flows into the right half-plane. It is a simple calculation to show that the net probability current in the left half-plane vanishes as well, as the reflected wave carries precisely the same current away from the interface as the incident wave carries toward the interface.

Case II. $E > V_0$

We now turn to the case in which the wave can propagate on both sides of the interface. As above, the wave function in the left half-space is assumed to be of the form of (2.8), which includes both an incident wave and a reflected wave. Similarly, the transmitted wave will be assumed to be of the form

$$\psi_2 = Ce^{ik_2 x}, \tag{2.20}$$

where k_2 is given by (2.10). The relationships between this wave vector and that for the region $x < 0$ are schematically described in Figure 2.2(*b*). Again, we will match both the wave function and its derivative at $x = 0$. This leads to

$$\begin{aligned} \psi_1(0) = \psi_2(0) &\rightarrow A + B = C \\ \left.\frac{d\psi_1}{dx}\right|_{x=0} = \left.\frac{d\psi_2}{dx}\right|_{x=0} &\rightarrow k(A - B) = k_2 C. \end{aligned} \tag{2.21}$$

These equations can now be solved to obtain the constants C and B in terms of A. One difference here from the previous treatment is that these will be real numbers now, rather than complex numbers. Indeed, adding and subtracting the two equations of (2.21) leads to

$$\frac{C}{A} = \frac{2k}{k + k_2}, \quad \frac{B}{A} = \frac{k - k_2}{k + k_2}. \tag{2.22}$$

Here, we see that if $V_0 \rightarrow 0$, $k_2 \rightarrow k$ and the amplitude of the reflected wave vanishes, and the amplitude of the transmitted wave is equal to the incident wave.

2.2 DENSITY AND CURRENT

The Schrödinger equation is a complex diffusion equation. The wave function ψ is a complex quantity. The potential energy $V(x,t)$, however, is usually a real quantity. Moreover, we discerned in Chapter 1 that the probabilities were real quantities, as they relate to the chance of finding the particle at a particular position. Thus, the probability density is just

$$P(x,t) = \psi^*(x,t)\,\psi(x,t) = \left|\psi(x,t)\right|^2. \tag{2.23}$$

This, of course, leads to the normalization of (1.22), which just expresses the fact that the sum of the probabilities must be unity. If (2.23) were multiplied by the electronic charge e, it would represent the charge density carried by the particle (described by the wave function).

One check of the extension of the Schrödinger equation to the classical limit lies in the continuity equation. That is, if we are to relate (2.23) to the local charge density, then there must be a corresponding current density $\mathbf{J}$, such that ($\rho = -eP$) we arrive at the continuity equation of the previous chapter, as

$$e\frac{\partial P}{\partial t} = \nabla \cdot \mathbf{J}, \tag{2.24}$$

although we use only the x-component here. Now, the complex conjugate of (2.3) is just

$$-i\hbar\frac{d\psi^*(x,t)}{dt} = -\frac{\hbar^2}{2m}\nabla^2\psi^*(x,t) + V(x)\psi^*(x,t). \tag{2.25}$$

We now postmultiply (2.25) by $\psi(x,t)$and premultiply (2.3) by $\psi^*(x,t)$ and subtract one from the other. This now gives us

$$i\hbar\frac{\partial P}{\partial t} = -\frac{\hbar^2}{2m}\left[\psi^*(\nabla^2\psi) - (\nabla^2\psi)\psi^*\right]. \tag{2.26}$$

Now, comparing this with (2.24), and dropping the charge, we find the probability current to be given by

$$\mathbf{J}_\psi = \frac{\hbar}{2mi}\left[\psi^*(\nabla\psi) - (\nabla\psi)\psi^*\right]. \tag{2.27}$$

If the wave function is to be a representation of a single electron, then this ‘current’ must be related to the velocity of that particle. On the other hand, if the wave function represents a large ensemble of particles, then the actual current (obtained by multiplying by e) represents some average velocity, with an average taken over that ensemble.

The probability current should be related to the momentum of the wave function, as discussed earlier. The gradient operator in (2.27) is, of course, related to the momentum operator, and the factors of the mass and Planck’s constant connect this to the velocity. In fact, we can rewrite (2.27) as

$$\mathbf{J}_\psi = \frac{1}{2m}2\mathrm{Re}\{p\}|\psi|^2. \tag{2.25}$$

In general, when the momentum is a ‘good’ operator, which means that it is measurable, the eigenvalue is a real quantity. Then, the imaginary part vanishes, and (2.28) is simply the product of the velocity and the probability, which yields the probability current.

The result (2.28) differs from the earlier form that appears in (1.35). If the expectation of the momentum is real, then the two forms agree, as the gradient of the action just gives the momentum. On the other hand, if the expectation of the momentum is not real, then the two results differ. For example, if the average momentum were entirely imaginary, then (2.28) would yield zero identically, while (1.35) would give a nonzero result. However, (1.35) was obtained by separating the real and imaginary parts of (1.31), and the result in this latter equation assumed that S was entirely real. An imaginary momentum would require that S be other than purely real. Thus, (1.35) was obtained for a very special form of the wave function. On the other hand, (2.28) results from a quite general wave function, and while the specific result depended upon a plane wave, the approach was not this limited. If (2.2) is used with the general wave function, then (2.28) is evaluated using the expectation values of the momentum and suggests that in fact these eigenvalues should be real, *if a real current is to be measured.*

By real eigenvalues, we simply recognize that if an operator A can be measured by a particular wave function, then this operator produces the eigenvalue a, which is a real quantity (we may assert without proof that one can only measure real numbers in a measurement). This puts certain requirements upon the operator A, as we note that, by eigenvalue, we require

$$A\psi(x,t) = a\psi(x,t) \tag{2.29}$$

for properly normalized wave functions. For the eigenvalue to be measureable, we require that the eigenvalue be real, and thus will satisfy

$$a^* = \langle A^\dagger \rangle = \int_{-\infty}^{\infty} \psi^* A^\dagger \psi dx = \int_{-\infty}^{\infty} A\psi^* \psi dx \equiv a. \tag{2.30}$$

with

$$\int_{-\infty}^{\infty} \psi^* A\psi(x,t)dx = \int_{-\infty}^{\infty} \psi^* a\psi(x,t)dx = a. \tag{2.31}$$

Not all operators are Hermitian, however, and the definition of the probability current allows for consideration of those cases in which the momentum may not be a real quantity and may not be measurable, as well as those more normal cases in which the momentum is measurable.

For an example, let us compute the current for case II of the preceeding section, where transport goes over the barrier. The probability current in the left-hand and right-hand spaces is found through the use of (2.28). For the incident and transmitted waves, these currents are simply

$$J_C = \frac{\hbar k_2}{m}\left(\frac{2k}{k+k_2}\right)^2, \qquad (2.32)$$
$$J_A = \frac{\hbar k}{m}$$

where we assume that $A = 1$ for convenience. Thus, we have the transmission coefficient as

$$T = \frac{J_C}{J_A} = \frac{4kk_2}{(k+k_2)^2}. \qquad (2.33)$$

Now, we note that the transmission is reciprocal, in that we can change the two momentum without changing the coefficient. The system is reversible, so that it doesn't matter from which direction the incident wave approaches—the same transmission is obtained. Similarly, we can compute the refection coefficient as

$$R = -\frac{J_B}{J_A}\left(\frac{k-k_2}{k+k_2}\right)^2. \qquad (2.34)$$

This leads to the result that

$$T + R = 1. \qquad (2.35)$$

For energies above the potential barrier height, the behavior of the wave at the interface is quite similar in nature to what occurs with an optical wave at a dielectric discontinuity. This is to be expected as we are using the wave representation of the particle and should expect to see optical analogs.

2.3 THE POTENTIAL WELL

If we now put two barriers together, we have a choice of making a potential in which there is a barrier between two points in space, or a well between two points in space. The former will be treated in the next chapter. Here, we want to consider the latter case, as shown in Figure 2.3. In this case, the two barriers are located at $|x| = a$. In general, the wave function will penetrate into the barriers a distance given roughly by the decay constant γ. As we know from Section 2.1.2, the size of the decay constant will depend upon the height of the barrier. One special case, however, is when the barrier is infinite in height, so that $\gamma \rightarrow \infty$, so that the wave does not penetrate the barrier. We deal with this case first, as it is considerably easier.

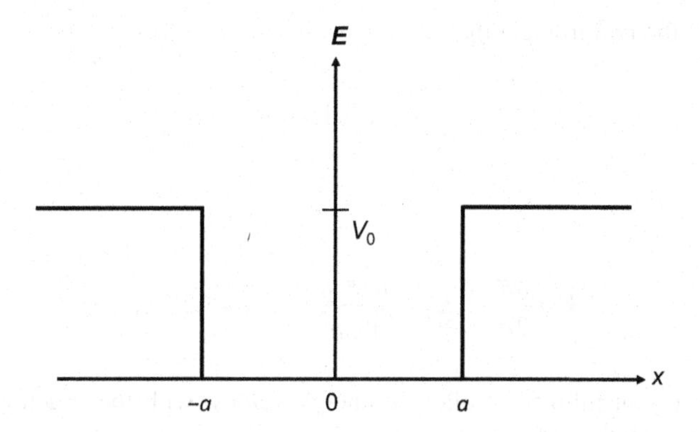

FIGURE 2.3 A potential well is formed by two barriers located at $|x| = a$.

2.3.1 The Infinite Potential Well

From the results obtained in Section 2.1, it is clear that the wave function decays infinitely rapidly under an infinite barrier. This leads to a boundary condition that requires the wave function to vanish at the barrier interfaces, that is $\Psi = 0$ at $|x| = a$. Within the central region, the potential vanishes, and the Schrödinger equation becomes just (2.4), with the wave vector defined by (2.6). The solution is now given, just as in the free-particle case, by (2.7). At the right-hand boundary, this leads to the situation

$$Ae^{ika} + Be^{-ika} = 0, \tag{2.36}$$

and at the left boundary

$$Ae^{-ika} + Be^{ika} = 0. \tag{2.37}$$

Here, we have two equations with two unknowns, apparently. However, one of the constants must be determined by normalization, so only A or B can be treated as unknown constants. The apparent dilemma is resolved by recognizing that the wave vector k cannot take just any value, and the allowed values of k are recognized as the second unknown. Since the two equations cannot give two solutions, they must be *degenerate*, and the determinant of coefficients must vanish, that is

$$\begin{vmatrix} e^{ika} & e^{-ika} \\ e^{-ika} & e^{ika} \end{vmatrix} = 0. \tag{2.38}$$

This leads to the requirement that

$$sin(2ka) = 0 \tag{2.39}$$

or

$$k = \frac{n\pi}{2a}, \quad E_n = \frac{n^2\pi^2\hbar^2}{8ma^2}, \quad n = 1,2,3,... \tag{2.40}$$

Thus, there are an infinity of allowed energy values, with the spacing increasing quadratically with the index n.

In order to find the wave function corresponding to each of the energy levels, we put the value for k back into one of the equations above for the boundary conditions; we chose to use (2.36). This leads to

$$\frac{B}{A} = -e^{in\pi} = (-1)^{n+1}. \tag{2.41}$$

Thus, as we move up the hierarchy of energy levels, the wave functions alternate between cosines and sines. This can be summarized as

$$\psi(x) = \begin{cases} A cos\left(\frac{n\pi x}{a}\right), & n\, odd \\ A sin\left(\frac{n\pi x}{a}\right), & n\, even \end{cases}. \tag{2.42}$$

We still have to normalize the wave functions. To do this, we can use either of the two solutions, and the general inner product with the range of integration now defined from $-a$ to a. This leads to

$$A^2\int_{-\infty}^{\infty}|\psi|^2\,dx = A^2\int_{-\infty}^{\infty}|\psi|^2\,dx \equiv 1. \tag{2.43}$$

Let us use the cosine form (the sine form will yield the same result) to give

$$\begin{aligned} A^2\int_{-\infty}^{\infty} cos^2\left(\frac{n\pi x}{a}\right)dx &= \frac{1}{2}A^2\int_{-\infty}^{\infty}\left[cos\left(\frac{2n\pi x}{a}\right)+1\right]dx \\ &= \frac{1}{2}A^2\left[\frac{a}{2n\pi}sin\left(\frac{2n\pi x}{a}\right)+x\right]\Bigg|_{-a}^{a} = A^2 a, \end{aligned} \tag{2.44}$$

or

$$A = \frac{1}{\sqrt{a}}. \tag{2.45}$$

If the particle resides exactly in a single energy level, we say that it is in a *pure* state. The more usual case is that it moves around between the levels and on the average many different levels contribute to the total wave function. Then the total wave function is a sum over the Fourier series, with coefficients related to the probability that each level is occupied.

It may be seen that the solutions to the Schrödinger equation in this situation were a set of odd wave functions and a set of even wave functions in (2.42), where by even and odd we refer to the symmetry when $x \to -x$. (The cosine function is then even, while the sine function is odd.) This is a general result when the potential is an even function; that is, $V(x) = V(-x)$. In the Schrödinger equation, the equation itself is unchanged when the substitution $x \to -x$ is made providing that the potential is an even function. Thus, for a bounded wave function, $\psi(-x)$ can differ from $\psi(x)$ by no more than a constant, say α. Repeated application of this variable replacement shows that $\alpha^2 = 1$, so α can only take on the values ± 1, which means that the wave function is either even or odd under the variable change. We note that this is only the case when the potential is even; no such symmetry exists when the potential is odd. Of course, if the wave function has an unbounded form, such as a plane wave, it is not required that the wave function have this symmetry, although both symmetries are allowed for viable solutions.

2.3.2 The Finite Potential Well

Now let us turn to the situation in which the potential is not infinite in amplitude and hence the wave function penetrates into the regions under the barriers. We continue to treat the potential as a symmetric potential centered about the point $x = 0$. However, it is clear that we will want to divide our treatment into two cases: one for energies that lie above the top of the barriers, and a second for energies that confine the particle into the potential well. In this regard, the system is precisely like the single finite barrier that was discussed in Section 2.3.1. When the energy is below the height of the barrier, the wave must decay into the region where the barrier exists, as shown in Figure 2.4. On the other hand, when the energy is greater than the barrier height, propagating waves exist in all regions, but there is a mismatch in the wave vectors, which leads to quasi-bound states and reflections from the interface. We begin with the case for the energy below the barrier height, which is the case shown in Figure 2.4.

Case I. $0 < E < V_0$

For energies below the potential, the particle has freely propagating characteristics only for the range $|x| < a$, for which the Schrödinger equation becomes the same as (2.5) with the energy and wave vector given in (2.6). It must be remembered that V_0

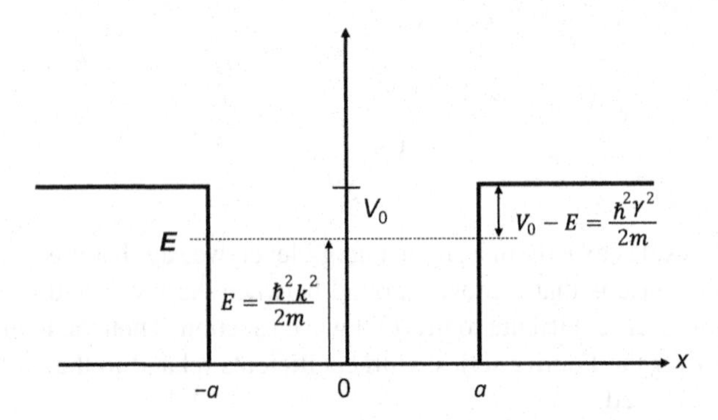

FIGURE 2.4 The various wave vectors are related to the energy of the wave for the case of $E < V_0$.

is the magnitude of the potential barrier and is a positive quantity. Similarly, in the range $|x| > a$, the Schrödinger equation becomes

$$\frac{d^2\psi}{dx^2} - \gamma^2\psi = 0\,, \quad \gamma^2 = \frac{2m(V_0 - E)}{\hbar^2}\,. \tag{2.46}$$

We saw at the end of the last section that with the potential being a symmetric quantity, the solutions for the Schrödinger equation would have either even or odd symmetry. The basic properties of the last section will carry over to the present case, and we expect the solutions in the well region to be either sines or cosines. Of course, these solutions have the desired symmetry properties, and will allow us to solve for the allowed energy levels somewhat more simply.

Thus, we can treat the even and odd solutions separately. In either case, the solutions of (2.46) for the damped region will be of the form $Ce^{-\gamma|x|}$, $|x| > a$. We can match this to the proper sine or cosine function. However, in the normal case, both the wave function and its derivative are matched at each boundary. If we attempt to do the same here, this will provide four equations. However, there are only two unknowns—the amplitude of C relative to that of either the sine or cosine wave and the allowed values of the wave vector k (and hence γ, since it is not independent of k) for the bound-state energy levels. We can get around this problem in one fashion, and that is to make the ratio of the derivative to the wave function itself continuous. That is, we make the logarithmic derivative ψ'/ψ continuous. (This is obviously called the logarithmic derivative since it is the derivative of the logarithm of ψ.) Of course, if we choose the solutions to have even or odd symmetry, the boundary condition at $-a$ is redundant, as it is the same as that at a by these symmetry relations.

To begin, let us consider the even-symmetry wave function (the cosine states) for which the logarithmic derivative at $x = a$ is given by

$$\frac{-k\sin(ka)}{\cos(ka)} = -k\tan(ka) = -\gamma. \tag{2.47}$$

On the far right-hand side, the logarithmic derivative of the damped function arises because of the magnitude in the argument of the exponent. We note that we can match the boundary condition at either a or $-a$, and the result is the same, a fact that gives rise to the even function that we are using. This transcendental equation now determines the allowed values of the energy for the bound states. If we define a new, reduced variable $\xi = ka$, then this equation becomes

$$tan(\xi) = \frac{\gamma}{k} = \sqrt{\left(\frac{\beta}{\xi}\right)^2 - 1}, \quad \beta^2 = \frac{2mV_0 a^2}{\hbar^2}. \tag{2.48}$$

The right-hand side of the transcendental equation is a decreasing function, and it is only those values for which the energy lies in the range (0, V_0) that constitute bound states. In general, the solution must be found graphically. This is shown in Figure 2.5, in which we plot the left-hand side of (2.48) and the right-hand side separately. The crossings (circled) are allowed energy levels.

As the potential amplitude is made smaller, or as the well width is made smaller, the value of β is reduced, and there is a smaller range of ξ that can be accommodated before the argument of the square root becomes negative. Variations in the width affect both parameters, so we should prefer to think of variations in the amplitude, which affects only β. We note, however, that the right-hand side varies from infinity (for $\xi = 0$) to zero (for $\xi = \beta$), regardless of the value of the potential. A similar

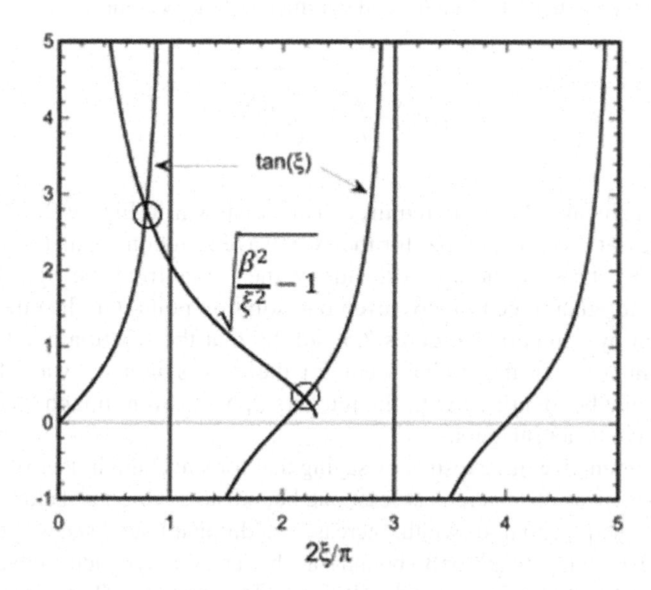

FIGURE 2.5 The graphical solution of (2.65) is indicated by the circled crossings. Here, we have used the values of a = 5 nm, V_0 = 0.3 eV, and $m = 0.067m_0$, appropriate to a GaAs quantum well between two layers of GaAlAs. The two circled crossings indicate that there are two even-symmetry solutions.

variation, in inverse range, occurs for the tangent function (that is, the tangent function goes to zero for $\xi = 0$ or $n\pi$, and the tangent diverges for ξ taking on odd values of $\pi/2$). Thus, there is always at least one crossing. However, there may only be the one. As the potential amplitude is reduced, the intercept β of the decreasing curve in Figure 2.5 moves toward the origin. Thus, the solution point approaches $\xi = 0$, or $k = 0$. By expanding the tangent function for small ξ, it is found that the solution is approximately $\beta\xi$. However, this requires $E \simeq V_0$, which means that the energy level is just at the top of the well. Thus, there is at least one crossing of the curves for $\xi < \pi/2$. For larger values of the amplitude of the potential, the zero point (β) moves to the right and more allowed energy levels appear for the even functions. It is clear from the construction of Figure 2.5, that at least one solution must occur, even if the width is the parameter made smaller, as the ξ-axis intersection cannot be reduced to a point where it does not cross the $\tan(\xi)$ axis at least once. The various allowed energy levels may be identified with the integers 1,3, 5… , just as is the case for the infinite well (it is a peculiarity that the even-symmetry wave functions have the odd integers) although the levels do not involve exact integers any more.

Let us now turn to the odd-symmetry wave functions in (2.42). Again, the logarithmic derivative of the propagating waves for $|x| < a$ may be found to be

$$\frac{k\cos(ka)}{\sin(ka)} = k\cot(ka) = -\gamma. \tag{2.49}$$

Using the same change of variables and parameters, this becomes

$$\cot(ka) - \frac{\gamma}{k} = -\sqrt{\left(\frac{\beta}{\xi}\right)^2 - 1}. \tag{2.50}$$

Again, a graphical solution is required. This is shown in Figure 2.6. The difference between this case and that for the even wave functions is that the left-hand side of (2.68) starts on the opposite side of the ξ-axis from the right-hand side and we are not guaranteed to have even one solution point. On the other hand, it may be seen by comparing Figures 2.5 and 2.6 that the solution point that does occur lies in between those that occur for the even-symmetry wave functions. Thus, this may be identified with the integers 2, 4, … even though the solutions do not involve exact integers.

We can summarize these results by saying that for small amplitudes of the potential, or for small widths, there is at least one bound state lying just below the top of the well. As the potential, or width, increases, additional bound states become possible. The first (and, perhaps, only) bound state has an even-symmetry wave function. The next level that becomes bound will have odd symmetry. Then a second even-symmetry wave function will be allowed, then an odd-symmetry one, and so on. In the limit of an infinite potential well, there are an infinite number of bound states whose energies are given by (2.40).

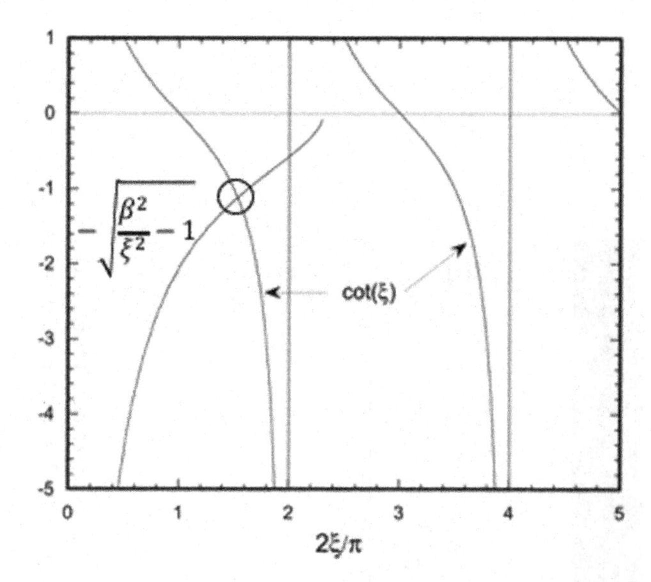

FIGURE 2.6 The graphical solution of (2.68). The parameters here are the same as those used in Figure 2.5. Only a single solution (circled) is found for the antisymmetric solution, and this energy lies between the two found in Figure 2.5. The parameters used are the same as for Figure 2.5.

Once the energy levels are determined for the finite potential well, the wave functions can be evaluated. We know the form of these functions, and the energy levels ensure the continuity of the logarithmic derivative, so we can generally easily match the partial wave functions in the well and in the barriers. One point that is obvious from the preceding discussion is that the energy levels lie below those of the infinite well. This is because the wave function penetrates into the barriers, which allows for example a sine function to *spread out* more, which means that the momentum wave vector k is slightly smaller, and hence corresponds to a lower energy level. Thus, the sinusoidal function does not vanish at the interface for the finite-barrier case, and in fact couples to the decaying exponential within the barrier. The typical sinusoid then adopts long exponential tails if the barrier is not infinite.

Now, all this plotting and graphing to find the solutions is rather slow given the technology available today. In fact, a site known as NanoHub.org has brought together a large set of tools for electronic science and given them all a common graphical user interface (accounts are free). We can examine the solutions here with the "bound-state laboratory" available in the AQME set of tools for quantum mechanics (Klimeck et al. 2019). In Figure 2.7, part of the entry panel is shown for this tool along with the paramters that are entered for the simulation. These parameters are those for GaAs and are listed in the caption of Figure 2.5 (a 10 nm width well has $a = 5$ nm). In Figure 2.8, the solutions for the energy levels are shown overlaid on the potential well itself. The three energy levels lie at 34.2 meV, 132 meV, and 271 meV, all measured from the bottom of the well, corresponding to Figure 2.4. In Figure 2.9,

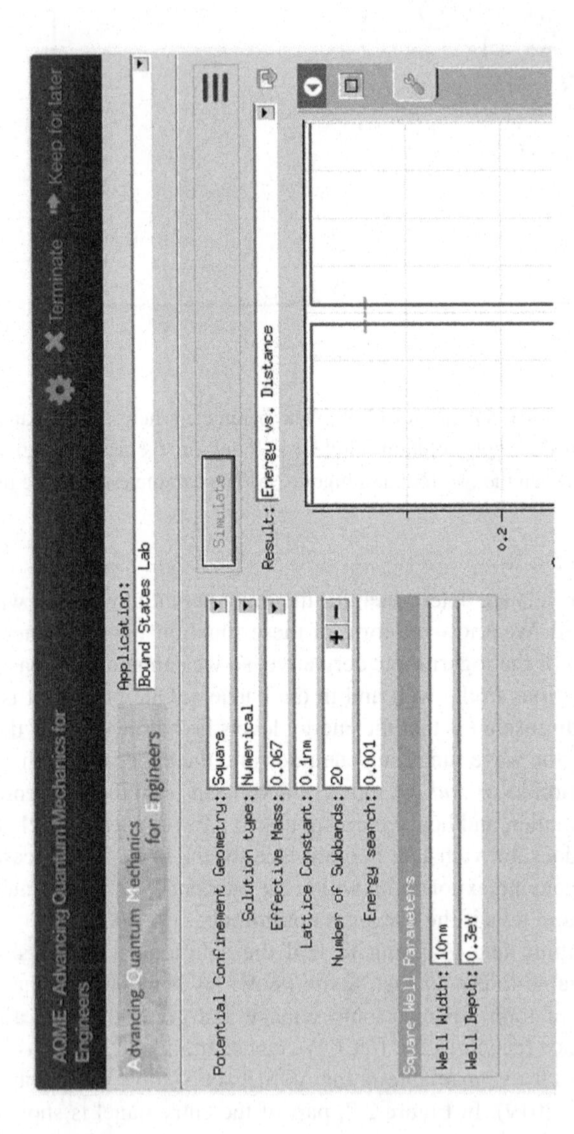

FIGURE 2.7 The front panel of AQME, showing that the "bound states lab" has been selected. The parameters from Figure 2.5 have been added at the appropriate spots. The results are partially obscured but are shown in Figure 2.8.

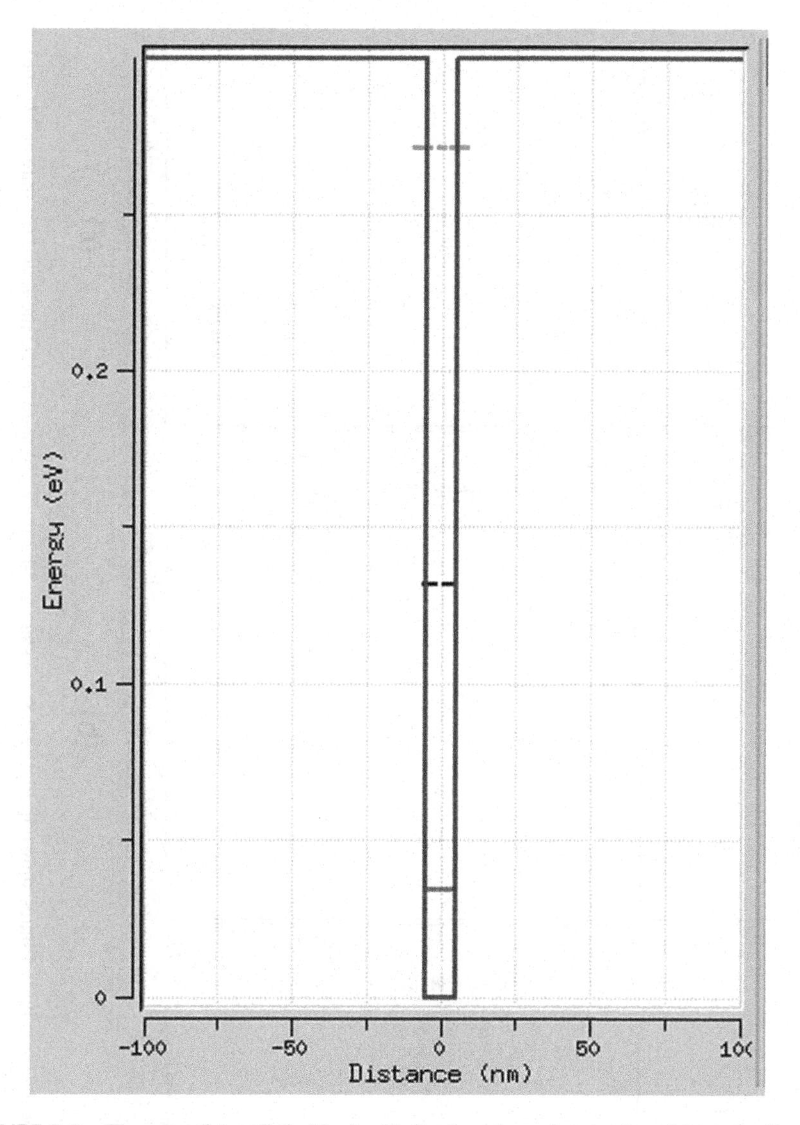

FIGURE 2.8 The potential well (in blue) with the three bound states found from the "bound states lab" are shown.

the wave functions for these three energy levels are plotted in panels (a), (b), and (c) respectively, and the clear even-odd-even parity can be seen.

Case II. $E > V_0$

Let us now turn our attention to the completely propagating waves that exist for energies above the potential well. It might be thought that these waves will show no effect of the quantum well, but this is not the case. Each interface is equivalent to a dielectric interface in electromagnetics, and the thin layer is equivalent to a thin dielectric

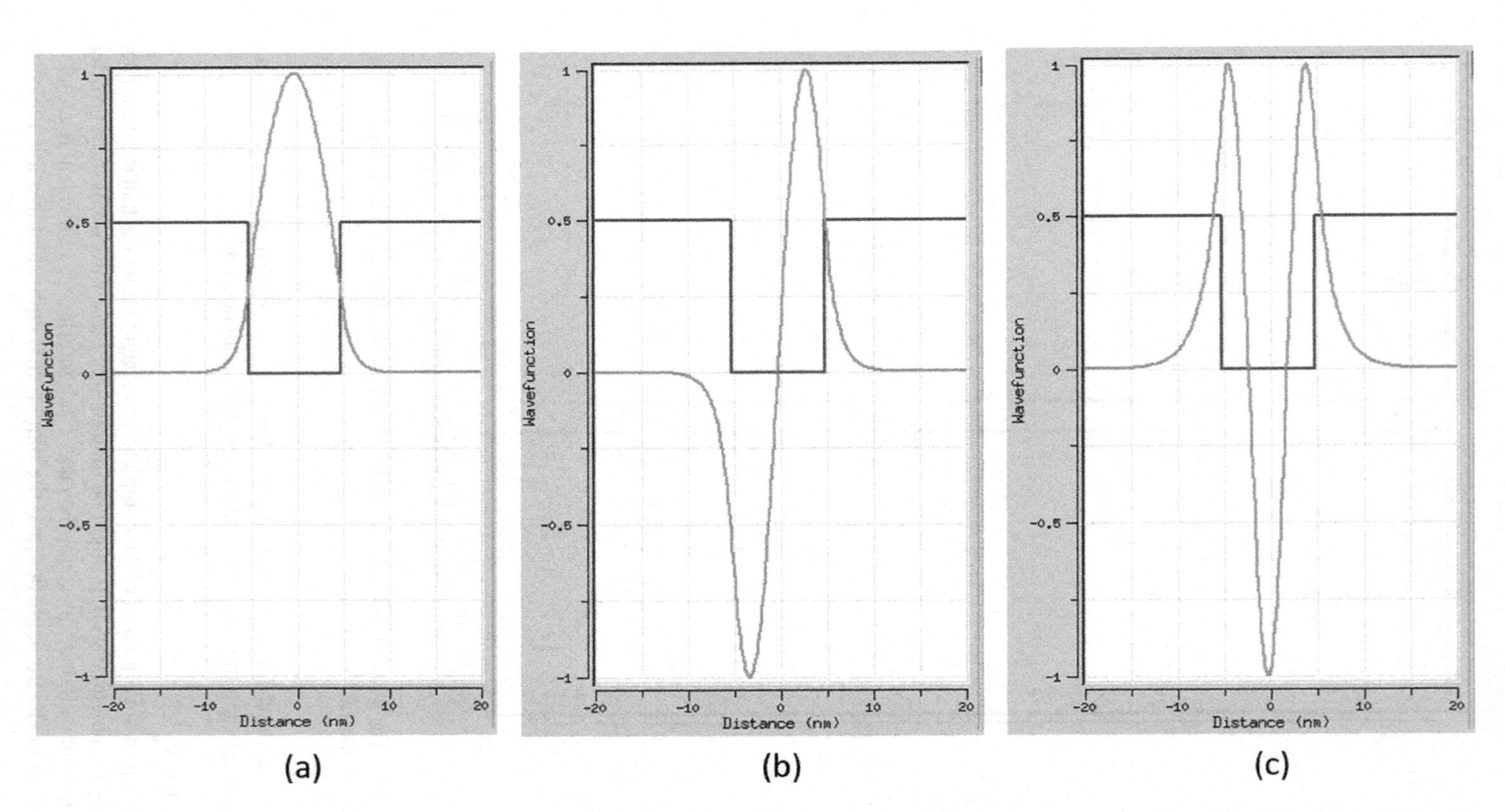

FIGURE 2.9 The wave functions for the three bound states shown in Figure 2.8: (a) state at 34.2 meV, (b) state at 132 meV, and (c) state at 271 meV.

layer in which interference phenomena can occur. The same is expected to occur here. We will make calculations for these phenomena by calculating the transmission coefficient for waves propagating from the left (negative x) to the right (positive x).

Throughout the entire space, the Schrödinger equation is given by the form (2.5), with different values of k in the various regions. The value of k given in (2.6) remains valid in the quantum well region, while for $|x| > a$ (see Figure 2.10)

$$k_0^2 = \frac{2m(E-V_0)}{\hbar^2}. \tag{2.51}$$

For $x > a$, we assume that the wave function propagates only in the outgoing direction, and is given by

$$\psi_{x>a} = Fe^{ik_0x}. \tag{2.52}$$

In the quantum well region, we need to have waves going in both directions, so the wave function is assumed to be the same as in the previous case, and is given by (2.5) and (2.6), but with coefficients C and D, rather than A and B. Similarly, in the incoming region, we need a reflected wave, which we take as (we assume A = 1, but is set by normalization)

$$e^{ik_0x} + Be^{-ik_0x}. \tag{2.53}$$

Now, let us define two reduced variables as

$$\begin{aligned} \omega &= e^{ika} \\ \omega_0 &= e^{ik_0a} \end{aligned}. \tag{2.54}$$

While this simplifies the solution to the four resulting equations for matching the wave function and the derivatives at the two boundaries, the result for F is independent of these two parameters, and

$$F = \frac{e^{-2ik_0a}}{\cos(2ka) - i\dfrac{k^2+k_0^2}{2kk_0}\sin(2ka)}, \tag{2.55}$$

which is symmetrical in the momenta wave numbers. Hence, the transmission is

$$T = |F|^2 = \frac{1}{1+\left[\dfrac{k^2+k_0^2}{2kk_0}\right]^2 sin^2(2ka)}. \tag{2.56}$$

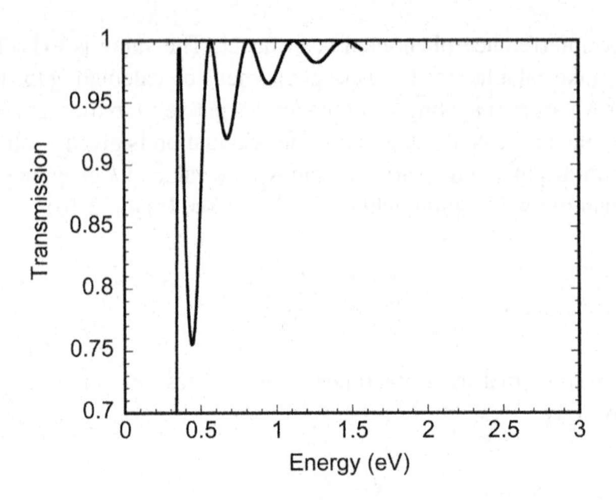

FIGURE 2.10 The transmission, given by (2.56), for a finite potential well. The parameters are those of Figure 2.5, appropriate to a GaAs quantum well situated between two GaAlAs layers.

Using (2.35), we can now write the reflection coefficient as

$$R = 1 - T = 1 - \left| \frac{1}{1 + \left[\frac{k^2 - k_0^2}{2kk_0} \right]^2 sin^2(2ka)} \right|^2 . \tag{2.57}$$

There are resonances, which occur when $2ka$ is equal to odd multiples of $\pi/2$, and for which the transmission is a minimum. The transmission rises to unity when $2ka$ is equal to even multiples of $\pi/2$, or just equal to $n\pi$. The reduction in transmission depends upon the amplitude of the potential well, and hence on the difference between k and k_0. We note that the transmission has minima that drop to small values only if the well is infinitely deep (and the energy of the wave is not infinite; i.e., $k_0 \gg k$). A deeper potential well causes a greater discontinuity in the wave vector, and this leads to a larger modulation of the transmission coefficient. An example is shown in Figure 2.10.

It seems strange that a wave function that lies above the quantum well should not be perfectly transmitting. It is simple enough to explain this via the idea of 'dielectric discontinuity', but is this really telling the whole truth of the physics? Yes and no. It explains the physics with the mathematics, but it does not convey the understanding of what is happening. In fact, it is perhaps easier to think about the incident wave as a particle. When it impinges upon the region where the potential well exists, it cannot be trapped there, as its energy lies above the top of the well. However, the well potential can *scatter* the particle as it arrives. In the present case, the particle is scattered back into the direction from which it came with a probability given by the reflection coefficient. It proceeds in an unscattered state with a probability given by the transmission coefficient. In general, the scattering probability is nonzero, but at special values of the incident energy, the scattering vanishes. In this case, the particle is transmitted with unit probability. This type of potential scattering is quite special, because only the direction of the momentum (this is a one-dimensional problem) is changed, and the energy of the particle remains unchanged. This type of scattering is termed *elastic*, as in elastic scattering of a billiard ball from a 'cushion' in three dimensions. We will see other examples of this in the following chapters.

2.4 THE TRIANGULAR WELL

Another type of potential well is quite common in everyday semiconductor devices, such as the common metal–oxide–semiconductor (MOS) transistor (Figure 2.11(a)). The latter is the workhorse in nearly all microprocessors and computers today, yet the presence of quantization has not really been highlighted in the operation of these devices. These devices depend upon capacitive control of the charge at the interface between the oxide and the semiconductor. If we consider a parallel-plate capacitor made of a metal plate, with an insulator made of silicon dioxide, and a second plate composed of the semiconductor silicon, we essentially have the MOS transistor. Voltage applied across the capacitor varies the amount of charge accumulated in the metal and in the semiconductor, in both cases at the interface with the insulator. On the semiconductor side, contacts (made of n-type regions embedded in a normally p-type material) allow one to pass current through the channel in which the charge resides in the semiconductor. Variation of the current, through variation of the charge via the capacitor voltage, is the heart of the transistor operation.

Consider the case in which the semiconductor is p-type, and hence the surface is in an 'inverted' condition (more electrons than holes) and mobile electrons are drawn to the interface by a positive voltage on the metal plate (the channel region is isolated from the bulk of the semiconductor by the inversion process). The surface charge in the semiconductor is composed of two parts: (i) the surface electrons, and (ii) ionized acceptors from which the holes have been pushed into the interior of the semiconductor. In both cases the charge that results is negative and serves to balance the positive charge on the metal gate. The electron charge is localized right at the interface with the insulator, while the ionized acceptor charge is distributed over a large region. In fact, it is the localized electron charge that is mobile in the direction along the interface, and that is quantized in the resulting potential well. The field in the oxide is then given by the total surface charge through Gauss's law (we use the approximation of an infinite two-dimensional plane) as

$$E_s = \frac{e}{\varepsilon_{ox}}\left(N_a W + n_s\right), \tag{2.58}$$

where W is the thickness of the layer of ionized acceptors N_a (normal to the surface), the surface electron density n_s is assumed to be a two-dimensional sheet charge, and

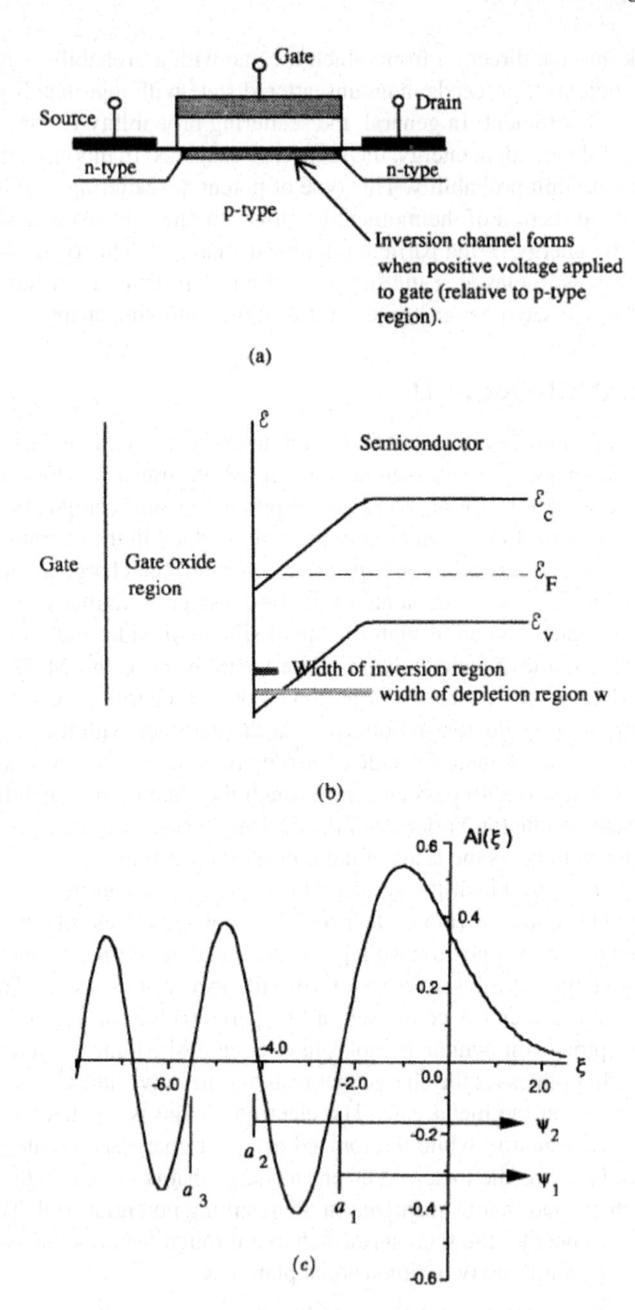

FIGURE 2.11 (*a*) A MOS field-effect transistor, (*b*) the triangular potential, and (*c*) the Airy function and the use of the zeros to match the boundary conditions.

the permittivity is that of the oxide. On the semiconductor side of the interface, the normal component of D is continuous, which means that E in (2.58) is discontinuous by the dielectric constant ratio. Thus, just inside the interface, (2.58) represents the field if the oxide permittivity is replaced by that of the semiconductor. However, just a short distance further into the semiconductor, the field drops by the amount produced by the surface electron density. Thus, the average field in the semiconductor, in the region where the electrons are located, is approximately

$$E_s = \frac{e}{\varepsilon_{ox}}\left(N_a W + \frac{n_s}{2}\right). \tag{2.59}$$

In this approximation, a constant electric field in this region gives rise to a linear potential in the Schrödinger equation (Figure 2.11(*b*)). We want to solve for just the region inside the semiconductor, near to the oxide interface. Here, we can write the Schrödinger equation in the form

$$-\frac{\hbar^2}{2m}\frac{\partial^2}{\partial x^2}\psi + eE_s x\psi = E\psi, \quad x > 0. \tag{2.60}$$

We assume that the potential barrier at the interface is infinitely high, so no electrons can get into the oxide, which leads to the boundary condition that $\psi(0) = 0$. The other boundary condition is merely that the wave function must remain finite, which means that it also tends to zero at large values of x.

While the previous discussion has been for the case of a MOS transistor, such as found in silicon integrated circuits, quite similar behavior arises in the GaAs–AlGaAs heterojunction high-electron-mobility transistor (HEMT). In this case, the AlGaAs plays a role similar to the oxide in the MOS transistor, with the exception that the dopant atoms are placed in this layer. The dopants near the interface are ionized, with the electrons falling into a potential well on the GaAs side of the interface, a process that is facilitated by the barrier created by the difference in the two conduction band edges, as shown in Figure 2.12. There are very few ionized dopants in the GaAs, although the interface electric field still satisfies (2.58). This field is created by the positively charged, ionized donors and the negatively charged electrons in the potential well. By placing a metal gate on the surface of the AlGaAs layer, bias applied to the gate can affect the density of the carriers in the quantum well, reducing them to zero. This is a *depletion* device, as opposed to the inversion device of the MOS transistor. That is, the gate *removes* carriers from the channel in the HEMT while the gate pulls carriers into the channel in the MOS transistor. Since the impurities are removed from the region where the carriers reside, the mobility is higher in the HEMT, hence its name. Such devices have found extensive use as analoge microwave amplifiers, either as power amplifiers or as low-noise amplifiers in receivers. Nevertheless, the potential variation near the interface still appears as approximately a triangular potential well, just as in the MOS transistor.

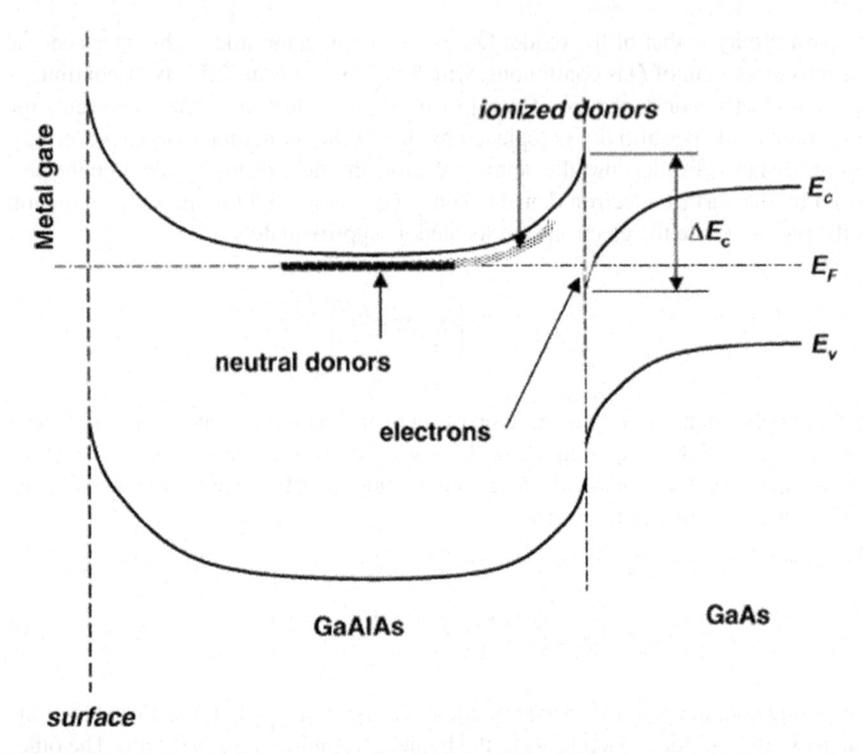

FIGURE 2.12 Band alignment for the AlGaAs–GaAs high-electron-mobility transistor. Ionized donors in the GaAlAs produce the electrons that reside in the triangular quantum well on the GaAs side of the interface.

Thus, the goal is to solve (2.60). To simplify the approach, we will make a change of variables in (2.60), which will put the equation into a standard form. One may recognize (2.60) as the Airy equation, which is one standard form in the Sturm-Liouville problem, and it guides our choice for the change of variables. For this, we redefine the position and energy variables as

$$z = \left(\frac{2meE_s}{\hbar^2}\right)^{1/3} x, \quad z_0 = \frac{2mE}{\hbar^2 z^2}. \tag{2.61}$$

We also introduce the shift $\xi = z - z_0$, so (2.60) becomes

$$\frac{\partial^2}{\partial \xi^2}\psi - \xi\psi = 0. \tag{2.62}$$

This is the final reduced Airy equation.

Airy functions are combinations of Bessel functions and modified Bessel functions. It is not important here to discuss their properties in excruciating detail. The important facts for us are that: (i) the Airy function $Ai(-\xi)$ decays as an exponential for positive ξ; and (ii) $Ai(\xi)$ behaves as a damped sinusoid with a period that also varies as ξ, as shown in Figure 2.11(*c*). For our purposes, this is all we need. The second solution of (2.62), the Airy functions $Bi(\xi)$, diverges in each direction and must be discarded in order to keep the probability function finite. The problem is in meeting the desired boundary conditions. The requirement that the wave function decay for large x is easy. This converts readily into the requirement that the wave function decay for large ξ, which is the case for $Ai(-\xi)$. However, the requirement that the wave function vanish at $x = 0$ is not arbitrarily satisfied for the Airy functions. On the other hand, the Airy functions are oscillatory. In the simple quantum well of the last two sections, we noted that the lowest bound state had a single peak in the wave function, while the second state had two, and so on. This suggests that we associate the vanishing of the wave function at $x = 0$ with the intrinsic zeros of the Airy function, which we will call a_s. Thus, choosing a wave function that puts the first zero a_1 at the point $x = 0$ would fit all the boundary conditions for the lowest energy level (Figure 2.11(*c*)). Similarly, putting the second zero a_2 at $x = 0$ fits the boundary conditions for the next level, corresponding to $n = 2$. By this technique, we build the set of wave functions, and also the energy levels, for the bound states in the wells.

We will treat only the lowest bound state as an example. For this, we require the first zero of the Airy function. Because the numerical evaluation of the Airy functions yields a complicated series, we cannot give exact values for the zeros. However, they are given approximately by the relation (Abramowitz and Stegun 1964)

$$a_s = -\left[\frac{3\pi(4s-1)}{8}\right]^{2/3}. \tag{2.63}$$

Thus, the first zero appears at approximately $-(9\pi/8)^{2/3}$. Now, this may be related to the required boundary condition at $x = z = 0$ through

$$\xi = -\left(\frac{9\pi}{8}\right)^{\frac{2}{3}} = -z_0 = \frac{2mE}{\hbar^2}\left(\frac{\hbar^2}{2mE_s}\right)^{2/3}, \tag{2.64}$$

or

$$E_1 = \frac{\hbar^2}{2m}\left(\frac{9\pi meE_s}{4\hbar^2}\right)^{\frac{2}{3}}, \tag{2.65}$$

remembering, of course, that this is an approximate value since we have only an approximate value for the zero of the Airy function.

In fact, we can return to NanoHub.org and once again use our AQME tool to find the solutions. In the "bound states lab," we are able to select a *triangular well*. Then, we select the Si heavy mass ($0.91m_0$) and a field of 0.33 MV/cm. (If we consider that we have 1 V on the gate of the MOS, and an oxide thickness of 10 nm, we find that the oxide field is 1 MV/cm.) The field in the semiconductor is smaller by the ratio of the permitivities.) Now, the bound-state lab gives the lowest energy level as 83.43 meV. In comparison, the lowest energy level in the Airy function approach, equation (2.65), gives 86.4 meV. The second and third levels are at 147 meV and 198 meV, respectively. The energy levels for this potential as determined by this tool are plotted in Figure 2.13. In Figure 2.14, the magnitude squared of the first three wave functions are plotted.

2.5 UNCERTAINTY

The description of the momentum operator in the position representation is that of a differential operator, as we discussed in Section 1.5.1. In dealing with this in the last chapter, we found that the position and momentum do not cummute, but satisfy a relation

$$\left[x, p\right] = xp - px \rightarrow i\hbar, \tag{2.66}$$

where, of course, this relationship yields the result on the right-hand side *when it operates upon a wave function*. If variables, or operators, do not commute, there is an implication that these quantities cannot be measured simultaneously. Here again, there is another and deeper meaning. In the previous chapter, we noted that the operation of the position operator x on the wave function in the position representation produced an expectation value $\langle x \rangle$. The momentum operator does not produce this simple result with the wave function of the position representation. Rather, the differential operator produces a more complex result. For example, if the differential operator were to produce a simple eigenvalue, then the wave function would be constrained to be of the form $\exp(ipx/\hbar)$ (which can be shown by assuming a simple eigenvalue form with the differential operator and solving the resulting equation). This form is not integrable (it does not fit our requirements on normalization), and thus the same wave function cannot simultaneously yield eigenvalues for both position and momentum. Since the eigenvalue relates to the expectation value, which corresponds to the most likely result of an experiment, these two quantities cannot be simultaneously measured. Thus, we will find that this leads to an uncertainty in the values of the two noncommuting operators. If the position is known, for example if we choose a delta function in space, then the Fourier transform has unit amplitude everywhere; that is, the momentum has equal probability of taking on any value. Another way of looking at this is to say that since the position of the particle is completely determined, it is impossible to say anything about the momentum, as

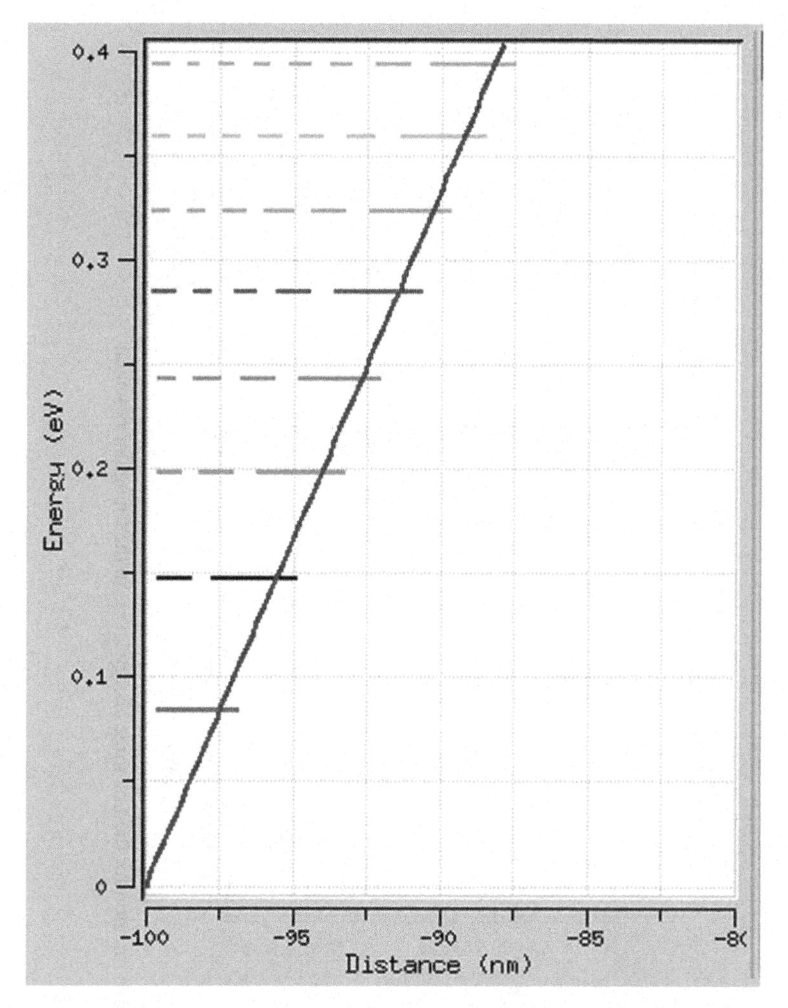

FIGURE 2.13 The triangular potential well and the ladder of bound states determined from the bound-state tool at Nanohub.org.

any value of the momentum is equally likely. Similarly, if a delta function is used to describe the momentum wave function, which implies that we know the value of the momentum exactly, then the position wave function has equal amplitude everywhere. This means that if the momentum is known, then it is impossible to say anything about the position, as all values of the latter are equally likely. As a consequence, if we want to describe both of these properties of the particle, the position wave function and its Fourier transform must be selected carefully to allow this to occur. Then there will be an uncertainty Δx in position, and there will be a corresponding uncertainty Δp in momentum. Heisenberg (1927) give an explicit value for the uncertainty in any two noncommuting operators in the form of

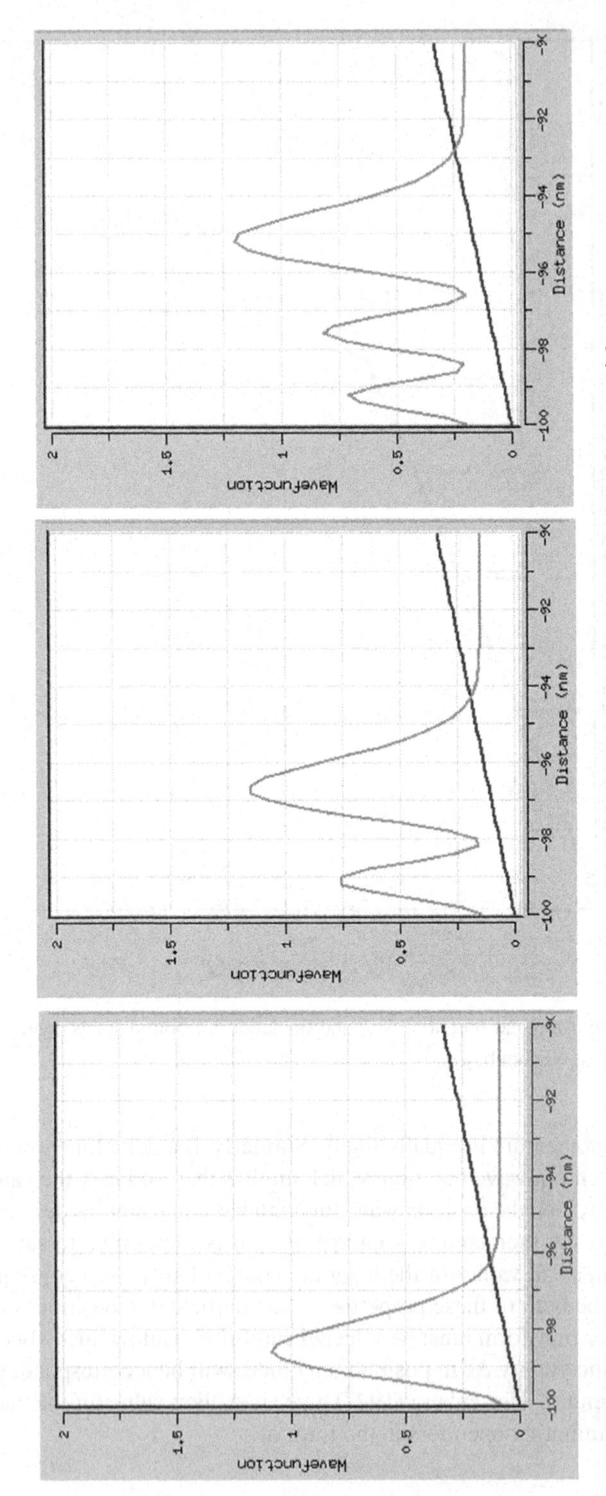

FIGURE 2.14 The squared magnitude of the wave function for the first three energy levels in the triangular potential.

$$\Delta x \Delta p \geq \frac{\hbar}{2}. \tag{2.67}$$

Suppose we take the Gaussian wave function shown in Figure 1.4. We can use this example wave function to illustrate the uncertainty. Let us write this wave function in the form

$$\psi(x) = \frac{1}{(2\pi)^{1/4}\sqrt{\sigma}} exp\left(-\frac{x^2}{4\sigma^2}\right), \tag{2.68}$$

where the prefactor assures the normalization of the probability and σ is the standard deviation of the probability. This wave function has its peak at $x = 0$, and symmetrical about this point. This assures us of

$$\langle x \rangle = \frac{1}{\sqrt{2\pi\sigma}} \int_{-\infty}^{\infty} exp\left(-\frac{x^2}{2\sigma^2}\right) x dx = 0. \tag{2.69}$$

Now, Heisenberg tells us that Δx is the rms values of various expectations, as

$$\Delta x = \sqrt{\langle x^2 \rangle - \langle x \rangle^2}. \tag{2.70}$$

Hence, we compute the expectation value of the square of x as

$$\langle x^2 \rangle = \frac{1}{\sqrt{2\pi\sigma}} \int_{-\infty}^{\infty} exp\left(-\frac{x^2}{2\sigma^2}\right) x^2 dx = \sigma^2. \tag{2.71}$$

This latter result is a well known property of a Gaussian, and gives us the meaning of the standard deviation. Now, we find that

$$\Delta x = \sqrt{\langle x^2 \rangle - \langle x \rangle^2} = \sqrt{\sigma^2 - 0} = \sigma. \tag{2.72}$$

The appropriate momentum wave function can now be found by Fourier transforming this position wave function. Using (1.25), this gives

$$\phi(k) = \left(\frac{2}{\pi}\right)^{1/4} \sqrt{\sigma} exp\left(-\sigma^2 k^2\right). \tag{2.73}$$

From this, we can determine Δp from

$$\langle \mathrm{p} \rangle = \left(\frac{2}{\pi}\right)^{1/2} \sigma \int_{-\infty}^{\infty} exp\left(-2\sigma^2 k^2\right) \hbar k dk = 0 \tag{2.74}$$

and

$$\langle \mathrm{p}^2 \rangle = \left(\frac{2}{\pi}\right)^{1/2} \sigma \int_{-\infty}^{\infty} exp\left(-2\sigma^2 k^2\right) \hbar^2 k^2 dk = \frac{\hbar^2}{4\sigma^2}. \tag{2.75}$$

Then, the uncertainty in momentum is given by

$$\Delta p = \sqrt{\langle \mathrm{p}^2 \rangle - \langle \mathrm{p} \rangle^2} = \sqrt{\frac{\hbar^2}{4\sigma^2} - 0} = \frac{\hbar}{2\sigma}. \tag{2.76}$$

We now find the uncertainty to be

$$\Delta x \Delta p = \sigma \frac{\hbar}{2\sigma} = \frac{\hbar}{2}. \tag{2.77}$$

Here we have achieved the minimum uncertainty expected by Heisenberg. Thus, the Gaussian wave packet is ofter referred to as the minimum uncertainty wave packet.

Let us now examine what this says about our infinite square well wave functions (2.42). Using these wave functions, we can examine the various quantities we need. Let us begin with the even-symmetry wave functions, as

$$\begin{aligned}
\langle x \rangle &= \frac{1}{a}\int_{-a}^{a} cos^2\left(\frac{n\pi x}{a}\right) x dx = \frac{1}{2a}\int_{-a}^{a}\left[1 + cos\left(\frac{2n\pi x}{a}\right)\right] x dx = 0 \\
\langle x^2 \rangle &= \frac{1}{a}\int_{-a}^{a} cos^2\left(\frac{n\pi x}{a}\right) x^2 dx = \frac{1}{2a}\int_{-a}^{a}\left[1 + cos\left(\frac{2n\pi x}{a}\right)\right] x^2 dx = \frac{a^2}{3} - \frac{a^2}{2n^2\pi^2} \\
\langle p \rangle &= \frac{-i\hbar}{a}\int_{-a}^{a} cos\left(\frac{n\pi x}{a}\right)\frac{n\pi}{a} sin\left(\frac{n\pi x}{a}\right) dx = 0 \\
\langle p^2 \rangle &= -\frac{\hbar^2}{2a}\left(\frac{n\pi}{a}\right)^2 \int_{-a}^{a} cos^2\left(\frac{n\pi x}{a}\right) dx = -\frac{1}{2}\left(\frac{\hbar n\pi}{a}\right)^2
\end{aligned} \tag{2.78}$$

We note here that the uncertainty in momentum is imaginary due to the negative sign on the last factor. This is because each wave function has a very precise energy level and therefore a very precise momentum. We do note, however, that as we move up in the ladder of allowed energy states, the spreading of these leads to an increase in the

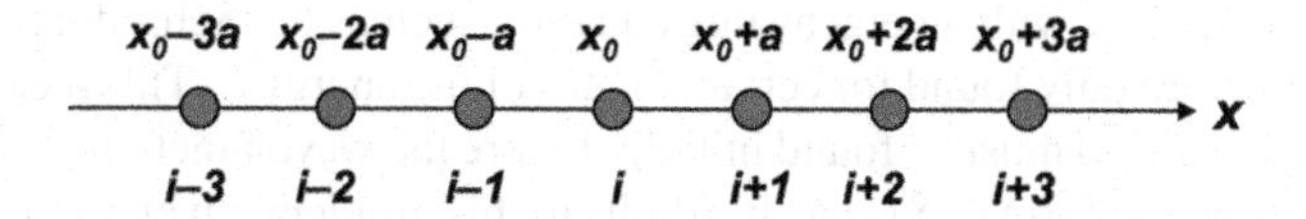

FIGURE 2.15 A one-dimensional, equidistant mesh for the evaluation of the Schrödinger equation by numerical methods

uncertainly in position. But the conclusion is that this is not a good system in which to discuss the Heisenberg uncertainty!

2.6 NUMERICAL SOLUTIONS OF THE SCHRÖDINGER EQUATION

The numerical solutions that we dealt with in the preceeding sections were largely done by programs available at NanoHub.org. To understand the limitations of these programs, and to gain appreciation for the assumptions and limitations inherent in their use, we want to look into the task of simulating the Schrödinger equation itself. If some arbitrary function is known at a series of equally spaced points, one can use a Taylor series to expand the function about these points and evaluate it in the regions between the known points. Consider the set of points shown in Figure 2.15 for example. Here, each point is separated from its neighbors by the distance a (in position). This allows us to develop a finite-difference scheme for the numerical evaluation of the Schrödinger equation. If we first expand the function in a Taylor series about the points on either side of x_0, we get

$$\begin{aligned} f(x_0+a) &= f(x_0)+a\left.\frac{\partial f}{\partial x}\right|_{x=x_0}+\frac{a^2}{2}\left.\frac{\partial^2 f}{\partial x^2}\right|_{x=x_0}+\dots \\ f(x_0-a) &= f(x_0)-a\left.\frac{\partial f}{\partial x}\right|_{x=x_0}+\frac{a^2}{2}\left.\frac{\partial^2 f}{\partial x^2}\right|_{x=x_0}+\dots \end{aligned} \quad (2.79)$$

By adding the two equations together, we arrive at

$$\left.\frac{\partial^2 f}{\partial x^2}\right|_{x=x_0} \approx \frac{1}{a^2}\left[f(x_0+a)+f(x_0-a)-2f(x_0)\right]. \quad (2.80)$$

Let us now introduce a short-hand notation. We can arbitrarily set $x_0 = x_i$, and then denote $f(x_0)$ as f_i. Then, we can write the time-independent Schrödinger equation as

$$-\frac{\hbar^2}{2ma^2}\left(\psi_{i+1}+\psi_{i-1}-2\psi_i\right)+V_i\psi_i = E\psi_i. \quad (2.81)$$

Of course, the trouble with an eigenvalue equation such as the Schrödinger equation is that solutions are only found for certain values of the energy E. This means that the energies (eigenvalues) must be found initially before the wave functions can be determined. However, the form (2.81) easily admits to this problem, given a reasonable set of computational routines on a modern computer. Equation (2.81) can be rewritten as a set of equations, one for each value of i. This then gives us a matrix equation that can be written as

$$[S][\psi]=0. \tag{2.82}$$

Here $[\psi]$ is a column matrix. If there are N nodes (see Figure 2.15), then the size of this column matrix is $N \times 1$; for example, a single column with N rows. The boundary conditions are imposed on the zeroth node (the supposed value of the node just below the first node) and the N+1 node. The matrix $[S]$ is now a tridiagonal matrix of size $N \times N$. The diagonal elements, and first off-diagonal elements, are given by

$$S_{ii} = \frac{\hbar^2}{ma^2} + V_i - E, \quad S_{i,i+1} = S_{i,i-1} = -\frac{\hbar^2}{2ma^2}. \tag{2.83}$$

The last term on the right is often called the "hopping energy" as it connects two adjacent sites on the grid. Now, immediately from (2.82), we know that the S matrix is singular; that is, its determinant must be zero

$$det[S]=0. \tag{2.84}$$

Expanding the determinant gives an Nth order equation whose solutions are the N values of the energy E. This latter equation is known as the eigenvalue equation and the solutions for the energy are the eigen values. For N nodes, we can find N solutions, each of which are representative of the various combinations of amplitudes for the nodes. These are the eigen functions.

Let us consider an example of an infinite potential well, whose width is 20 nm (we use free electrons). The exact solutions are given in (2.40) in units of the total width of the well (we set $2a = W$) as

$$E = \frac{n^2\pi^2\hbar^2}{2mW^2}. \tag{2.85}$$

where n is an integer. To simulate the wave functions, we take the discretization variable (in Figure 2.15) a = 1.0 nm, so that N = 20. Since the end points i = 0 and i = 20 both have the wave function zero because of the infinite height of the barrier, so we only need to consider the N – 1 interior grid points and $[S]$ is a 19 × 19 matrix. To compare the computed values, we will give the energies in units of $\pi^2\hbar^2/2mW^2$ so that the actual energy levels are given by n^2. We compare the results with the numerically

TABLE 2.1
Comparison of numerically computed energies with the actual values

n	Energy	Estimate
1	1	0.99795
2	4	3.96721
3	9	8.83468
4	16	15.4805
5	25	23.74103
6	36	33.41287

determined values of the lowest six energies in Table 2.1. It may be seen that while the results are close, the numerical estimates are somewhat below the actual values.

We now turn to computing the wave functions. Since the wave function has only been evaluated on the nodes, using the energy values E_n in the determinant of $[S]$ gives the set of values the node functions take for each energy. Only $N-2$ values can be found for each energy level, with the last value needing to be determined from the normalization condition. It is important to note that the boundary conditions at $i = 0$, and $i = N$ must be imposed at the beginning in establishing (2.84). In Figure 2.16, we plot the first and fourth wave functions for this example, showing the exact solutions, with the numerically computed discrete eigenfunctions. The solid curves are the exact values from (2.42) while the solid points are the numerically computed values at the grid points. It may be seen that these appear to be fairly accurate given the large scale of the figure.

When the problem is time-varying, the numerical approach is more complicated. The Schrödinger equation is a diffusion equation, but with complex coefficients. This becomes more apparent if we rewrite (2.3) as

$$\frac{\partial \psi}{\partial t} = \frac{i\hbar}{2m}\frac{\partial^2 \psi}{\partial x^2} + \frac{1}{i\hbar}V(x,t)\,\psi. \tag{2.86}$$

The quantity $\hbar/2m$ has the units of a diffusion constant (cm^2s^{-1}). Nevertheless, we can use a Taylor series, exactly as we did for (2.79), for the time derivative as

$$f(t+\Delta t) = f(t) + \Delta t \left.\frac{\partial f}{\partial t}\right|_{\Delta t=0} + \ldots. \tag{2.87}$$

We still use the subscript i to denote the position at which the function is evaluated and denote the time evolution by $t = n\Delta t$. Hence, the explicit first-order evaluation of (2.86) is given by

$$\psi_i^{n=1} = \psi_i^n + \frac{i\hbar\Delta t}{2ma^2}\left(\psi_{i+1}^n + \psi_{i-1}^n - 2\psi_i^n\right) - \frac{i\Delta t}{\hbar}V_i\psi_i^n. \tag{2.88}$$

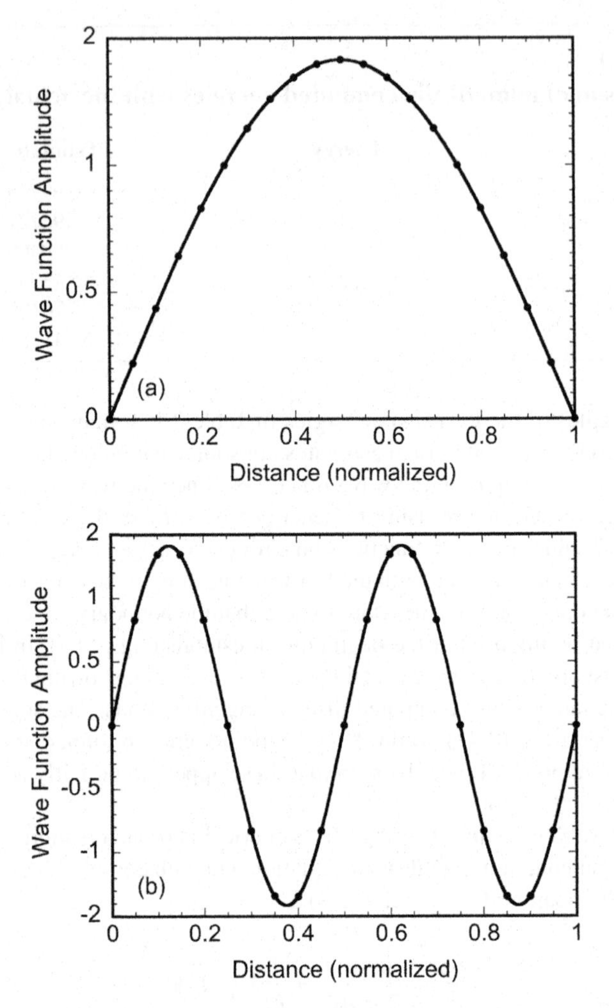

FIGURE 2.16 A comparison of the exact wave function with a numerically computed estimate for an infinite quantum well. In (*a*), ψ_1 is shown, while in (*b*), ψ_4 is plotted.

This is basically the simplest approach and develops the value at grid point i and new time $t + \Delta t$ in terms of the function evaluated at the preceding time step.

There are clearly errors associated with the time step Δt and the space step a. In fact, if these are too big, then an instability develops in the solution. We can check the linear stability by assuming that the wave function is composed of a set of Fourier modes

$$\psi = \tilde{\psi} e^{iqx}, \tag{2.89}$$

for which (2.88) becomes (we ignore the potential for the moment)

$$\tilde{\psi}^{n+1}e^{iqx} = \tilde{\psi}^{n}e^{iqx} + \frac{i\hbar\Delta t}{2ma^2}\left(\tilde{\psi}^{n}e^{iq(x+a)} + \tilde{\psi}^{n}e^{iq(x-a)} - 2\tilde{\psi}^{n}e^{iqx}\right), \tag{2.90}$$

or

$$\tilde{\psi}^{n+1} = \tilde{\psi}^{n}\left[1 - \frac{2i\hbar\Delta t}{ma^2} sin^2\left(\frac{qa}{2}\right)\right]. \tag{2.91}$$

For stability, the term in brackets must be less than 1 for all values of q. This can only occur if

$$-\frac{2i\hbar\Delta t}{ma^2} \geq -2 \tag{2.92}$$

or

$$\Delta t \leq \frac{ma^2}{2\hbar}. \tag{2.93}$$

For example, for the previous example, in which $a = 1$ nm, the time step must be smaller than 4.3×10^{-15} s for a free electron and less than 2.9×10^{-16} s for an electron in GaAs. This is a severe limitation for many applications.

An improvement is to go to a second-order method, due to Crank and Nicholson (1947) (see also Richtmyer and Morton 1967). In this, a two-step approach is utilized in which an estimate for ψ_i^{n+1} is obtained using (2.88). Once this estimate is obtained, an improved value is found by including this in the update of the final value of the wave function via

$$\begin{aligned}\psi_i^{n=1} = \psi_i^n &+ \frac{i\hbar\Delta t}{4ma^2}\left(\psi_{i+1}^n + \psi_{i-1}^n - 2\psi_i^n\right) - \frac{i\Delta t}{2\hbar}V_i\psi_i^n \\ &+ \frac{i\hbar\Delta t}{4ma^2}\left(\psi_{i+1}^{n+1} + \psi_{i-1}^{n+1} - 2\psi_i^{n+1}\right) - \frac{i\Delta t}{2\hbar}V_i\psi_i^{n+1}.\end{aligned} \tag{2.94}$$

In essence, this is a form of a predictor–corrector algorithm to estimate the time variation. The intrinsic advantage is that the growth factor (the bracketed term in (2.91)) becomes

$$\left[\psi_i^{n=1}\right] \rightarrow \frac{1 - \frac{2i\hbar\Delta t}{ma^2} sin^2\left(\frac{qa}{2}\right)}{1 + \frac{2i\hbar\Delta t}{ma^2} sin^2\left(\frac{qa}{2}\right)} \tag{2.95}$$

and the magnitude is less than unity for all values of the coefficients. This means that any combination of a and Δt can be used, and the solution is, in principle, stable. This is, of course, an oversimplification, and the solutions must be checked for stability (convergence) with variable values of, for example, Δt for a given value of a.

REFERENCES

Abramowitz, M. and Stegun, I. A. 1964. *Handbook of Mathematical Functions*. Washington, DC: US National Bureau of Standards. p. 450

Crank, D. C. and Nicholson, P. 1947. Proc. *Cambridge Phil. Soc.* 43: 50–67

Heisenberg, W. 1925. Über quantentheoreticsche undeutung Kinematicsher und mechanischer bezeihungen. *Z. Phyzik* 33: 879–893

Heisenberg, W. 1927. Über den anschaulichen Inhalt der quantentheoretischen Kinematik und Mechanik. *Z. Physik.* 41: 172–198; Trans. J. A. Wheeler and W. H. Zurek. 1983. *Quantum Theory of Measurement*. Princeton: Princeton University Press. pp. 62–84

Klimeck, G., Wang, X., and Vasileska, D. 2019. *AQME—Advancing Quantum Mechanics for Engineers*. Nanohub.org. https://nanohub.org/resources/aqme. DOI: 10.21981/4PKZ-Q312

Richtmyer, R. D. and Morton, K. W. 1967. *Difference Methods for Initial-Value Problems*. New York: Interscience

Schrödinger, E. 1926. Quantizierung als eigenwertprobleme. *Ann. Phyzik* 79: 361–379

PROBLEMS

1 For the wave packet defined by $\phi(k)$, shown below, find $\psi(x)$. What are Δx and Δk?

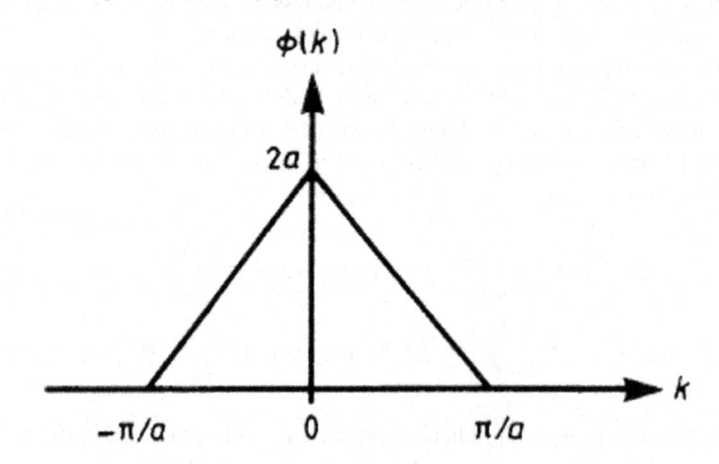

2 If a Gaussian wave packet approaches a potential step ($V > 0$ for $x > 0$, $k_0 > 0$), it is found that it becomes broader for the region $x > 0$. Why?

3 Assume that $\psi(x)$ are the eigenfunctions in an infinite square well ($V \to \infty$ for $|x| > d/2$). Calculate the overlap integrals

$$\int_{-d/2}^{d/2} \psi_n(x)\,\psi_m(x)\,dx.$$

4 Suppose that electrons are confined in an infinite potential well of width 0.5 nm. What spectral frequencies will result from transitions between the lowest four energy levels? Use the free-electron mass in your computations.

5 A particle confined to an infinite potential well has an uncertainty that is of the order of the well width, $\Delta x \simeq a$. The momentum can be estimated as its uncertainty value as well. Using these simple assumptions, estimate the energy of the lowest level. Compare with the actual value.

6 In terms of the momentum operator $p = -i\hbar\nabla$, and

$$H = \frac{p^2}{2m} + \frac{m\omega^2}{2}x^2$$

and using the fact that $p = x = 0$ in a bound state, with

$$\left\langle p^2 \right\rangle = (\Delta p)^2 + \left\langle p^2 \right\rangle = (\Delta p)^2$$

$$\left\langle x^2 \right\rangle = (\Delta x)^2 + \left\langle x \right\rangle^2 = (\Delta x)^2$$

use the uncertainty principle to estimate the lowest bound-state energy. (Hint: recall the classical relation between the average kinetic and potential energies.)

7 Consider a potential well with $V = -0.3$ eV for $|x| < a/2$, and $V = 0$ for $|x| > a/2$, with $a = 7.5$ nm. Write a computer program that computes the energy levels for $E < 0$ (use a mass appropriate for GaAs, $m \simeq 6.0 \times 10^{-32}$ kg). How many levels are bound in the well, and what are their energy eigenvalues? Using a simple wave-function-matching technique, plot the wave functions for each bound state. Plot the transmission coefficient for $E > 0$.

8 For the situation in which a linear potential is imposed on a system, compute the momentum wave functions. Show that these wave functions form a normalized set.

9 Using the continuity of the wave function and its derivative at each interior interface, verify (2.65).

10 Consider an infinite potential well that is 10 nm wide. At time zero, a Gaussian wave packet, with half-width of 1 nm, is placed 2 nm from the center of the well. Plot the evolving wave functions for several times up to the stable steady state. How does the steady state differ from the initial state, and why does this occur?

11 For an infinite potential well of width 20 nm, compute the energies associated with the transitions between the five lowest levels. Give the answers in eV.

12 For a finite potential well of width 20 nm and height 0.3 eV (use an effective mass appropriate to GaAs electrons, $m = 0.067\text{m}_0$), compute the energy needed to ionize an electron (move it from the lowest energy level to the top of the well). If a 0.4 eV photon excites the electron out of the well, what is its kinetic energy? What is its wavelength?

13 For a finite potential well of width 20 nm and height 0.3 eV (use an effective mass appropriate to GaAs electrons, $m = 0.067m_0$), carry out a numerical evaluation of the energy levels. Use enough grid points in the 'forbidden' region to assure that the wave function is thoroughly damped. How many bound states are contained in the well?

14 Consider a potential well with $\text{V}_0 \to \infty$ for $x < 0$, and $V(x) = Fx$ for $x > 0$, with $F = 0.02$ eV nm^{-1}. Using a numerical procedure, compute the ten lowest energy levels in the quantum well.

3 Tunneling

When we dealt in the last chapter with the finite potential well, it was observed that the wave function penetrated into the barriers. This process does not occur in classical mechanics, since a particle will in all cases bounce off the barrier. However, when we treat the particle as a wave, then the wave nature of barrier penetration can occur. This is familiar in electromagnetic waves, where the decaying wave (as opposed to a propagating wave) is termed an *evanescent* wave. For energies below the top of the barrier, the wave is attenuated, and it decays exponentially. Yet, it takes a significant distance for this decay to eliminate the wave completely. If the barrier is thinner than this critical distance, the evanescent wave can excite a propagating wave in the region beyond the barrier. In optics, this effect is termed *frustrated total internal reflection*. We are familiar with total internal reflection just from a swimming pool. Water has a higher dielectric constant than free space, so when we are under water, there is critical angle beyond which we can no longer see out of the pool. In optics, we can use a prism, and there will be reflection from the back surface for angles greater than this critical angle. The wave in back of the prism is attenuated and decaying. But, if we place another high dielectric constant material close to the prism, the decaying wave can excite a wave in this new material. Thus, the wave can penetrate the barrier, and continue to propagate, with an attenuated amplitude, in the trans-barrier region. For particles, this process is termed *tunneling*, with analogy to the miners who burrow through a mountain in order to get to the other side! This process is quite important in modern semiconductor devices, and Leo Esaki received the Nobel prize for first recognizing that tunneling was important in degenerately doped *p–n* junction diodes. Nevertheless, tunneling arises *when the particle is treated as a wave*. The behavior is fully consistent with optics in that regard.

Since Esaki's discovery of the tunnel diode, tunneling has become important in a variety of situations. In reverse biased *p–n* junctions, Zener breakdown occurs when electrons in the valence band can tunnel across the gap into the conduction band if a sufficiently high electric field is applied to bring these two bands to the same energy levels (on opposite sides of the junction). Similarly, resonant tunneling diodes have been fabricated in heterostructures such as GaAs–AlGaAs, and we will discuss these in some detail in a later section. Finally, as semiconductor devices become smaller, particularly the metal–oxide–semiconductor field-effect transistor (MOSFET), where a thin layer of insulator (usually SiO_2 of HfO_2) is used as the gate insulator, this thin oxide becomes susceptible to leakage currents via tunneling through the oxide.

In this chapter, we will address this tunneling process. First, we will treat those few cases in which the tunneling probability can be obtained exactly. Then we will discuss its use in solid-state electronics. Following this, we will move to approximate treatments suitable for those cases in which the solution is not readily obtainable in an exact manner. Finally, in the next chapter, we turn to periodic tunneling structures, which give rise for example to the band structure discussed in semiconductors.

3.1 THE TUNNEL BARRIER

The general problem is that posed in Figure 3.1. Here, we have a barrier, whose height is taken to be V_0, that exists in the region $|x| < a$. To the left and to the right of this barrier, the particle can exist as a freely propagating wave, but, in the region $|x| < a$, and for energies $E < V_0$, the wave is heavily attenuated and is characterized by a decaying exponential 'wave'. Our interest is in determining just what the transmission probability through the barrier is for an incident particle. We are also interested in the transmission behavior for energies above the top of the barrier. To solve for these factors, we proceed in precisely the same fashion as we did for the examples of the last chapter. That is, we assume waves with the appropriate propagation characteristics in each of the regions of interest, but with unknown coefficients. We then apply boundary conditions, in this case the continuity of the wave function and its derivative at each interface, in order to evaluate the unknown coefficients. We consider first a simple barrier.

3.1.1 THE SIMPLE RECTANGULAR BARRIER

The simple barrier is shown in Figure 3.1. Here the potential is defined to exist only between $-a$ and a, and the zero of potential for the propagating waves on either side is the same. We can therefore define the wave vector k in the region $|x| > a$, and the decaying wave vector γ in the region $|x| < a$, by the equations ($E < V_0$) as

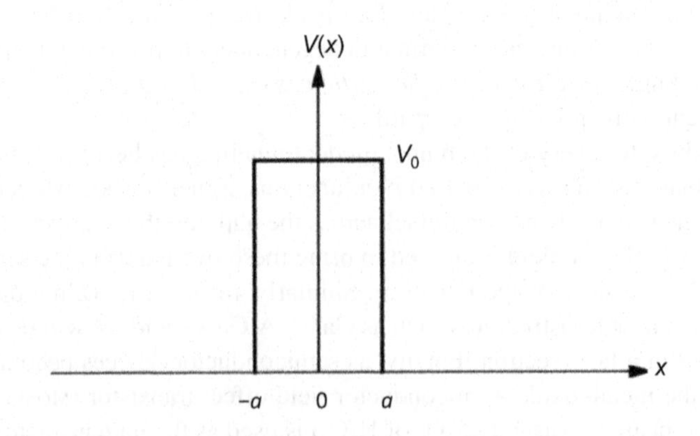

FIGURE 3.1 The simple rectangular tunneling barrier.

$$k = \sqrt{\frac{2mE}{\hbar^2}}, \quad \gamma = \sqrt{\frac{2m(V_0 - E)}{\hbar^2}}, \tag{3.1}$$

respectively. These equations, of course, are exactly (2.6) and (2.12), as they should be. To the right and left of the barrier, the wave is described by propagating waves, while in the barrier region, the wave is attenuated. Thus, we can write the wave function quite generally as was done in the last chapter via

$$\psi(x) = \begin{cases} Ae^{ikx} + Be^{-ikx}, & x < -a, \\ Ce^{\gamma x} + De^{-\gamma x}, & |x| < a, \\ Fe^{ikx}, & x > a. \end{cases} \tag{3.2}$$

This assumes that the wave is incident from the left. We now have five unknown coefficients to evaluate. However, we can get only four equations from the two boundary conditions, and we get a fifth from normalizing the incoming wave (A) from one side or the other. For other applications, we can count on eventually using the principle of superposition, as the Schrödinger equation is linear; thus, our approach is perfectly general.

The boundary conditions are applied by asserting continuity of the wave function and its derivative at each interface. Thus, at the interface $x = -a$, continuity of these two quantities gives rise to

$$\begin{aligned} Ae^{ikx} + Be^{-ikx} &= Ce^{\gamma x} + De^{-\gamma x} \\ ik\left(Ae^{ikx} - Be^{-ikx}\right) &= \gamma\left(Ce^{\gamma x} - De^{-\gamma x}\right) \end{aligned}. \tag{3.3}$$

As in the last chapter, we can now solve for two of these coefficients in terms of the other two coefficients. For the moment, we seek A and B in terms of C and D. This leads to the matrix equation

$$\begin{bmatrix} A \\ B \end{bmatrix} = \begin{bmatrix} \left(\frac{ik+\gamma}{2ik}\right)e^{(ik-\gamma)a} & \left(\frac{ik-\gamma}{2ik}\right)e^{(ik+\gamma)a} \\ \left(\frac{ik-\gamma}{2ik}\right)e^{-(ik+\gamma)a} & \left(\frac{ik+\gamma}{2ik}\right)e^{-(ik-\gamma)a} \end{bmatrix} \begin{bmatrix} C \\ D \end{bmatrix}. \tag{3.4}$$

Now, we turn to the other boundary interface. The continuity of the wave function and its derivative at $x = a$ leads to

$$\begin{aligned} Ce^{\gamma x} + De^{-\gamma x} &= Fe^{ikx} \\ \gamma\left(Ce^{\gamma x} - De^{-\gamma x}\right) &= ikFe^{ikx} \end{aligned}. \tag{3.5}$$

Again, we can solve for two of these coefficients in terms of the one on the right. Here, we seek to find C and D in terms of F (we will eliminate the former two through the use of (3.4)). This leads to the matrix equation

$$\begin{bmatrix} C \\ D \end{bmatrix} = \begin{bmatrix} \left(\frac{ik+\gamma}{2\gamma}\right) e^{(ik-\gamma)a} & -\left(\frac{ik-\gamma}{2\gamma}\right) e^{-(ik+\gamma)a} \\ -\left(\frac{ik-\gamma}{2\gamma}\right) e^{(ik+\gamma)a} & \left(\frac{ik+\gamma}{2\gamma}\right) e^{-(ik-\gamma)a} \end{bmatrix} \begin{bmatrix} F \\ 0 \end{bmatrix}. \tag{3.6}$$

From the pair of equations, (3.4) and (3.6), the two propagating coefficients on the left of the barrier, A and B, can be related directly to those on the right of the barrier, here only F, with the two under the barrier dropping out of consideration. This leads to the matrix equation

$$\begin{bmatrix} A \\ B \end{bmatrix} = \begin{bmatrix} M_{11} & M_{12} \\ M_{21} & M_{22} \end{bmatrix} \begin{bmatrix} F \\ 0 \end{bmatrix}. \tag{3.7}$$

After a little algebra, the various elements are determined to be

$$M_{11} = \left[cosh(2\gamma a) - \frac{i}{2}\left(\frac{k^2-\gamma^2}{k\gamma}\right) sinh(2\gamma a) \right] e^{2ika} = M_{22}^*, \tag{3.8}$$

$$M_{21} = -\frac{i}{2}\left(\frac{k^2+\gamma^2}{k\gamma}\right) sinh(2\gamma a) = M_{12}^*. \tag{3.9}$$

It is a simple algebraic exercise to show that, for the present case, the determinant of the matrix M is unity, so this matrix has quite interesting properties. It is *not* a unitary matrix, because the diagonal elements are complex. In the simple case, the transmission coefficient is simply given by the reciprocal of $|M_{11}|^2$, since the momentum is the same on either side of the barrier and hence the current does not involve any momentum adjustments on the two sides. The transmission probability is the ratio of the currents on the two sides of the barrier, directed in the same direction of course, so

$$\begin{aligned} T = \frac{1}{{M_{11}}^2} &= \left[cosh^2(2\gamma a) + \left(\frac{k^2-\gamma^2}{k\gamma}\right)^2 sinh^2(2\gamma a) \right]^{-1} \\ &= \left[1 + \left(\frac{k^2+\gamma^2}{k\gamma}\right)^2 sinh^2(2\gamma a) \right]^{-1} \end{aligned}. \tag{3.10}$$

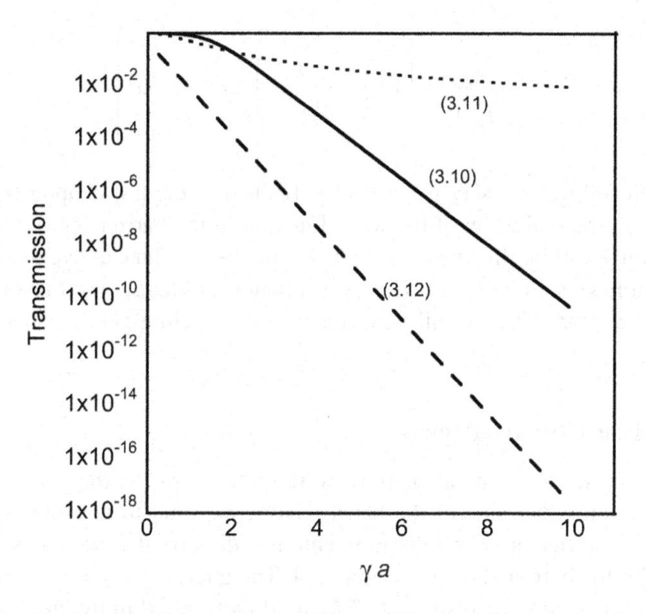

FIGURE 3.2 Comparison of the exact tunneling coefficient with two approximations.

There are a number of limiting cases that are of interest. First, for a very weak barrier, in which $2\gamma a \ll 1$, the transmission coefficient becomes

$$T \to \frac{1}{1+(ka)^2}. \tag{3.11}$$

On the other hand, when the potential is very strong, where $2\gamma a \gg 1$, the transmission coefficient falls off exponentially as

$$T \to \left(\frac{4k\gamma}{k^2+\gamma^2}\right)^2 e^{-\gamma a}. \tag{3.12}$$

It is important to understand that these approximations have existed for some time, but they are not very useful. In Figure 3.2, we plot the above three equations by varying the parameter γa and set the energy to be one-half the barrier height. It is clear that neither approximation is very good, although the small γa approximation (3.11) is not too band. However, the large γa approximation (3.12) is very bad, being often orders of magnitude away from the proper value!

It is important to note that the result (3.10) is valid only for a weak potential for which the energy is actually *below the top of the barrier*. If we consider an incident energy above the barrier, we expect the barrier region to act as a thin dielectric and cause interference fringes. We can see this by making the simple substitution suggested by (3.1) through $\gamma \to -ik'$. This changes (3.10) into

$$T=\frac{1}{\left[M_{11}\right]^{2}}=\left[1+\left(\frac{k^{2}-k'^{2}}{2kk'}\right)^{2}sin^{2}\left(2k'a\right)\right]^{-1}, \tag{3.13}$$

which would follow precisely from (2.22) obtained in the last chapter (with a suitable change in the definition of the wave function in the barrier region) if we made the barrier only a finite thickness as we have done here. Thus, above the barrier, the transmission has oscillatory behavior as a function of energy, with resonances that occur for $2k'a = n\pi$. The overall behavior of the tunneling coefficient is shown in Figure 3.3.

3.1.2 A More Complex Barrier

In the previous section, the calculations were quite simple as the wave momentum was the same on either side of the barrier. Now, we want to consider a somewhat more realistic barrier in which the momentum differs on the two sides of the barrier. Consider the barrier shown in Figure 3.4. The interface at $x = -a$ is the same as treated previously, and the results of (3.4) are directly used in the present problem. However, the propagating wave on the right-hand side of the barrier $(x > a)$ is characterized by a different wave vector through

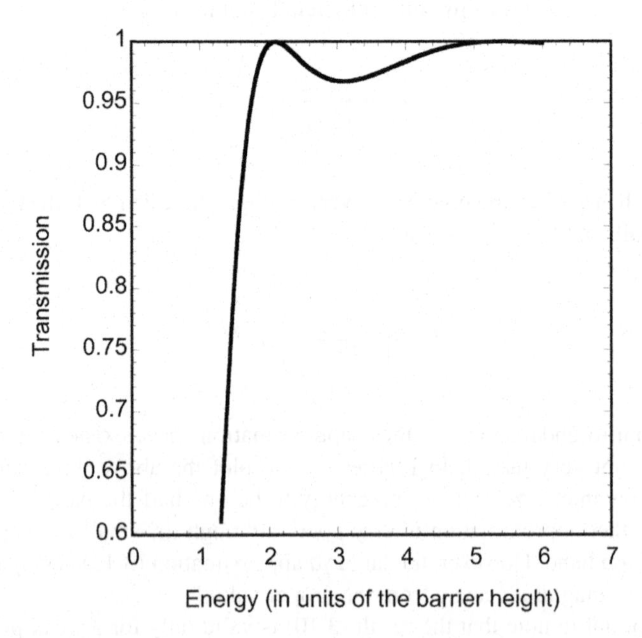

FIGURE 3.3 Tunneling (transmission) probability for a simple barrier. The energy is normalized to the barrier height and the quantity $2\sqrt{2mV_0/\hbar^2}\,a = 3$.

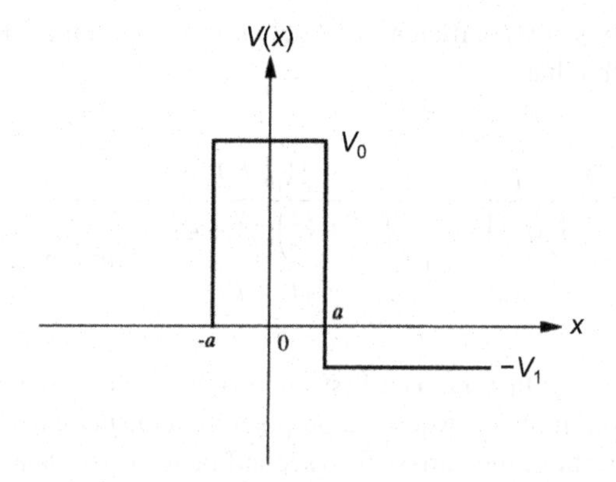

FIGURE 3.4 A more complex tunneling barrier.

$$k_1 = \sqrt{\frac{2m(E+V_1)}{\hbar^2}}. \tag{3.14}$$

Matching the wave function and its derivative at $x = a$ leads to

$$\begin{bmatrix} C \\ D \end{bmatrix} = \begin{bmatrix} \left(\frac{ik_1+\gamma}{2\gamma}\right)e^{(ik_1-\gamma)a} & -\left(\frac{ik_1-\gamma}{2\gamma}\right)e^{-(ik_1+\gamma)a} \\ -\left(\frac{ik_1-\gamma}{2\gamma}\right)e^{(ik_1+\gamma)a} & \left(\frac{ik_1+\gamma}{2\gamma}\right)e^{-(ik_1-\gamma)a} \end{bmatrix} \begin{bmatrix} F \\ 0 \end{bmatrix}, \tag{3.15}$$

which, of course, differs from (3.6) by the change in the wave momentum. But now we find that the entries in the M matrix are changed. These now become

$$M_{11} = \left[\frac{1}{2}\left(1+\frac{k_1}{k}\right)cosh(2\gamma a) - \frac{i}{2}\left(\frac{kk_1-\gamma^2}{k\gamma}\right)sinh(2\gamma a)\right]e^{i(k+k_1)a} = M_{22}^*, \tag{3.16}$$

$$M_{21} = -\left[\frac{i}{2}\left(\frac{kk_1+\gamma^2}{k\gamma}\right)sinh(2\gamma a) + \frac{1}{2}\left(\frac{k_1}{k}-1\right)cosh(2\gamma a)\right]e^{i(k-k_1)a} = M_{12}^*. \tag{3.17}$$

The determinant of the matrix M is also no longer unity, but is given by the ratio k_1/k. This determinant also reminds us that we must be careful in calculating the transmission coefficient as well, due to the differences in the momenta, at a given energy, on the two sides of the barrier. We proceed as in the previous section to compute the transmission coefficient. The actual transmission coefficient relates the currents, and we find that

$$T = \frac{k_1}{k}\frac{1}{\left[M_{11}\right]^2} = \frac{4k_1 k / (k_1 + k)^2}{1 + \frac{(\gamma^2 + k^2)(\gamma^2 + k_1^2)}{\gamma^2 (k_1 + k)^2} sinh^2(2\gamma a)}. \tag{3.18}$$

In (3.18), there are two factors. The first factor is the one in the numerator, which describes the discontinuity between the propagation constants in the two regions to the left and to the right of the barrier. The second factor is the denominator, which is the actual tunneling coefficient describing the *transparency* of the barrier. It is these two factors together that describe the total transmission of waves from one side to the other. It should be noted that if we take the limiting case of $k_1 = k$, we recover the previous result (3.10).

There is an important relationship that is apparent in (3.18). The result represented by (3.18) is reciprocal in the two wave vectors. They appear symmetrical in the transmission coefficient *T*. This is a natural and important result of the symmetry. Even though the barrier and energy structure of Figure 3.4 does not appear symmetrical, the barrier is a linear structure that is passive (there is no active gain in the system). Therefore, the electrical properties should satisfy the principle of reciprocity, and the transmission should be the same regardless of from which direction one approaches the barrier. This is evident in the form of the transmission coefficient (3.18) that is obtained from these calculations.

3.2 THE DOUBLE BARRIER

We now want to put together two tunnel barriers separated by a quantum well. The quantum well (that is, the region between the two barriers) will have discrete energy levels because of the confinement quantization, just as in Section 2.3. We will find that, when the incident wave energy corresponds to one of these resonant energy states of the quantum well, the transmission through the double barrier will rise to a value that is unity (for equal barriers). This resonant tunneling, in which the transmission is unity, is quite useful as an energy filter.

There are two approaches to solving for the composite tunneling transmission coefficient. In one, we resolve the entire problem from first principles, matching the wave function and its derivative at four different interfaces (two for each of the two barriers). The second approach, which we will pursue here, uses the results of the previous sections, and we merely seek knowledge as to how to put together the transmission matrices that we already have found. The reason we can pursue this latter approach effectively is that the actual transmission matrices found in the previous sections depend only upon the wave vectors (the *k*'s and γ), and the thickness of the barrier, $2a$. They do *not* depend upon the position of the barrier, so the barrier may be placed at an arbitrary point in space without modifying the transmission properties. Thus, we consider the generic problem of Figure 3.5, where we have indicated

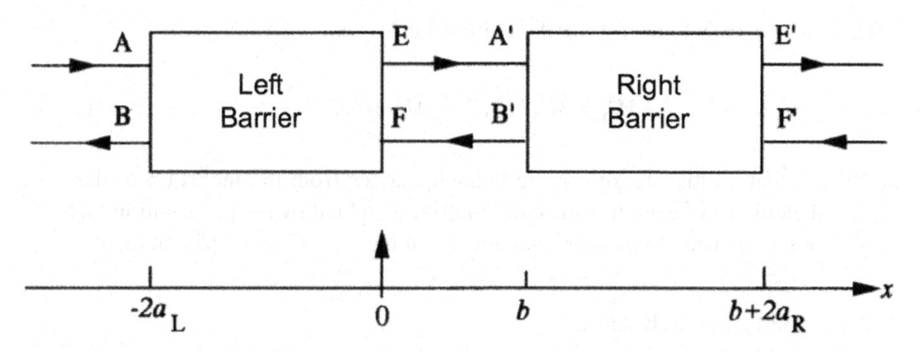

FIGURE 3.5 Two generic barriers are put together to form a double-barrier structure.

the coefficients in the same manner as that in which they were defined in the earlier sections. To differentiate between the two barriers, we have used primes on the coefficients of the right-hand barrier. Our task is to now relate the coefficients of the left-hand barrier to those of the right-hand barrier.

We note that both E and A' describe a wave propagating to the right. Denoting the definition of the thickness of the well region as b, we can simply relate these two coefficients via

$$A' = Ee^{ikb}, \tag{3.19}$$

where k is the propagation constant *in the well region.* Similarly, F and B' relate the same wave propagating in the opposite direction. These two can thus be related by

$$B' = Fe^{-ikb}. \tag{3.20}$$

These definitions now allow us to write the connection as a matrix in the following manner:

$$\begin{bmatrix} E \\ F \end{bmatrix} = \begin{bmatrix} e^{-ikb} & 0 \\ 0 & e^{ikb} \end{bmatrix} \begin{bmatrix} A' \\ B' \end{bmatrix}. \tag{3.21}$$

Equation (3.21) now defines a matrix $\boldsymbol{M}_{\mathrm{W}}$, where the subscript indicates the well region. This means that we can now take the matrices defined in Section 3.1 for the left-hand and right-hand regions and write the overall tunneling matrix as

$$\begin{bmatrix} A \\ B \end{bmatrix} = [M_L][M_W][M_R] \begin{bmatrix} E' \\ F' \end{bmatrix}. \tag{3.22}$$

From this, it is easy to now write the composite M_{11} as

$$M_{11}^{T} = M_{11}^{L} M_{11}^{R} e^{ikb} + M_{12}^{L} M_{21}^{R} e^{-ikb}, \tag{3.23}$$

and it is apparent that the resonance behavior arises from the inclusion of the off-diagonal elements of each transmission matrix, weighted by the propagation factors. At this point, we need to be more specific about the individual matrix elements.

3.2.1 Simple, Equal Barriers

For the first case, we use the results of Section 3.1, where a simple rectangular barrier was considered. Here, we assume that the two barriers are exactly equal, so the same propagation wave vector k exists in the well and in the regions to the left and right of the composite structure. By the same token, each of the two barriers has the same potential height and therefore the same γ. We note that this leads to a magnitude-squared factor in the second term of (3.23), *but not in the first term* with one notable exception. The factor of e^{i2ka} does cancel since we are to the left of the right-hand barrier (-a-direction) but to the right of the left-hand barrier (+a-direction). Thus, the right-hand barrier contributes a factor of e^{-i2ka}, and the left-hand barrier contributes a factor of e^{i2ka}, so the two cancel each other. In order to simplify the mathematical details, we write the remainder of (3.23) as

$$\begin{aligned} M_{11}^{T} &= \left[cosh(2\gamma a) - \frac{i}{2}\left(\frac{k^2-\gamma^2}{k\gamma}\right) sinh(2\gamma a)\right]^2 e^{ikb} \\ &\quad - \frac{1}{4}\left[\left(\frac{k^2+\gamma^2}{k\gamma}\right) sinh(2\gamma a)\right]^2 e^{-ikb}. \end{aligned} \tag{3.24}$$

While this latter form shows the explicit elements in the total transmission, it is easier both to understand and to manipulate, if we leave the explicit terms out for the present.

To do the bit a little better, let us write the terms as a phasor; that is, let us write each of the various matrix elements as

$$M_{11} = m_{11} e^{i\vartheta}, \tag{3.25}$$

where

$$m_{11} = \sqrt{cosh^2(2\gamma a) + \left(\frac{k^2-\gamma^2}{k\gamma}\right)^2 sinh^2(2\gamma a)} \tag{3.26}$$

and

$$\vartheta = atan\left[\left(\frac{k^2 - \gamma^2}{k\gamma}\right)tanh(2\gamma a)\right]. \tag{3.27}$$

Then, (3.23) can be rewritten as

$$\begin{aligned} \left|M_{11}^T\right|^2 &= |M_{11}|^4 + |M_{12}|^4 + 2|M_{11}|^2|M_{12}|^2 cos\left[2(kb+\vartheta)\right] \\ &= \left(|M_{11}|^2 - |M_{12}|^2\right)^2 + 4|M_{11}|^2|M_{12}|^2 cos^2(kb+\vartheta). \end{aligned} \tag{3.28}$$

The first term, the combination within the parentheses, is just the determinant of the individual barrier matrix, and is unity for the simple rectangular barrier. Thus, the overall transmission is now

$$\left|M_{11}^T\right|^2 = 1 + 4|M_{11}|^2|M_{12}|^2 cos^2(kb+\vartheta). \tag{3.29}$$

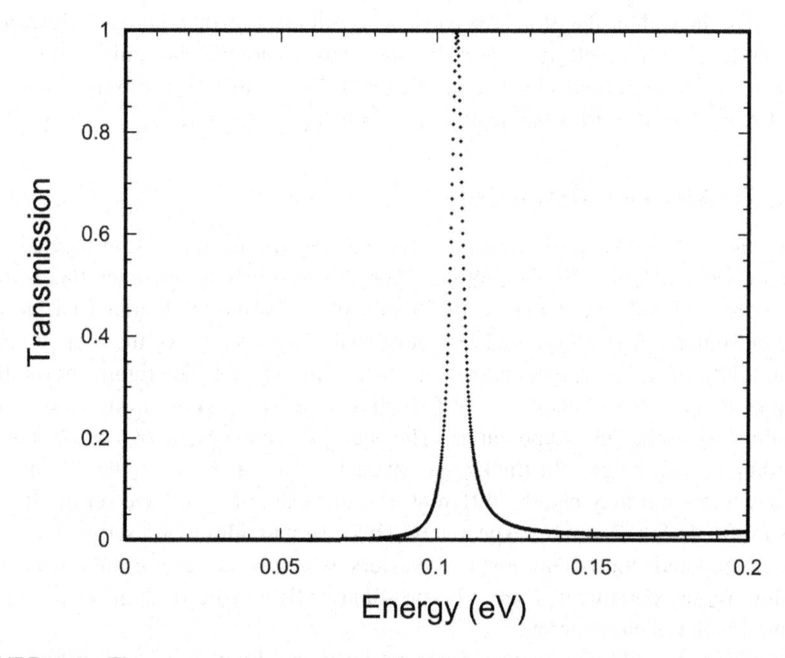

FIGURE 3.6 The transmission through a double-barrier system composed of AlGaAs barriers and a GaAs well. Here, the barrier height is 0.25 eV and the thicknesses are 4 nm. The well is taken to be 2 nm thick so that only a single resonant level is present in the well.

In general, the cosine function is nonzero, and the composite term of (3.29) is actually larger than that for the single barrier T_1, with

$$T_{total} \sim \frac{1}{4|M_{12}|^2}, \quad \textit{off resonance}. \tag{3.30}$$

However, for particular values of the wave vector, the cosine term vanishes, and

$$T_{total} = 1, \quad kb + \vartheta = (2n+1)\frac{\pi}{2}. \tag{3.31}$$

These values of the wave vector correspond to the resonant levels of a finite- depth quantum well (the finite-well values are shifted in phase from the infinite-well values by θ, which takes the value $-\pi/2$ in the latter case). Hence, as we supposed, the transmission rises to unity at values of the incident wave vector that correspond to resonant levels of the quantum well. In essence, the two barriers act like mirrors, and a resonant structure is created just as in electromagnetics. The incoming wave excites the occupancy of the resonance level until an equilibrium is reached in which the incoming wave is balanced by an outgoing wave and the overall transmission is unity. This perfect match is broken up if the two barriers differ, as we see below. In Figure 3.6, we plot the transmission for a double quantum well in which the barriers are 0.25 eV high and 4 nm thick. The well is taken to be 2 nm thick. Here, it is assumed that the well is GaAs and the barriers are AlGaAs, so that there is only a single resonant level in the well. It is clear that the transmission rises to unity at the value of this level. We will return later to the shape of the resonant transmission peak, and how it may be probed in some detail experimentally.

3.2.2 The Unequal-Barrier Case

In the case where the two barriers differ, the results are more complicated, and greater care is necessary in the mathematics. The best way to approach this is to go once more to Nanohub.org and use the "Piece-Wise Constant Potential Tool," as pictured in Figure 3.7(a). Here, we have individual wave vectors for the regions to the left and right of the composite barrier, as well as in the well. The figure shows these two potentials as 0.3 eV and 0.4 eV (which are set by clicking on the colored line and adjusting to the desired potential). The decay constants of the two barriers differ, according to their height. the thicknesses of each layer may be set in the "Geometry" pull-down tab and may also be different. For our example, we have set the five dimensions to 4, 2, 4, 2, 4 nm, respectively. This tool will allow for a series of barriers to be created and studied, not just two barriers, which is the case of interest here. In fact, almost any complicated set of barriers and wells may be chosen, so long as the potential is flat in each region.

There are several important points to be discussed. First, there will still be a resonance in the quantum well, but its position will be affected by the unequal barriers.

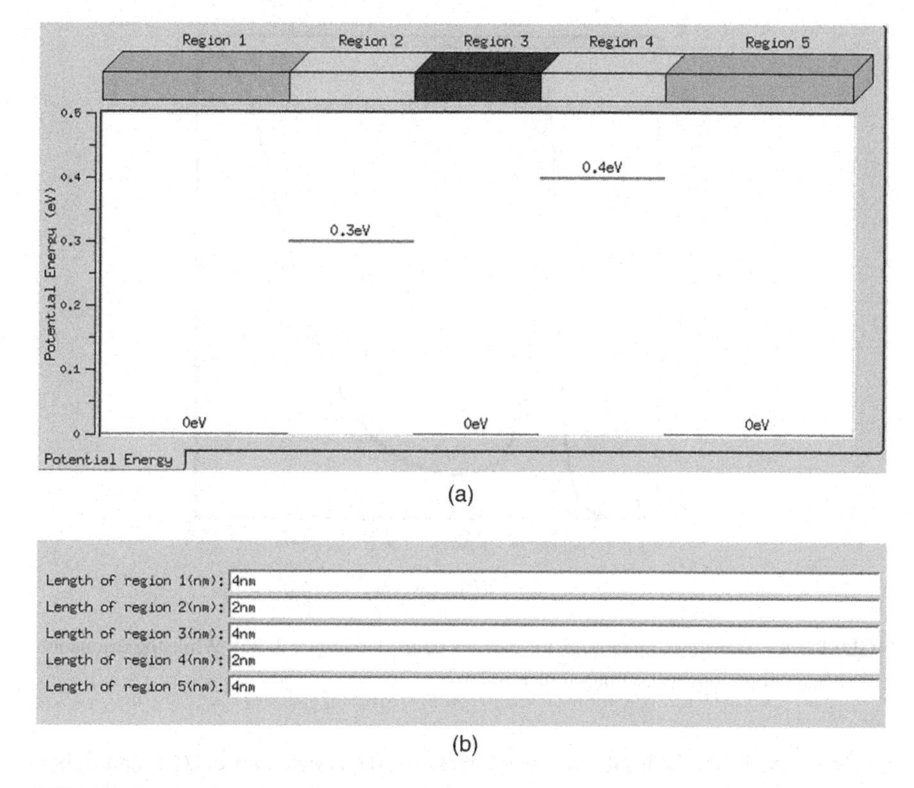

(a)

(b)

FIGURE 3.7 (a) Two barriers set in the piece-wise constant potential tool. (b) The dimensions set in the "geometry" section.

In addition, the peak of the resonance will no longer be unity, because of the unequal barriers. In the optics analog, the reflectivity of the two mirrors is no longer perfect for each of the mirrors. Thus, we expect a dimunition in the height of the resonance. But the resonance is not exponentially reduced, but linearly reduced; that is, it is attenuated, but not exponentially attenuated by the unequal barriers in the problem. We see this in Figure 3.8, where we plot the transmission for the barrier defined in Figure 3.7(a). Here, we have utilized the ability to download the data from the tool as a text file and replotted it with another graphics package. There is a large peak at about 0.13 eV, which appears at the energy of the resonant level in the structure. But, the height is only about 0.88–0.9, which is less than the unity peak in Figure 3.6, and arises from the factors discussed above—unequal barriers and lack of complete phase matching at the reflection points of the barriers. One can give an approximate form for this reduction through the relationship

$$T = \frac{4T_L T_R}{\left(T_L + T_R\right)^2}, \tag{3.32}$$

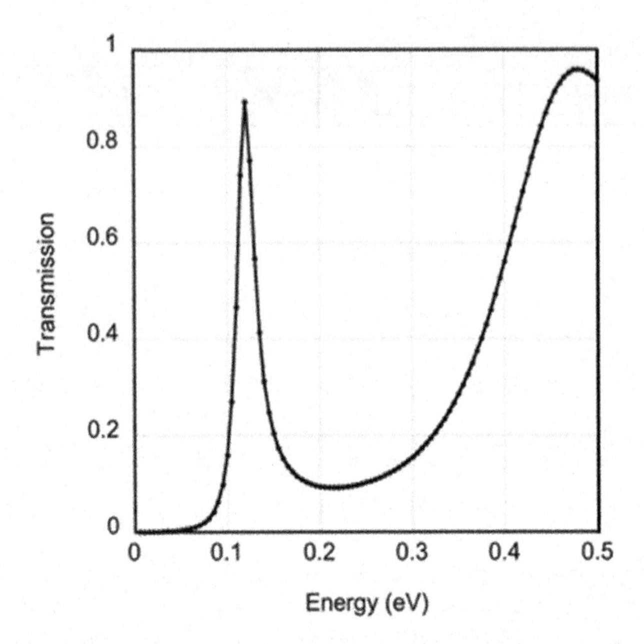

FIGURE 3.8 Transmission through the barriers defined in Figure 3.7(a). While the resonance still appears, it no longer has unity transmission.

where T_L and T_R are the transmissions of the two independent barriers (left and right). It is clear that, if the two are equal, then (3.32) gives a value of unity. But, if they are not equal, the peak transmission begins to be attenuated. If one is significantly larger than the other, it is thought that the result will be

$$T \sim 4\frac{T_{min}}{T_{max}}, \tag{3.33}$$

where again the two values are appropriate for the individual barriers acting alone.

The opposite extreme is reached when we are away from one of the resonant levels. In this case, the maximum attenuation is achieved when the cosine function has the value of unity, and the resulting minimum in overall transmission is given approximately by the value

$$T = \frac{T_L T_R}{2}. \tag{3.34}$$

This is for the limit of low transmission. It is clear from these discussions that if we are to maximize the transmission on resonance in any device application, it is important to have the transmission of the two barriers equal under any bias conditions that cause the resonance to appear.

3.2.3 Shape of the Resonance

It is apparent from the shape of the transmission curve in Figure 3.8 that there is a very sharp resonance that coincides with the resonant energy level in the quantum well. We can analyze this resonance more by considering it to have a Lorentzian shape around the maximum, that is

$$\frac{1}{T(E)} = 1 + \left(\frac{E - E_0}{\Delta E}\right)^2, \tag{3.35}$$

where E_0 is the resonance energy and ΔE is the half-width of the resonance at half-height. In fact, this form holds quite accurately over orders of magnitude in transmission (Price, 1999). In essence, this means that we can write $|M_{11}|^2 = PQ$, with $P = Q^*$. From (3.25), we identify

$$M_{11} = Q = Q_0\left[1 - i\left(\frac{E - E_0}{\Delta E}\right)\right], \quad M_{11}^* = P. \tag{3.36}$$

Hence, PQ must have two zeros located at $E = E_0 \pm \Delta E$. One of these is a zero of P and the other is a zero of Q. Thus, when the energy E is swept through the resonance energy E_0, the phase angle arg(P) is swept through an amount approaching π. Price (1999) has pointed out that this has an analogy with formal wave propagation theory, so that one can assign a *transit time* for passage of an electron through the double-barrier structure as

$$t_{transit} \sim \hbar\frac{d\left[arg(P)\right]}{dE} \rightarrow \frac{\hbar / \Delta E}{1 + \left(\frac{E - E_0}{\Delta E}\right)^2}. \tag{3.37}$$

While it is quite difficult to measure the transit time of the electron through the resonant structure, creative methods have been found to measure the phase shift. In Chapter 1, we discussed the Aharonov–Bohm effect in a semiconductor nanostructure. Indeed, experiments have embedded a semiconductor quantum dot in one leg of the AB ring (Yacoby et al., 1995). The quantum dot is a two-dimensional electron gas at the interface of a GaAlAs/GaAs heterostructure, which is further confined by lateral in-plane gates. We will discuss these two-dimensional structures in Chapter 9. Propagation through the quantum dot is facilitated by two *tunneling* quantum point contacts which, along with the dot itself, are located in one leg of the Aharonov–Bohm (A–B) loop. Measurements of the transmitted current through the A–B ring give the total phase through each branch in terms of the interference between the two legs. This phase interference is measured by varying the magnetic field through the ring. When the size of the dot is tuned by a *plunger* gate, the resonant energy is swept through the Fermi energy of the remaining parts of the system. By this gate voltage tuning the resonance shape is swept through the transmission energy, and a change in phase by a factor of π is found, as expected.

3.3 APPROXIMATION METHODS—THE WKB METHOD

So far, the barriers that we have been treating are simple barriers in the sense that the potential $V(x)$ has always been piece-wise constant. The reason for this lies in the fact that if the barrier height is a function of position, then the Schrödinger equation is a complicated equation that has solutions that are special functions. The example we treated in the last chapter merely had a linear variation of the potential—a constant electric field—and the result was solutions that were identified as Airy functions which already are quite complicated. What are we to do with more complicated potential variations? In some cases, the solutions can be achieved as well-known special functions—we treat Hermite polynomials in Chapter 5—but in general these solutions are quite complicated. On the other hand, nearly all of the solution techniques that we have used involve propagating waves or decaying waves, and the rest of the problem lay in matching boundary conditions. This latter, quite simple, observation suggests an approximation technique to find solutions, the Wentzel-Kramers-Brillouin (WKB) approach (Wentzel 1926, Kramers 1926, Brillouin 1926).

Consider Figure 3.8, in which we illustrate a general spatially varying potential. At a particular energy level, there is a position (shown as a) at which the wave changes from propagating to decaying. This position is known as a *turning point*. The description arises from the simple fact that the wave (particle) would be reflected from this point in a classical system. In fact, we can generally extend the earlier arguments and definitions of this chapter to say that

$$k(x)=\sqrt{\frac{2m}{\hbar^2}\left[E-V(x)\right]}, \quad E>V(x) \tag{3.38}$$

and:

$$\gamma(x)=\sqrt{\frac{2m}{\hbar^2}\left[V(x)-E\right]}, \quad E<V(x). \tag{3.39}$$

These solutions suggest that, at least to zero order, the solutions can be taken as simple exponentials that correspond either to propagating waves or to decaying waves. We then adopt the results (3.38) and (3.39) as the lowest approximation but seek higher approximations. To proceed, we assume that the wave function is generically definable as

$$\psi(x)\sim e^{iu(x)}, \tag{3.40}$$

and we now need to determine just what form *u(x)* takes. This, of course, is closely related to the formulation adopted in Section 2.1, and the differential equation for *u(x)* is just (2.8) when the variation of the prefactor of the exponent is ignored. This gives:

$$i\frac{\partial^2 u}{\partial x^2}-\left(\frac{\partial u}{\partial x}\right)^2+k^2(x)=0, \tag{3.41}$$

and equivalently for the decaying solution (we treat only the propagating one, and the decaying one will follow easily via a sign change). If we had a true free particle, the last two terms would cancel ($u = kx$) and we would be left with

$$i\frac{\partial^2 u}{\partial x^2}=0. \tag{3.42}$$

This suggests that we approximate *u(x)* by making this latter equality an initial assumption for the lowest-order approximation to *u(x)*. To carry this further, we can then write the *i*th iteration of the solution as the solution of

$$\left(\frac{\partial u_i}{\partial x}\right)^2=k^2(x)+i\frac{\partial^2 u_{i-1}}{\partial x^2}. \tag{3.43}$$

We will only concern ourselves here with the first-order correction and approximation. The insertion of the zero-order approximation (which neglects the last term in (3.43)) into the equation for the first-order approximation leads to

$$\frac{\partial u_1}{\partial x}=\sqrt{k^2(x)+i\frac{\partial k}{\partial x}}=\pm k(x)+i\frac{1}{2k(x)}\frac{\partial k}{\partial x}. \tag{3.44}$$

In arriving at this last expression, we have assumed, in keeping with the approximations discussed, that the second term on the right-hand side in (3.44) is much smaller than the first term on the right. This implies that, in keeping with the discussion of Section 2.1, the potential is slowly varying on the scale of the wavelength of the wave packet.

The result (3.44) can now be integrated over the position, with an arbitrary initial position as the reference point. This gives

$$u_1 \approx \pm\int^x k(x')dx'+\frac{1}{2}ln\left[k(x)\right]+lnC_1, \tag{3.45}$$

which leads to

$$\psi(x) \sim \frac{C_1}{\sqrt{k(x)}} exp\left[\pm i\int^{x} k(x')dx'\right]. \tag{3.46}$$

The equivalent solution for the decaying wave function is

$$\psi(x) \sim \frac{C_1}{\sqrt{\gamma(x)}} exp\left[\pm \int^{x} \gamma(x')dx'\right]. \tag{3.47}$$

It may be noted that these results automatically are equivalent to the requirement of making the current continuous at the turning point, which is achieved via the square-root prefactors.

To see how this occurs, we remind ourselves of (2.13) and (2.14). There, it was necessary to define a current which was continuous in the time-independent situation. Using (2.17), we then require the continuity of

$$|J| = \left|p\psi^2\right|. \tag{3.48}$$

This continuity of current was used both in the last chapter, and in this chapter, to determine the transmission coefficient. Now, if we are to infer a connection formula for the wave function, then we must use a generalization of (3.48) to say that

$$\frac{1}{\sqrt{|p|}}\psi \tag{3.49}$$

must be continuous. It is the recognition of this fact that leads to the forms (3.46) and (3.47) for the wave functions on either side of the interface of Figure 3.9. These connection formulas are the heart of the WKB technique, but they have their source in the decomposition of the wave function discussed in Section 2.2.

The remaining problem lies in connecting the waves of one type with those of the other at the turning point. The way this is done is through a method called the method of stationary phase. The details are beyond the present treatment but are actually quite intuitive. In general, the connection formulas are written in terms of sines and cosines, rather than as propagating exponentials, and this will insert a factor of two, but only in the even functions of the propagating waves. In addition, the cosine waves always couple to the decaying solution, and a factor of $\pi/4$ is always subtracted from the phase of the propagating waves (this is a result of the details of the stationary phase relationship and arises from the need to include a factor that is the square root of i). In Figure 3.9, the turning point is to the right of the classical region (where $E > V$). For this case, the connection formulas are given by

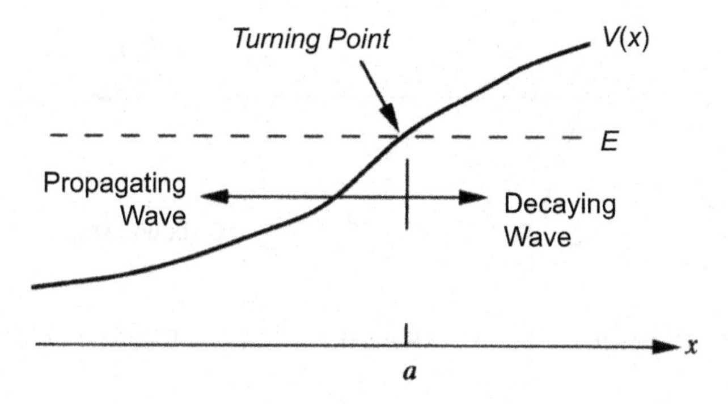

FIGURE 3.9 A simple variation of potential and the corresponding energy surface.

$$\frac{2}{\sqrt{k}} \cos\left(\int_x^a k dx' - \frac{\pi}{4}\right) \leftrightarrow \frac{1}{\sqrt{\gamma}} \exp\left(-\int_a^x \gamma dx'\right), \tag{3.50}$$

$$\frac{2}{\sqrt{k}} \sin\left(\int_x^a k dx' - \frac{\pi}{4}\right) \leftrightarrow \frac{1}{\sqrt{\gamma}} \exp\left(+\int_a^x \gamma dx'\right). \tag{3.51}$$

The alternative case is for the mirror image of Figure 3.9, in which the turning point is to the left of the classical region (in which the potential would be a decreasing function of x rather than an increasing function). For this case, the matching formulas are given as (the turning point is taken as $x = b$ in this case)

$$\frac{1}{\sqrt{\gamma}} \exp\left(-\int_x^b \gamma dx'\right) \leftrightarrow \frac{2}{\sqrt{k}} \cos\left(\int_b^x k dx' - \frac{\pi}{4}\right) \tag{3.52}$$

$$-\frac{1}{\sqrt{\gamma}} \exp\left(\int_x^b \gamma dx'\right) \leftrightarrow \frac{2}{\sqrt{k}} \sin\left(\int_b^x k dx' - \frac{\pi}{4}\right). \tag{3.53}$$

To illustrate the application of these matching formulas, we consider some simple examples.

3.3.1 Bound States of a General Potential

As a first example of the WKB technique, and the matching formulas, let us consider the general potential shown in Figure 3.10. Our aim is to find the bound states, or the energy levels to be more exact. It is assumed that the energy level of interest is such that the turning points are as indicated; that is, the points $x = a$ and $x = b$ correspond to the turning points. Now, in region 1, to the left of $x = b$, we know that the

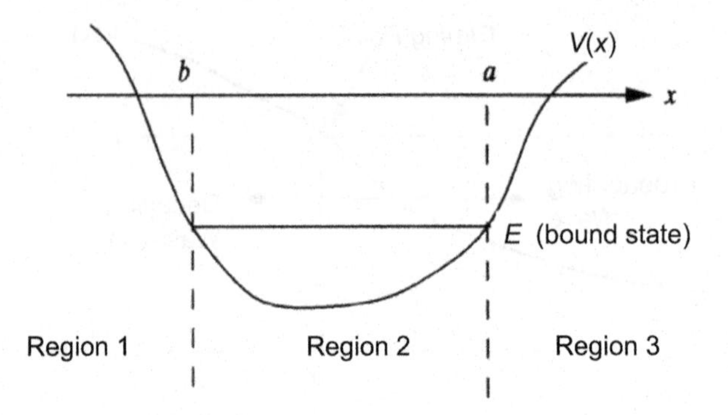

FIGURE 3.10 An arbitrary potential well in which to apply the WKB method.

solution has to be a decaying exponential as we move away from *b*. This means that we require that

$$\psi_1(x) = \frac{1}{\sqrt{\gamma}} exp\left(-\int_x^b \gamma dx'\right), \quad x < b. \tag{3.54}$$

At $x = b$, this must match to the cosine wave if we use (3.52). Thus, we know that in region 2, the wave function is given by

$$\psi_2(x) = \frac{2}{\sqrt{k}} cos\left(\int_b^x k dx' - \frac{\pi}{4}\right), \quad b < x < a. \tag{3.55}$$

We now want to work our way across to $x = a$, and this is done quite simply with simple manipulations of (3.55), as

$$\begin{aligned}\psi_2(x) &= \frac{2}{\sqrt{k}} cos\left(\int_b^x k dx' + \frac{\pi}{4} - \frac{\pi}{2}\right) = \frac{2}{\sqrt{k}} sin\left(\int_b^x k dx' + \frac{\pi}{4}\right) \\ &= \frac{2}{\sqrt{k}} sin\left(\int_b^a k dx' - \int_x^a k dx' + \frac{\pi}{4}\right),\end{aligned} \tag{3.56}$$

We now expand the sine function to give

$$\begin{aligned}\psi_2(x) = &-\frac{2}{\sqrt{k}} cos\left(\int_b^a k dx'\right) sin\left(\int_x^a k dx' - \frac{\pi}{4}\right) \\ &+ \frac{2}{\sqrt{k}} sin\left(\int_b^a k dx'\right) cos\left(\int_x^a k dx' - \frac{\pi}{4}\right).\end{aligned} \tag{3.57}$$

We also know that the solution for the matching at the interface $x = a$ must satisfy (3.50), as the wave function in region 3 must be a decaying wave function. This means that at this interface, $\Psi_2(a)$ must be given *only* by the second term of (3.57). In this second term, the sine function is just a constant. The desired result can *only* be achieved by requiring that the first term must vanish, and this leads to

$$cos\left(\int_b^a kdx'\right) = 0, \tag{3.58}$$

or

$$\int_b^a kdx' = (2n+1)\frac{\pi}{2}. \tag{3.59}$$

This equation now determines the energy eigenvalues of the potential well, at least within the WKB approximation.

If we compare (3.59) with the result for a sharp potential such as the infinite quantum well of Chapter 2, with $b = -a$, we see that there is an additional phase shift of $\pi/2$ on the left-hand side. While one might think that this is an error inherent in the WKB approach, we note that the sharp potentials of the last chapter violate the assumptions of the WKB approach (slowly varying potentials). The extra factor of $\pi/2$ arises from the soft variation of the potentials. Without exactly solving the true potential case, one cannot say whether or not this extra factor is an error, but this factor is a general result of the WKB approach.

3.3.2 Tunneling

It is not necessary to work out the complete tunneling problem here, since we are interested only in the decay of the wave function from one side of the barrier to the other (recall that the input wave was always normalized to unity). It suffices to say that the spirit of the WKB approximation lies in the propagation (or decaying) wave vector, and the computation of the argument of the exponential decay function. The result (3.59) is that it is only the combination of forward and reverse waves that matter. For a barrier in which the attenuation is relatively large, only the decaying forward wave is important, and the tunneling probability is approximately

$$T \sim exp\left[-2\int_b^a \gamma(x')dx'\right], \tag{3.60}$$

which implies that it is only the numerical coefficients (which involve the propagating and decaying wave vectors) that are lost in the WKB method. This tell us that we can use the limiting form of (3.12) $(b = -a)$, or the equivalent limit of (3.18), with the argument of the exponential replaced with that of (3.60).

3.4 TUNNELING DEVICES

One of the attractions of tunneling devices is that it is possible to apply textbook quantum mechanics to gain an understanding of their operation, and still achieve a reasonable degree of success in actually getting quantitative agreement with experimental results. The concept of the "tunnel diode" goes back several decades and is usually implemented in heavily doped *p–n* junctions. In this case, the tunneling is through the forbidden energy gap, as we will see below. Here, the tunneling electrons make a transition from the valence band, on one side of the junction, to the conduction band on the other side. More recently, effort has centered on resonant tunneling devices which can occur in a material with a single carrier type. Each of these will be discussed below, but first we need to formulate a current representation for the general tunneling device, and this will lead us to the Landauer formula.

3.4.1 The Landauer Formulation

In the treatment of the tunneling problem that we have encountered in the preceding sections, the tunneling process is that of a single plane-wave energy state from one side of the barrier to the other. The tunneling process, in this view, is an energy-conserving process, since the energy at the output side is the same as that at the input side. In many real devices, the tunneling process can be more complex, but we will follow this simple approach and treat a general tunneling structure, such as that shown in Figure 3.11. In the 'real' device, the tunneling electrons are those within a narrow energy range near the Fermi energy, where the range is defined by the applied voltage as indicated in the figure. For this simple view, the device is treated in the linear-response regime, even though the resulting current is a nonlinear function of the applied voltage. The general barrier can be a simple square barrier, or a multitude of individual barriers, just so long as the total tunneling probability through the entire structure is *coherent*. By coherent here, we mean that the tunneling through the entire barrier is an energy- and momentum-conserving process, so no further complications are necessary. Hence, the properties of the barrier are completely described by the quantity $T(k)$.

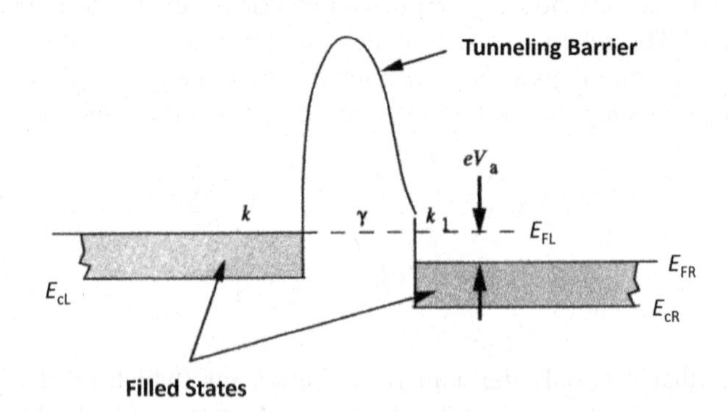

FIGURE 3.11 Tunneling occurs from filled states on one side of the barrier to the empty states on the opposite side. The current is the net flow of particles from one side to the other.

In equilibrium, where there is no applied bias, the left-going and right-going waves are equivalent and there is no net current. By requiring that the energy be conserved during the process, we can write the z-component of energy as (we take the z-direction as that of the tunneling current)

$$E = \frac{\hbar^2 k_z^2}{2m} = \frac{\hbar^2 k_{1z}^2}{2m} + constant, \tag{3.61}$$

where the constant accounts for the bias and is negative for a positive potential applied to the right of the barrier. The two wave vectors are easily related to one another by this equation, and we note that the derivative of this equation allows us to relate the velocities on the two sides. In particular, we note that

$$v_z(k_z)dk_z = v_z(k_{1z})dk_{1z}. \tag{3.62}$$

The current flow through the barrier is related to the tunneling probability and to the total number of electrons that are available for tunneling. Thus, the flow from the left to the right is given by

$$J_{LR} = 2e\int \frac{d^3k}{(2\pi)^3} v_z(k_z)T(k_z)f(E_L), \tag{3.63}$$

where the factor of 2 is for spin degeneracy of the electron states, the $(2\pi)^3$ is the normalization on the number of k states (related to the density of states in k space), and $f(E_L)$is the electron distribution function *at the barrier.* Similarly, the current flow from the right to the left is given by

$$J_{RL} = 2e\int \frac{d^3k_1}{(2\pi)^3} v_z(k_{1z})T(k_{1z})f(E_R). \tag{3.64}$$

we know that the tunneling probability is reciprocal at the same energy, regardless of the direction of approach, and these two equations can be combined, using (3.62), as

$$J = 2e\int \frac{d^3k}{(2\pi)^3} v_z(k_z)T(k_z)\left[f(E_L) - f(E_L + eV_a)\right], \tag{3.65}$$

where we have related the energy on the left to that on the right through the bias, as shown in Figure 3.11, and expressed in (3.61). In the following, we will drop the subscript 'L' on the energy, but care must be used to ensure that it is evaluated on the left of the barrier.

Before proceeding, we want to simplify some of the relationships in (3.65). First, we note that the energy is a scalar quantity and can therefore be decomposed into its z-component and its transverse component, as

$$E = E_z + E_\perp \tag{3.66}$$

and

$$d^3k = d^2k_\perp dk_z. \tag{3.67}$$

We would like to change the differential dk_z to one on dE_z. This is done through

$$dk_z = \left(\frac{dE}{dk_z}\right)^{-1} \frac{dE}{dE_z} dE_z. \tag{3.68}$$

The second term on the right-hand side is unity, so it drops out. The first term may be evaluated from (3.61) as

$$\frac{dE}{dk_z} = \frac{\hbar^2 k_z}{m} = \hbar v_z. \tag{3.69}$$

The velocity term here will cancel that in (3.65), and we can write the final representation of the current as

$$J = \frac{e}{\pi\hbar}\int \frac{d^2k_\perp}{(2\pi)^2} \int dE_z T(E_z)\left[f(E_L) - f(E_L + eV_a)\right]. \tag{3.70}$$

At this point in the theory, we really do not know the form of the distributions themselves, other than some form of simplifying assumption such as saying that they are Fermi–Dirac distributions. If we make this assumption, and expand the second distribution as the first two terms of a Taylor series, we obtain

$$J = -\frac{e}{\pi\hbar}\int \frac{d^2k_\perp}{(2\pi)^2} \int dE_z T(E_z) eV_a \frac{\partial f(E_L)}{\partial E}. \tag{3.71}$$

At low temperatures, the derivative of the Fermi–Dirac distribution is a delta function in the negative direction, which allows us to simplify (3.71) further to get

$$J = \frac{e^2}{\pi\hbar}\int \frac{d^2k_\perp}{(2\pi)^2} T(E_{FL}) V_a. \tag{3.72}$$

The integration over the transverse momentum just gives us the number of two-dimensional transverse modes in the device. This may be a small number or a large number (approaching infinity in a large cross-section device). If we denote this number of modes as *N/A*, then (3.72) can be written as

$$I = AJ = \frac{e^2}{\pi\hbar} NTV_a, \tag{3.73}$$

and the conductance becomes

$$G = \frac{I}{V_a} = \frac{2e^2}{h} \sum_{i=1}^{N} T_i, \tag{3.74}$$

where we have replaced the product *NT* by the summation over the transmission of each of the modes. This is known as the Landauer (1957) equation.

In fact, in metals, the distribution functions are well approximated by Fermi–Dirac distributions. In semiconductors, however, the electric field and the current flow work to perturb the distributions significantly from their equilibrium forms, and this will introduce some additional complications. Additionally, the amount of charge in semiconductors is much smaller and charge fluctuations near the barriers can occur. Hence, there is a deviation of the distribution from its normal form as one approaches the barrier. This is quite simply understood. Electrons see a barrier in which the tunneling is rather small. Thus, the wave function tries to have a value near zero at the interface with the barrier. The wave function then peaks at a distance of approximately $\lambda/2$ from the barrier. But this leads to a charge depletion right at the barrier, and the self-consistent potential will try to pull more charge toward the barrier. Electrons with a higher momentum will have their peaks closer to the barrier, so this charging effect leads to a distribution function with more high-energy electrons close to the barrier. In essence, this is a result of the Bohm potential of (1.41), as quantum mechanics does not really like to have a strongly varying density. In metals, where the number of electrons is quite high, this effect is easily screened out, but in semiconductors it can be significant. Whether or not it affects the total current is questionable, depending upon the size of the tunneling coefficient. Nevertheless, we need to account for the distribution function being somewhat different from the normal Fermi–Dirac function. Generally, all of this results in the need for a self-consistent solution of the local potential and the distribution function, which is a large topic in its own right.

3.4.2 The Esaki Diode

Tunneling in a diode essentially arises in a very heavily doped *p–n* junction, so the built-in potential of the junction is larger than the band gap. This is shown in Figure 3.12(*a*). When a small bias is applied, as shown in Figure 3.12(b), the filled states on one side of the junction overlap empty, allowed states on the other side, which allows current to flow (assuming the actual barrier is thin so tunneling is

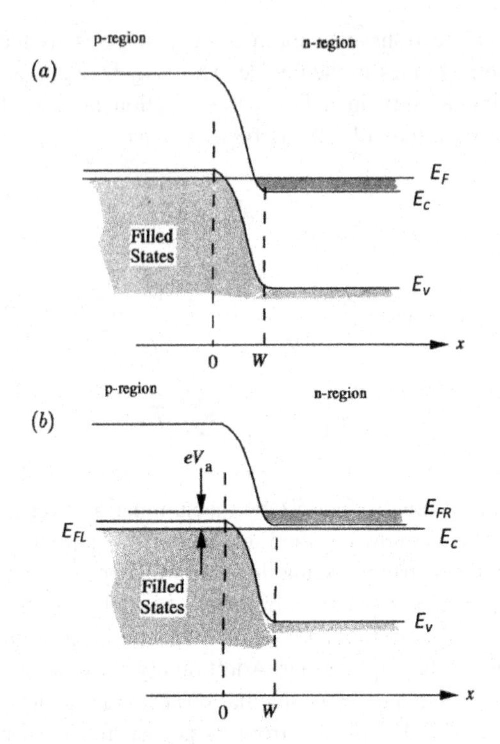

FIGURE 3.12 The band line-up for degenerately doped *p–n* junctions (a), and the possible tunneling transitions for small forward bias (b).

allowed). So far, this is no different from a normal junction diode, other than the fact that the carriers tunnel across the forbidden gap at the junction rather than being injected. However, it may be noted from Figure 3.12(b) that continuing to increase the forward bias (the polarity shown) causes the filled states to begin to overlap states in the band gap, which are forbidden. Thus, the forward current returns to zero with increasing forward bias, and a negative differential conductance is observed. When combined with the normal *p–n* junction injection currents, an *N*-shaped conductance curve is obtained, which leads to the possibility of the use of the device for many novel electronic applications. In the reverse bias direction, the overlap of filled and empty (allowed) states continues to increase with all bias levels, so no negative conductance is observed in this direction of the current.

When the electric field in the barrier region is sufficiently large, the probability of tunneling through the gap region is nonzero; for example, tunneling can occur when the depletion width W is sufficiently small. One view of the tunneling barrier is that it is a triangular potential, whose height is approximately equal to the band gap, and whose width at the tunneling energy is the depletion width W. In Section 2.4, we found that a triangular-potential region gave rise to wave functions that were Airy functions. The complications of these functions provide a strong argument for the use of the WKB approximation, and we do so here. We can take the decay coefficient as

$$\gamma(x) = \begin{cases} \sqrt{\dfrac{2mE_G}{\hbar^2}\left(1 - \dfrac{x}{W} + \dfrac{E_\perp}{E_G}\right)}, & 0 < x < W, \\ 0 & elsewhere. \end{cases} \tag{3.75}$$

where we have factored the energy gap out of the potential term and evaluated the electric field as E_G/eW. The last term in the square root accounts for the transverse energy, since the tunneling coefficient depends upon only the z-component of momentum (the z-component of energy must be reduced below the total energy by the transverse energy). This expression must now be integrated according to (3.60) over the tunneling region, which produces

$$T \approx exp\left[-\frac{4W}{3}\sqrt{\frac{2mE_G}{\hbar^2}}\left(1 + \frac{3E_\perp}{2E_G}\right)\right], \tag{3.76}$$

where we have expanded the radical to lowest order and retained only the leading term in the transverse energy since it is considerably smaller than the band gap. It turns out that the result (3.76) is not sensitive to the actual details of the potential, since it is measuring the area under the V–E curve. Different shapes give the same result if the areas are equal. Recognizing this assures us that the approximation (3.76) is probably as good as any other. We can rewrite (3.76) as

$$T \approx T_0 exp\left(-\frac{E_\perp}{E_0}\right), \quad E_0 = \frac{E_G}{W}\sqrt{\frac{\hbar^2}{2mE_G}}. \tag{3.77}$$

This can now be used in (3.70) to find the current.

We first will tackle the transverse energy integral. To lowest order, we note that the term involving the Fermi–Dirac functions is mainly a function of the longitudinal z-component of the energy, which we will show below, so the transverse terms are given by

$$\int_0^{E_F - eV_a} \frac{d^2k_\perp}{(2\pi)^2} exp\left(-\frac{E_\perp}{E_0}\right) = \frac{mE_0}{2\pi\hbar^2}\left[1 - exp\left(-\frac{E_{FL} - eV_a}{E_0}\right)\right]. \tag{3.78}$$

The limits on the previous integral are set by the fact that the transverse energy can only increase up to the sum of the Fermi energies on the two sides of the junction (measured from the band edges) reduced by the longitudinal energy.

The longitudinal contribution may be found by evaluating the energies in the Fermi–Dirac integrals, through shifting the energy on one side by the applied voltage eV_a. This leads to the result, in the linear-response limit, that

$$\begin{aligned}\left[f\left(E_z+E_\perp\right)-f\left(E_z+E_\perp+eV_a\right)\right]&=\frac{eV_a}{k_BT}f\left(1-f\right)\\ &\approx -eV_a\frac{\partial f}{\partial E_z}\approx eV_a\delta\left(E_z-E_{FL}\right).\end{aligned} \tag{3.79}$$

The last approximation is for strongly degenerate material (or equivalently, very low temperature). Then the integration over E_z gives just eV_a times the tunneling probability T_0. We can now put (3.78) and (3.79) in the general equation to obtain the total current density

$$J_z=\frac{e^2T_0}{\pi\hbar}\left(\frac{mE_{FL}}{2\pi\hbar^2}\right)\left(1-\frac{eV_a}{E_{FL}}\right)V_a. \tag{3.80}$$

As we discussed at the beginning of this section, the current rises linearly with applied bias, but then decreases as the electron states on the right-hand side begin to overlap the forbidden states in the energy gap, which cuts off the current. We show the tunneling current in Figure 3.13, along with the normal *p–n* junction current due to injection and diffusion.

3.4.3 The Resonant Tunneling Diode

The resonant tunneling diode is one in which a double barrier, such as that in Section 3.2, is inserted into, say, a conduction band, and the current through the structure is metered via the resonant level. The latter corresponds to the energy at which the transmission rises to a value near unity. The structure of such a system, in the GaAs/AlGaAs/GaAs/AlGaAs/GaAs system with the AlGaAs forming the barriers, is shown in Figure 3.14. Typically, the barriers are 3–5 nm thick and about 0.3 eV high, and the well is also 3–5 nm thick.

To proceed, we will use the same approximations as used for the *p–n* junction diode, at least for the distribution function. The difference beween the Fermi–Dirac distributions on the left-hand and right-hand sides, in the limit of very low temperature ($T \to 0$ K) gives

$$\left[f\left(E_z+E_\perp\right)-f\left(E_z+E_\perp+eV_a\right)\right]=eV_a\delta\left(E_z+E_\perp-E_{FL}\right). \tag{3.81}$$

We retain the transverse energy in this treatment, since we must be slightly more careful in the integrations in this model. The tunneling probability can also be taken as approximately a delta function, but with a finite width describing the nature of the

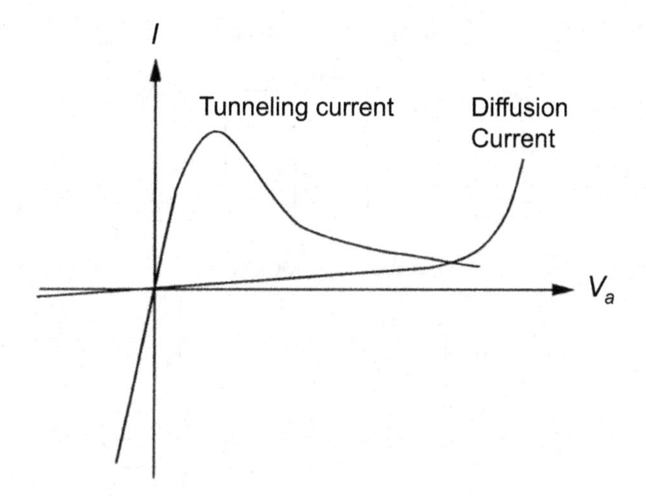

FIGURE 3.13 The contribution of the tunneling current to the overall current of a tunnel diode.

actual lineshape (an alternative is to use something like a Lorentzian line, but this does not change the physics). Thus, we write (we note that the transmission will be less than unity)

$$T(E) \cong E_W \delta\left(E_z + \frac{eV_a}{2} - E_0\right), \tag{3.82}$$

where we have assumed that the width of the transmission is E_W, and that the resonant level E_0 is shifted downward by an amount equal to half the bias voltage (everything is with reference to the Fermi energy on the left-hand side of the barrier, as indicated in the figure). Thus, the current can be written from (3.70) as

$$\begin{aligned} J &= \frac{e^2 V_a}{\pi\hbar}\left(\frac{mE_W}{2\pi\hbar^2}\right)\int dE_z \int_0^{E_{FL}} \delta(E_z + E_\perp - E_{FL})\,\delta\left(E_z + \frac{eV_a}{2} - E_0\right) dE_\perp \\ &= \frac{e^2 V_a}{\pi\hbar}\left(\frac{mE_W}{2\pi\hbar^2}\right)\int_0^{E_{FL}} \delta\left(E_z + E_\perp + \frac{eV_a}{2} - E_0\right) dE_\perp \\ &= \frac{e^2 V_a}{\pi\hbar}\left(\frac{mE_W}{2\pi\hbar^2}\right), \qquad 2(E_0 - E_{FL}) < eV_a < 2E_0. \end{aligned} \tag{3.83}$$

Outside the indicated range of applied bias, the current is zero. At finite temperature (or if a Lorentzian line shape for T is used), the current rises more smoothly and drops more smoothly. Essentially, the current begins to flow as soon as the resonant level E_0 is pulled down to the Fermi energy on the left-hand side (positive bias is to

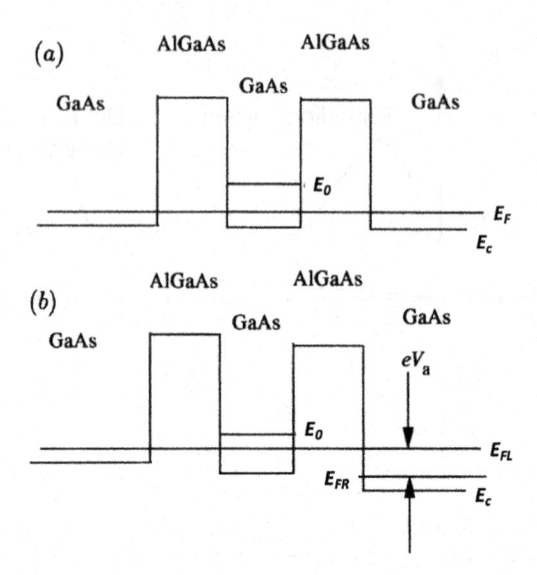

FIGURE 3.14 A typical double-barrier resonant tunneling diode potential system, grown by heteroepitaxy in the GaAs–AlGaAs system. In (a), the basic structure is shown for an n-type GaAs well and cladding. In (b), the shape under bias is shown.

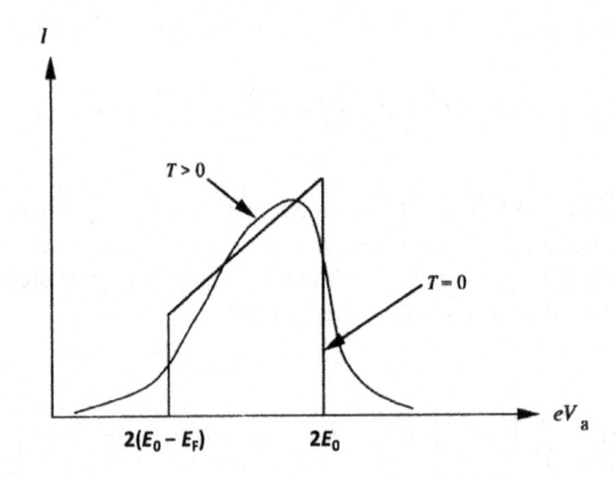

FIGURE 3.15 The theoretical curves for the simple model of (3.83) are shown for zero temperature and finite temperature.

the right), and current ceases to flow when the resonant level passes the bottom of the conduction band. This is shown in Figure 3.15.

3.4.4 Single-Electron Tunneling

Let us consider as an example tunneling through the insulator of a capacitor (which we take to be an oxide such as SiO_2 found in MOS structures). The tunneling through the capacitor oxide is an example of a very simple physical system that can exhibit quite complicated behavior when it is made small. The capacitor is formed by placing the insulator between two metals, or between a metal (gate) and a doped semiconductor in an MOS structure. What if the area of the capacitor is made small, so that the capacitance is also quite small? It turns out that this can affect the operation of tunneling through the oxide significantly. When an electron tunnels through the oxide, it lowers the energy stored in the capacitor by the amount

$$\delta E = \frac{e^2}{2C}. \tag{3.84}$$

This energy change leads to a voltage change of

$$\delta V = \frac{e}{C}. \tag{3.85}$$

What this means is that the tunneling current cannot occur until a voltage equivalent to (3.85) is actually applied across the capacitor. If the voltage on the capacitor is less than this, no tunneling current occurs because there is not sufficient energy stored in the capacitor to provide the tunneling transition. When the capacitance is large, say > 10^{-12} F, this voltage is immeasurably small in comparison with the thermally induced voltages ($k_B T/e$). On the other hand, suppose that the capacitance is defined with a lateral dimension of only 50 nm. This area is 2.5×10^{-15} m^2, and the capacitor has a capacitance of 2.8×10^{-17} F. The required voltage of (3.85) is 5.7 mV. These capacitors are easily made, and the effects easily measured at low temperatures. In Figure 3.16, we show measurements by Fulton and Dolan (1987) on such structures. The retardation of the tunneling current until a voltage according to (3.85) is reached is termed the *Coulomb blockade*. The name arises from the need to have sufficient Coulomb energy before the tunneling transition can occur. The Coulomb blockade causes the offset of the current in the small-(S) capacitor case. This offset scales with area, as shown in the inset to the figure, and hence with C as expected in (3.85).

To understand the Coulomb blockade somewhat more quantitatively, we consider the double-barrier structure shown in Figure 3.17, as it may be created with extremely small lateral dimensions so that the capacitance of each barrier satisfies

$$\delta E = \frac{e^2}{2C} \gg k_B T, \tag{3.86}$$

as required for Coulomb blockade, so that thermal fluctuations cannot induce an electron to cros the capacitor. The region between the two capacitors may be termed a *quantum dot*. In general, each of the two capacitors (barriers) will have some leakage, but the equivalent circuit is as shown in Figure 3.17(a). The leakage is indicated as the

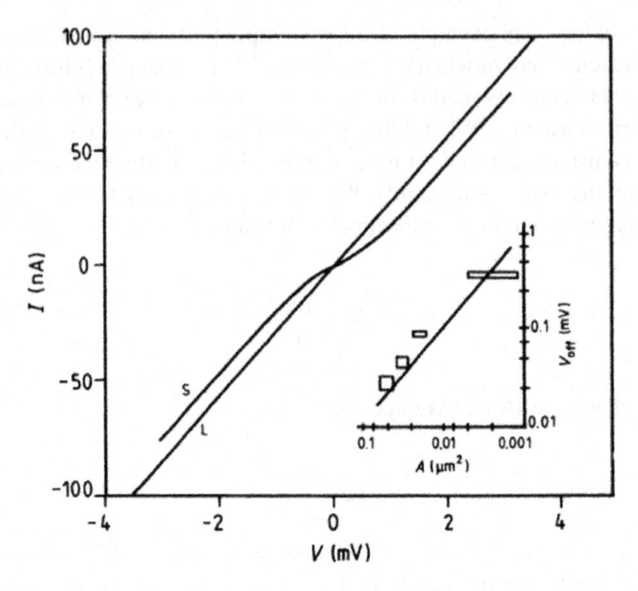

FIGURE 3.16 Single-electron tunneling currents in small capacitors. The voltage offset is due to the Coulomb blockade. (Reproduced with permission from Fulton and Dolan (1987), copyright 1987 by the American Physical Society.)

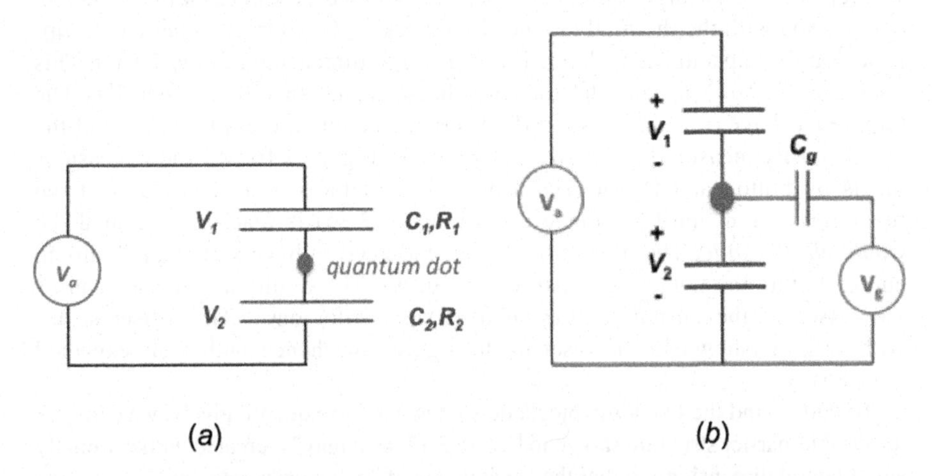

FIGURE 3.17 (a) Equivalent circuit for a quantum dot connected through two tunneling capacitors. The indicated resistance is the tunneling resistance, as discussed in the text. (b) The circuit when a gate bias is applied to directly vary the potential at the dot.

shunt resistance through each capacitor. This leakage, or tunneling resistance, represents the current flow when the electrons tunnel through the barrier. In this sense, these resistances are different from ordinary resistances. In the latter, charge and current flow is essentially continuous. Here, however, the current flow only occurs when the electrons are tunneling, and this is for a very short time. The charge on the two capacitors is related to the two voltages, and is given by

$$Q_1 = C_1 V_1, \quad Q_2 = C_2 V_2. \tag{3.87}$$

The net charge on the quantum dot island is given by

$$Q_{dot} = Q_2 - Q_1 = -ne, \tag{3.88}$$

where n is the net number of electrons on the island. The sum of the junction voltages is the applied voltage V_a, so that combining the two equations above, the voltage drops across the two junctions are just

$$\begin{aligned} V_1 &= \frac{1}{C_1 + C_2}\left(C_2 V_a + ne\right) \\ V_2 &= \frac{1}{C_1 + C_2}\left(C_1 V_a - ne\right) \end{aligned}. \tag{3.89}$$

The electrostatic energy can be written as

$$E_s = \frac{Q_1^2}{2C_1} + \frac{Q_2^2}{2C_2} = \frac{1}{2\left(C_1 + C_2\right)}\left(C_1 C_2 V_a^2 + Q_{dot}^2\right). \tag{3.90}$$

In addition, we must consider the work done by the voltage source in charging the two capacitors. This is an integral over the power delivered to the tunnel junctions during the charging process, or

$$E_a = \int dt V_a\left(t\right) = V_a \Delta Q, \tag{3.91}$$

where ΔQ is the total charge transferred from the voltage source to the capacitors. This includes the integer number of electrons on the quantum dot as well as the continuous polarization charge that builds up in response to the voltages on the capacitors (resulting in an electric field in the island). A change in the charge on the island due to one electron tunneling through C_2 changes the charge on the island to $Q' = Q + e$, and $n' = n - 1$. From (3.90), the voltage change on C_1 results in $V_1' = V_1 - e / C_{eq}$, where $C_{eq} = C_1 + C_2$. Therefore, a polarization charge flows in from the voltage source $\Delta Q = -eC_1 / C_{eq}$ to compensate. The total work done to pass n_2 charges through C_2 is then

$$E_a\left(n_2\right) = -n_2 e V_a C_1 / C_{eq}. \tag{3.92}$$

A similar argument is used to generate the total work done to pass n_1 charges through C_1, which is

$$E_a(n_1) = -n_1 e V_a C_2 / C_{eq}. \tag{3.93}$$

Combining these two results with (3.90), the total energy of the complete circuit, including that provided by the voltage source, is given by

$$E(n_1, n_2) = E_s - E_a = \frac{1}{2C_{eq}}\left(C_1 C_2 V_a^2 + Q_{dot}^2\right) + \frac{eV_a}{C_{eq}}\left(C_1 n_2 + C_2 n_1\right). \tag{3.94}$$

With this description of the total energy in terms of the charge on each of the capacitors (given by the factors n_i), we can now look at the *change* in the energy when a particle tunnels through either capacitor. At low temperature, the tunneling transition must take the system from a state of higher energy to a state of lower energy. The change in energy for a particle tunneling through C_2 is

$$\begin{aligned} \Delta E_2^{\pm} &= E(n_1, n_2) - E(n_1, n_2 \pm 1) \\ &= \frac{e}{C_{eq}}\left[-\frac{e}{2} \pm \left(Q_{dot} - V_a C_1\right)\right]. \end{aligned} \tag{3.95}$$

The value of the charge on the dot Q_{dot} is *prior* to the tunneling process. Similarly, the change in energy for a particle tunneling through C i is given by

$$\begin{aligned} \Delta E_1^{\pm} &= E(n_1, n_2) - E(n_1 \pm 1, n_2) \\ &= \frac{e}{C_{eq}}\left[-\frac{e}{2} \mp \left(Q_{dot} - V_a C_2\right)\right]. \end{aligned} \tag{3.96}$$

According to this discussion, only those transitions are allowed for which $\Delta E_i > 0$; e.g., the initial state is at a higher energy than the final state.

If we consider a system in which the dot is initially uncharged, $Q_{dot} = 0$, then (3.95) and (3.96) reduce to

$$\Delta E_{1,2}^{\pm} = -\frac{e^2}{2C_{eq}} \pm \frac{eV_a C_{2,1}}{C_{eq}} > 0. \tag{3.97}$$

Initially, at low voltage, the leading term on the right-hand side makes $\Delta E < 0$. Hence, no tunneling can occur until the voltage reaches a threshold that depends upon the lesser of the two capacitors. For the case in which $C_1 = C_2 = C$, the requirement becomes $|V_a| > e / C_{eq}$. Tunneling is prohibited and no current flows below this threshold voltage. This region of *Coulomb blockade* is a direct result of the additional Coulomb energy, which is required for an electron to tunnel through one of the capacitors. This leads to the offset shown in Figure 3.16.

Now, consider the situation when we make an additional contact, through another capacitor, to the quantum dot as shown in Figure 3.17(b). A new voltage source, V_g, is coupled to the quantum dot through an *ideal* capacitor C_g (no tunneling is possible through this capacitor). This additional voltage source modifies the charge balance on the quantum dot, so that

$$Q_g = C_g\left(V_g - V_2\right), \tag{3.98}$$

and the charge on the dot is modified to be

$$Q_{dot} = Q_2 - Q_1 - Q_g = -ne. \tag{3.99}$$

We also have to rewrite the two voltages as

$$\begin{aligned} V_1 &= \frac{1}{C_T}\left[\left(C_g + C_2\right)V_a - C_g V_2 + ne\right] \\ V_2 &= \frac{1}{C_T}\left(C_1 V_a + C_g V_2 - ne\right), \end{aligned} \tag{3.100}$$

where $C_T = C_{eq} + C_g$. The electrostatic energy (3.90) now becomes

$$E_s = \frac{1}{2C_T}\left(C_1 C_2 V_a^2 + C_g C_2 V_g^2 + C_g C_1\left(V_a - V_g\right)^2 + Q_{dot}^2\right). \tag{3.101}$$

The work done by the voltage sources during the tunneling now includes the work done by the gate voltage and the additional charge flowing onto the gate capacitor. Equations (3.92) and (3.9) now become

$$\begin{aligned} E_a\left(n_1\right) &= -n_1\left[\frac{eV_a C_2}{C_T} + e\left(V_a - V_g\right)\frac{C_g}{C_T}\right] \\ E_a\left(n_2\right) &= -n_2\left[\frac{eV_a C_1}{C_T} + \frac{eV_g C_g}{C_T}\right] \end{aligned} . \tag{3.102}$$

With these equations we can now rewrite the total energy (3.94) as

$$E(n_1,n_2)=\frac{1}{2C_T}\Big[\left(C_1C_2V_a^2+Q_{dot}^2\right)+C_gC_2V_g^2+C_gC_1\left(V_a-V_g\right)^2 \\ +2eV_a\left(C_1n_2+\left(C_2+C_g\right)n_1\right)-2n_1C_gV_g\Big]. \tag{3.103}$$

For the tunneling of an electron across C_1, the energy change is now given by

$$\begin{aligned}\Delta E_1^{\pm} &= E(n_1,n_2)-E(n_1\pm1,n_2)\\ &=\frac{e}{C_T}\left[-\frac{e}{2}\mp\left(Q_{dot}-V_a\left(C_2+C_g\right)-C_gV_g\right)\right].\end{aligned} \tag{3.104}$$

Similarly, the change in energy for an electron tunneling through C_2 is

$$\begin{aligned}\Delta E_2^{\pm} &= E(n_1,n_2)-E(n_1,n_2\pm1)\\ &=\frac{e}{C_T}\left[-\frac{e}{2}\pm\left(Q_{dot}-V_aC_1-C_gV_g\right)\right].\end{aligned} \tag{3.105}$$

When we compare these results with those of (3.95) and (3.96), it is apparent that the gate voltage allows us to change the effective charge on the quantum dot, and therefore to shift the region of Coulomb blockade with V_g. As before, the condition for tunneling at low temperature is that the change in energy must be negative and the tunneling must take the system to a lower energy state. We now have two conditions that exist for forward and backward tunneling as

$$\begin{aligned}&-\frac{e}{2}\mp\left(ne-V_a\left(C_2+C_g\right)-C_gV_g\right)>0\\ &-\frac{e}{2}\pm\left(ne-V_aC_1-C_gV_g\right)>0\end{aligned}. \tag{3.106}$$

The four equations (3.106) may be used to generate a stability plot in the (V_a,V_g) plane, which shows stable regions corresponding to each value of n, and for which no tunneling can occur. This diagram is shown in Figure 3.18 for the case in which $C_g = C_2 = C$ and $C_1 = 2C$ ($C_T = 4C$). The lines represent the boundaries given by (3.106). The trapezoidal shaded areas correspond to regions where no solution satisfies (3.106) and hence where Coulomb blockade exists. Each of the regions corresponds to a different number of electrons on the quantum dot, which is stable in the sense that this charge state does not change easily. The gate voltage allows us to 'tune' between different stable regimes, essentially adding or subtracting one electron

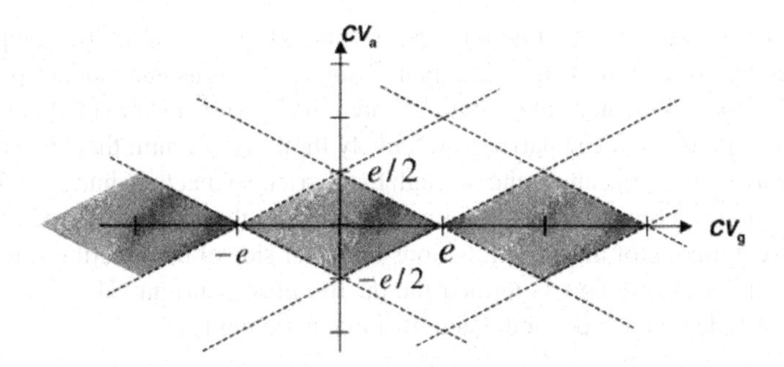

FIGURE 3.18 A stability diagram for the single-electron quantum dot. The parameters are discussed in the text, and it is assumed that the temperature is $T = 0$. The shaded regions are where the dot electron number is stable and Coulomb blockade exists.

at a time to the dot region. For a given gate bias, the range of V_a over which Coulomb blockade occurs is given by the vertical extent of the shaded region. The width of this blockaded region approaches zero as the gate charge approaches half-integer values of a single electron charge, and here tunneling occurs easily. It is at these gate voltages that tunneling current peaks can be observed as the gate bias is varied.

In Figure 3.18, the distance between the current peaks for tunneling through the structure of $\Delta V_g = e/C = e/C_g$. Between these peaks the number of electrons on the quantum dot remains constant. We will return to the quantum dot in Chapter 9, where we begin to worry about the quantum levels within the dot itself. To this point, we have ignored the fact that the quantum dot may be so small that the energy levels within the dot are quantized. Yet, this can occur, and the behavior of this section is modified. As remarked, we return to this in Chapter 9, where we deal with spectroscopy of just these quantum levels.

3.4.5 Josephson Tunneling

The Josephson junction (1962) is a tunnel junction in which the materials on either side of the insulator are superconductors. Superconductors will be discussed in Chapter 4, but it is enough to note that these are certain low conductivity metals whose resistivity goes to zero at low temperature, typically a few, or a few tens of, degrees Kelvin. In the Josephson junction, these superconducting leads produce zero loss in dc operation. With superconductors, the electrons condense in to Cooper pairs—two electrons with opposite spin (again, we will discuss the spin in Chapter 4 and especially in Chapter 10). The formation of the Cooper pairs lowers the energy of the entire electron gas below the Fermi energy, and this leads to a gap opening at the Fermi energy—the gap lies between the full states of Cooper pairs (at T = 0 K) and the empty single-electron states. This gap is 2Δ, where Δ is the condensation energy per electron for the pair.

Normally, with Josephson junctions, the tunneling is carried out by these Cooper pairs, but both the Cooper pair tunneling and single-electron tunneling can occur under the appropriate conditions. The Josephson junction itself operates as shown in

Figure 3.18. As can be seen, no tunneling occurs until $eV_a > 2\Delta$ at low temperature. As the temperature increases, both unpaired single electrons and Cooper pairs exist below the gap, and the normal (unpaired) can tunnel giving a very small current.

As with tunneling, if the barrier is relatively thin, say 1–2 nm, then the superconducting wave functions can extend through the barrier, so that they interact with those wave functions on the other side. An unusual effect in the Josephson junction is that this coherent mixing of the wave functions on either side of the junction can produce a current at zero bias! This is termed the dc Josephson current. This current has a fixed magnitude that can be modulated by a magnetic field as

$$I_J(B) = I_{J0}\cos\left(\frac{eBA}{h}\right), \tag{3.107}$$

where $BA = \Phi$ is the flux flowing through the area of the junction. As before, h/e is the quantum unit of flux, so that the flux can be quantized in a quantum system. The interesting aspect is that the current peak I_{J0} is proportional to the single-particle tunneling coefficient through the junction (Kuper 1968).

We note that if a bias is applied, nothing much happens to the normal curve of Figure 3.18(a), except a few unpaired electrons may tunnel through. But this voltage

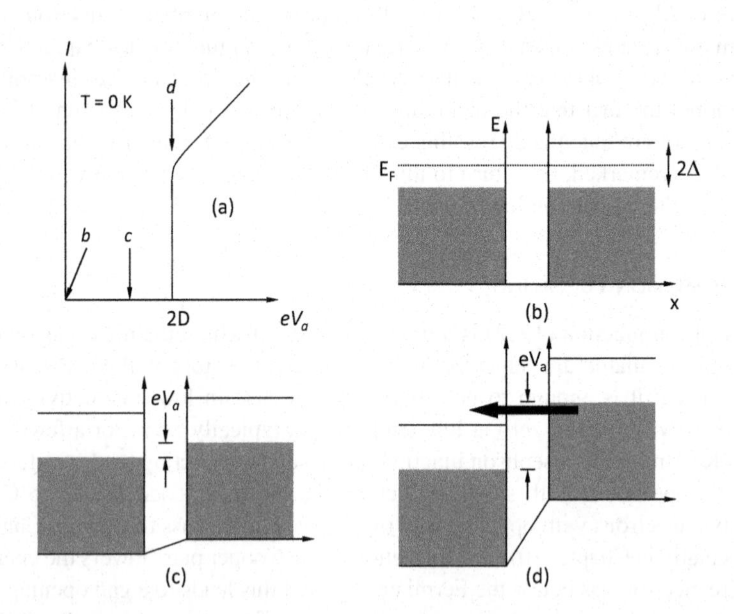

FIGURE 3.19 (a) The current-voltage curve of the Josephson junction at T = 0 K. The letters refer to the other pictures in the figure. (b) At zero bias, no tunneling can occur because the full Cooper pair states see only full Cooper pair states across the junction. (c) At small bias, the full states see only forbidden in the energy on the other side of the junction, and no current can flow. (d) When $eV_a > 2\Delta$, Cooper pairs can tunnel, but the pair breaks up into two free electrons above the gap. Coherent Josephson pair tunneling at zero bias is not shown.

has a significant effect on the coherent flux of the Josephson current (3.107). This follows from the relationship

$$\hbar \frac{d\Phi}{dt} = 2eV_a, \tag{3.108}$$

where the factor of 2 arises from the two electrons in the Cooper pair. Now, what this means is the associated flux is changed by the voltage and this produces an ac signal, known as the ac Josephson effect. This leads to the important relation

$$\hbar\omega = 2eV_a. \tag{3.109}$$

Hence, the Josephson junction can be a microwave source. It has also been noticed that there is a peak in the spectrum at $\omega = 4\Delta / \hbar$, which is called a Riedel peak in the response of the junction (Riedel 1964).

REFERENCES

Brillouin, L. 1926. The ondulatory mechanics of Schrödinger; A general method of resolution by successive approximations. *C. R. Acad. Sci.* 183: 24–27

Fulton, T. A. and Dolan, G. J. 1987. Observation of single-electron charging effects in small tunnel junctions. *Phys. Rev. Lett.* 59: 109–112

Kramers, H. A. 1926. Wellenmechnik und halbzahlige Quantisierung. *Z. Phys.* 39: 828–840

Josephson, B. D. 1962. Possible new effect in superconductive tunneling. *Physics Lett.* 1: 251

Kuper, C. G. 1968. *An Introduction to the Theory of Superconductivity* (Oxford: Clarendon Press) Sec. 8.2

Landauer, R. 1957. Spatial variation of the currents and fields due to localized scatterers in metallic conduction. *IBM J. Res. Develop.* 1: 223–231

Price, P. J. 1999. Quasi-bound states and resonances in heterostructures. *Microelectron. J.* 30: 925

Riedel, E. 1964. Zum Tunneleffeckt bei Supraleitern im Mikrowellenfeld. Z. *Naturforsch.* 19a: 1634

Wentzel, G. 1926. Eine Verallgemeinerung der Quantenbedingungen für die Zwecke der Wellenmechanik. *Z. Phys.* 38: 518–529

Yacoby, A., Heiblum, M., Mahalu, D., and Shtrikman, H. 1995. Coherence and phase sensitive measurements in a quantum dot. *Phys. Rev. Lett.* 74: 4047–4050

PROBLEMS

1. For a potential barrier with $V(x) = 0$ for $x > |a/2|$, and $V(x) = 0.3$ eV for $x < |a/2|$, plot the tunneling probability for E in the range 0–0.5 eV. Take the value $a = 5$ nm and use the effective mass of GaAs, $m^* = 6.0 \times 10^{-32}$ kg.
2. For a potential barrier with $V(x) = 0$ for $x > |a/2|$, and $V(x) = 0.4$ eV for $x < |a/2|$, plot the tunneling probability for E in the range 0–0.5 eV. Take the value $a = 5$ nm and use the effective mass of GaAs, $m^* = 6.0 \times 10^{-32}$ kg.

3 Consider the potential barrier discussed in problem 1. Suppose that there are two of these barriers forming a double-barrier structure. If they are separated by 4 nm, what are the resonant energy levels in the well? Compute the tunneling probability for transmission through the entire structure over the energy range 0–0.5 eV.

4 Suppose that we create a double-barrier resonant tunneling structure by combining the barriers of problems 1 and 2. Let the barrier with $V_0 = 0.3$ eV be on the left, and the barrier with $V_0 = 0.4$ eV be on the right, with the two barriers separated by a well of 4 nm width. What are the resonant energies in the well? Compute the tunneling probability through the entire structure over the energy range 0–0.5 eV. At an energy of 0.25 eV, compare the tunneling coefficient with the ratio of the tunneling coefficients (at this energy) for the barrier of problem 2 over that of problem 1 (i.e. the ratio T_{min}/T_{max}).

5 Let us consider a trapezoidal potential well, such as that shown in the figure below. Using the WKB method, find the bound states within the well. If $V_1 = 0.3$ eV, $V_2 = 0.4$ eV, and $a = 5$ nm, what are the bound-state energies?

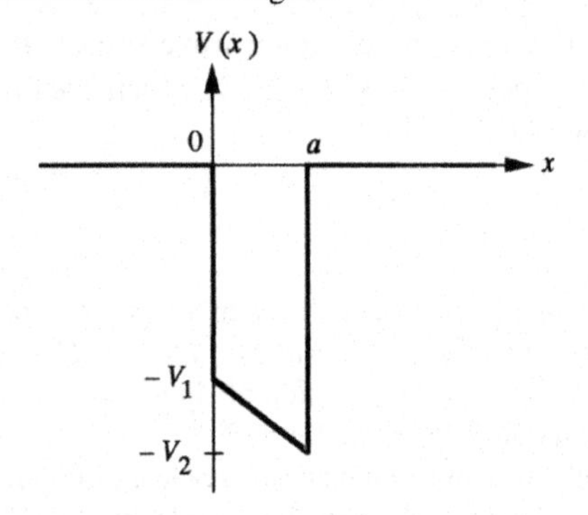

6 A particle is contained within a potential well defined by $V(x) \to \infty$ for $x < 0$ and $V(x) = ax$ for $x > 0$. Using the WKB formula, compute the bound-state energies. How does the lowest energy level compare to that found in (3.59)?

7 Consider the tunneling barrier shown below. Using the WKB form for the tunneling probability $T(E)$, calculate the tunneling coefficient for $E = V_0/2$.

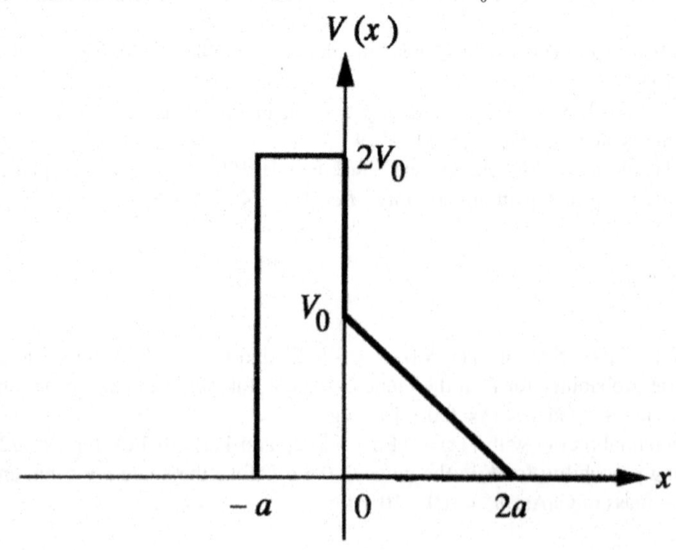

8 A particle moves in the potential well $V(x) = ax^4$. Calculate the bound states with the WKB approximation.

9 In the WKB approximation, show that the tunneling probability for a double barrier (well of width b, barriers of width $2a$, as shown in Figure 3.5, and a height of each barrier of V_0) is given by

$$T = \frac{4}{\left(4\theta^2 + 1/\left(4\theta^2\right)\right)\cos^2 L + 4\sin^2 L}$$

where

$$\theta = \exp\left(\int_b^{b+2a} \gamma(x)\, dx\right)$$

and

$$L = \int_0^b k(x) dx.$$

What value must b have so that only a single resonant level exists in the well?

10 Consider a single rectangular barrier in which the barrier height is 2.5 eV. A free electron of energy 0.5 eV is incident from the left. If the barrier is 0.2 nm thick, what are the tunneling and reflection coefficients?

11 In an infinite potential of width 15 nm, an electron is initialized with the wave function $\psi(x) = x(1 - x/a)e^{-x}$, where a is the width of the well. Develop a time-dependent solution of the Schrödinger equation to show how this wave function evolves in time.

12 Solve the two-dimensional Schrödinger equation for a two-dimensional infinite potential well with $a_x = 5$ nm and $a_y = 7$ nm. Determine the seven lowest energy levels.

13 Consider a potential barrier which is described by $V(x) = -Ax^2$. Using the WKB method, compute the tunneling coefficient as a function of energy for $E < 0$.

4 Periodic Potentials

The atoms in nearly all carbon compounds and nearly all semiconductors are covalently bonded. Covalent bonding involves the sharing of electrons among different atoms. In the hydrogen molecule (H_2), for example, the covalent bond contains two electrons. One electron is contributed from each atom of the molecule. Each of these two electrons moves back and forth between the two atoms (this is what the chemists call the resonating bond), and on average are situated between the atoms. In this special case, the two electrons come from the $1s$ level of the atoms. When we move into solids, in particular crystalline solids, the nature of the sharing becomes a little bit more complicated. This is because the potential energy in which the atom sits becomes periodic throughout the lattice. If we focus on an atom in say Si at point A and examine the potential landscape in which this atom sits, it will be exactly the same as any other atom in the crystal. When we decide to solve the Schrödinger equation for this system, we now need to involve a very great number of atoms and electrons in the solution. One can easily understand that this is an intractable problem. But we make some approximations through which the problem can be attacked in a manner to gather some results. The most important one is the adiabatic approximation, in which we realize that the atoms are heavy while the electrons are light. Thus, we can conceive that the atom moves around rather slowly, while the electrons are zipping around with a greater speed. For all practical purposes then, we can make the assumption that the atoms are frozen in their atomic positions when we solve for the electron quantum mechanics. Similarly, we may assume that the electrons instantaneously respond to the atomic motion (infinitely fast) when we solve the latter quantum mechanics. Here, we will not deal with the motion of the atoms themselves, but deal with the electrons and the periodic potential that they see. There is of course one other part, and that is the interaction between an atom and the electrons that are nearby. This is typically handled with perturbation theory, which is the topic of Chapters 7 and 8.

When we say that a potential is periodic, we mean that two atoms, separated by a distance a, will see exactly the same potential. That is, we say that the potential satisfies

$$V(x+a)=V(x), \tag{4.1}$$

in one dimension (which we will use throughout this chapter). The time independent Schrödinger equation for these two atoms may be written as

$$\begin{aligned}
-\frac{\hbar^2}{2m}\nabla^2\psi(x)+V(x)\psi(x)&=E\psi(x)\\
-\frac{\hbar^2}{2m}\nabla^2\psi(x+a)+V(x+a)\psi(x+a)&=E\psi(x+a).
\end{aligned} \tag{4.2}$$

If we now use (4.1) in the second of Equation (4.2), we may write it as

$$-\frac{\hbar^2}{2m}\nabla^2\psi(x+a)+V(x)\psi(x+a)=E\psi(x+a). \tag{4.3}$$

This means that $\psi(x+a)$ satisfies the same equation as $\psi(x)$. Since the probability that arises from the squared magnitude of the wave functions must be the same at each atom, these two wave functions can differ by at most a phase factor

$$\psi(x+a)=e^{i\varphi}\psi(x). \tag{4.4}$$

How do we determine this phase factor? What other constraints does the periodic potential impose upon the wave function? These two questions are at the heart of the matter, and we will address them in this chapter. The answers to these two questions are the basic tenets of what we call condensed matter physics, and this includes the properties of semiconductors that are important to our electronic world of semiconductor devices. We will begin with the phase factor by examining periodicity on at least two levels.

4.1 ATOMS ON A LATTICE

We begin by assuming a linear chain of atoms uniformly spaced by the lattice constant a. This is shown in Figure 4.1. Below this chain of atoms, the periodic potential is plotted as a guide to the reader to understand how each atom sits in a potential minimum. At this point it is stable, although of course it has thermal vibrations around the minimum. But those are not our concern here. We assume that each atom has a single electron. As we move from one atom to the next in the chain, the wave function for the electron at the next atom is related to the wave function at the current atom by (4.4). Hence, the periodicity of the potential has introduced one level of periodicity into our problem. But we now have to face the problem of boundary conditions. We handle this by introducing the concept of periodic boundary conditions. With periodic boundary conditions, atom N in the figure folds back onto atom 0, so that *these are the same atom*. That is, we wrap the linear array of atoms into a ring structure, so that the last becomes the first. This means that the wave function must adopt the phase in such a way that

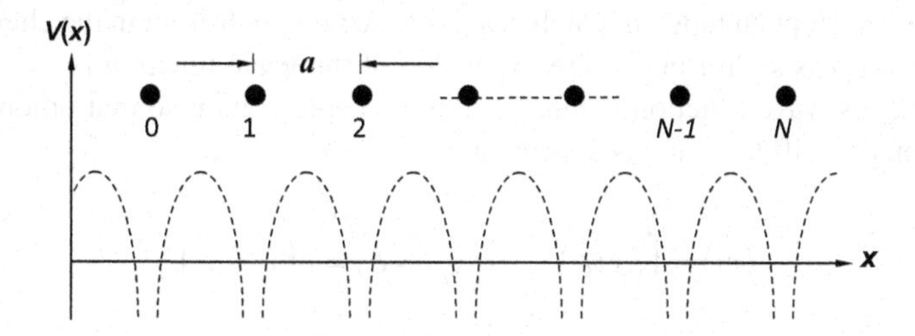

FIGURE 4.1 A one-dimensional chain of atoms and the corresponding periodic potential (dashed) for this system. We have assumed N atoms and will close the line as a ring so that atom N is folded back onto atom 0.

$$\psi(x+Na)=\psi(x+L)=\psi(x). \tag{4.5}$$

If we now combine (4.4) and (4.5), we see that

$$N\varphi=2\pi\rightarrow\varphi=\frac{2\pi}{N}. \tag{4.6}$$

Of course, the value given on the right of (4.6) is the smallest possible value. We could have taken many multiples of 2π. Because of this, we can allow φ to have many values, such as

$$\varphi=\frac{2\pi n}{N},\quad n=1,2,\ldots,N. \tag{4.7}$$

We cannot allow n to go past N, because we then begin to get repetitions of each phase value. So, the smallest allowed value for φ is $2\pi/N$, and there are n independent values. (Note that we don't need a value of 0, because $n = N$ produces 2π, which is the equivalent of 0 phase.)

In (4.5), we also introduced another notation, with $L = Na$. Let us replace N with the corresponding value given by this relation, so that (4.7) becomes

$$\varphi=\frac{2\pi n}{L}a\ ,\quad n=1,2,\ldots,N. \tag{4.8}$$

The units on the prefactor $2\pi n/L$ are inverse distance (1/cm). In electromagnetics, this inverse distance is called the wave number k. This is the same k we encountered in Chapter 1 when dealing with light and particle waves. And it is the same k we used in Chapters 2 and 3. This is the wave number of the electron waves we are studying in the periodic potential. Just as in Chapter 2, the periodic potential allows just a fixed number of values for this wave number. If we have N atoms in the chain, we can have only N distinct values for k.

Now let us adopt an index j, which designates with which atom in the chain we are dealing. The basis set for our future expansion of the wave function is one in which each electron's wave function is localized upon a single atom so that othonormality for electrons on different atoms appears as

$$\langle i|j\rangle = \int dx \psi^*(x_i)\psi(x_j) = \delta_{ij} = \delta(x_i - x_j). \tag{4.9}$$

In addition, since each atom and its electron are the same at each lattice site they have the same electronic energy. Hence, in the Hamiltonian for the system, we may write (in the absence of the potential)

$$H_0|j\rangle = H_0\psi(x_j) = E_1|j\rangle = E_1\,\psi(x_j). \tag{4.10}$$

Here, and in the previous equation, we have introduced the Dirac short-hand notation of using the index in the ket |j> to represent the wave function $\psi(x_j)$. The periodic potential will now produce some interactions between the electrons on adjacent sites. We represent this interaction by a potential V_P. We consider this interaction to occur between nearest neighbors, and we can write the Schrödinger equation as

$$H_0|j\rangle + V_P|j+1\rangle + V_P|j-1\rangle = E|j\rangle. \tag{4.11}$$

We now operate with the adjoint operator <j|, which indicates integration over space, and

$$E_1 + \langle j|V_P|j+1\rangle + \langle j|V_P|j\text{-}1> = E\rangle. \tag{4.12}$$

We now use (4.4) and the description of k following (4.8) to write

$$|j \pm 1\rangle = e^{\pm ika}|j\rangle, \tag{4.13}$$

where the exponential is exactly the real-space *displacement operator* in quantum mechanics and shifts the wave function by one atomic site. We are then left with the integration of the onsite pseudo-potential, which we define to be

$$\langle j|V_P|j\rangle = -A. \tag{4.14}$$

Using the above expansions and evaluation of the overlap integrals that appear in the equations, we may write (4.12) as

$$\begin{aligned} E &= E_1 - A\left(e^{ika} + e^{-ika}\right) \\ &= E_1 - 2A cos(ka). \end{aligned} \tag{4.15}$$

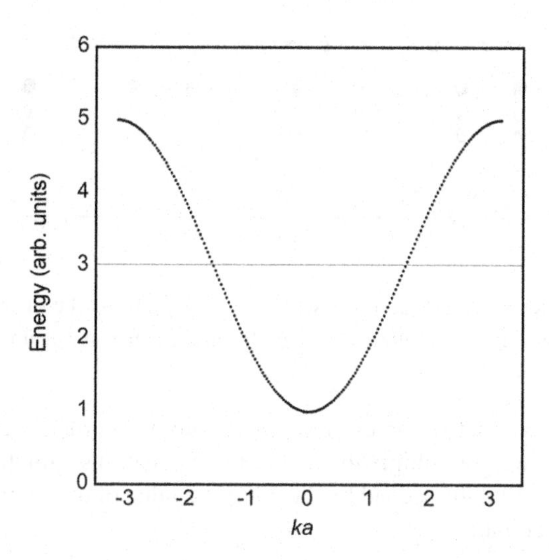

FIGURE 4.2 Plot of the energy for the linear atomic chain from (4.15). Here, arbitrary values of $E_1 = 3$ and $A = 1$ have been chosen for the plot.

This energy structure is plotted in Figure 4.2. The band is $2A$ wide (from lowest to highest energy), and is centered about the single site energy E_1, which says that the band forms by spreading around this single atom energy level. The band contains N values of k, as there are N atoms in the chain, and this is the level of quantization of the momentum variable, as was shown earlier. The minimum of the band appears at $ka = 0$ and maxima appear at $ka = \pi$.

Suppose we now add a second atom per unit cell, so that we have a diatomic basis. This is illustrated in Figure 4.3. Here, each unit cell contains the two atoms, and the need for a diatomic lattice arises from the unequal spacing. The lattice can be defined with either of the two atoms, but each lattice site contains a *basis* of two atoms. This will significantly change the electronic structure. We will have two wave functions, one for each of the two atoms in the basis.

This, in turn, leads us to have to write two equations to account for the two atoms, which we will call the A and B atoms. And there are two distances involved in the atom interactions. There is one interaction in which the two atoms are separated by b and one interaction in which the two atoms are separated by a–b. The lattice constant, however, remains a. To proceed, we need to adopt a slightly more complicated notation. We will assume that the A atoms will be indexed by the site j, when j is an even number (including 0). Similarly, we will assume that the B atoms will be indexed by the site j when j is an odd number. We have to write two equations, one for when the central site is an A atom and one for when the central site is a B atom. It really doesn't matter which sites we pick, but we take the adjacent sites so that these equations become

$$\begin{aligned} &H_0|j\rangle + V_P|j+1\rangle + V_P|j-1\rangle = E|j\rangle \\ &H_0|j+1\rangle + V_P|j+2\rangle + V_P|j\rangle = E|j+1\rangle. \end{aligned} \tag{4.16}$$

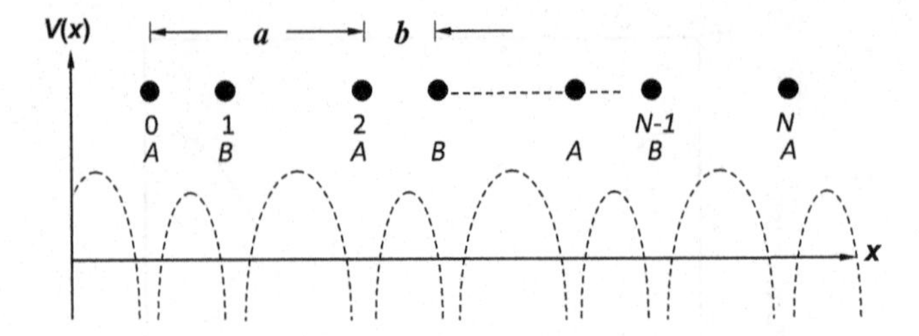

FIGURE 4.3 A linear chain of atoms, in which the spacing is not the same. Now, we require a basis of one A atom and one B atom, although the lattice constant remains a.

We will assume here that j is an even integer (A atom) in order to evaluate the integrals. Once again, we premultiply the first of these equations with the complex value of the central site j, and the second by the complex value of the central site $j + 1$, and integrate, we then obtain

$$\begin{aligned} E_1 + \langle j|V_P|j+1\rangle + \langle j|V_P|j-1\rangle &= E \\ E_1 + \langle j+1|V_P|j+2\rangle + \langle j+1|V_P|j\rangle &= E. \end{aligned} \tag{4.17}$$

Now, we have four integrals to evaluate, and these are given by the parameters

$$\begin{aligned} \langle j|V_P|j+1\rangle &= e^{ikb}\langle j|V_P|j\rangle \equiv e^{ikb}A_1 \\ \langle j V_P|j-1\rangle &= e^{ik(b-a)}\langle j|V_P|j\rangle \equiv -e^{ik(b-a)}A_2 \\ \langle j+1|V_P|j+2\rangle &= e^{-ik(b-a)}\langle j|V_P|j\rangle \equiv -e^{-ik(b-a)}A_2 \\ \langle j+1|V_P|j\rangle &= e^{-ikb}\langle j|V_P|j\rangle \equiv e^{-ikb}A_1. \end{aligned} \tag{4.18}$$

Here, we have taken the overlap integrals to be different for the two different atoms. The choice of the sign on the A_2 terms assures that the gap will occur at $k = 0$. The choice of in which direction to translate the wave functions is made to ensure that the Hamiltonian is Hermitian. The two equations now give us a secular determinant which must be solved, as

$$\begin{vmatrix} (E_1 - E) & \left(A_1 e^{ikb} - A_2 e^{ik(b-a)}\right) \\ \left(A_1 e^{-ikb} - A_2 e^{-ik(b-a)}\right) & (E_1 - E) \end{vmatrix} = 0. \tag{4.19}$$

It is clear that the two off-diagonal elements are complex conjugates of each other, and this assures that the Hamiltonian is Hermitian and the energy solutions are real. This basic requirement must be satisfied, no matter how many dimensions we have in

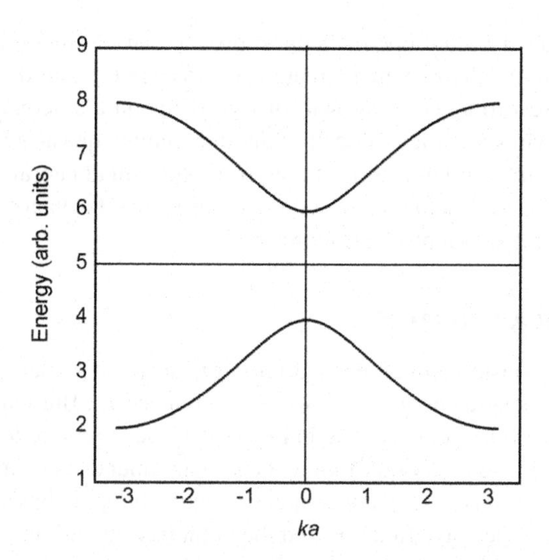

FIGURE 4.4 The resulting energy bands for the one-dimensional chain with two atoms per unit cell. Here, (4.20) is plotted for the values $E_1 = 5, A_1 = 2, A_2 = 1$. The two bands are mirrored around the value 5, and each has a band width of 2.

the lattice and is a basic property of quantum mechanics—in order to have real measurable eigen-values, the Hamiltonian must be Hermitian. We now find that two bands are formed from this diatomic lattice, and these are mirror images around an energy midway between the lowest energy of the upper band and the highest energy of the lower band. The energy is given by

$$E = E_1 \pm \sqrt{A_1^2 + A_2^2 - 2A_1A_2\cos(ka)} \tag{4.20}$$

The two bands are shown in Figure 4.4 for the values $E_1 = 5, A_1 = 2, A_2 = 1$. The two bands are mirrored around the value 5, and each has a band width of 2.

As in the monatomic lattice, each band contains $N/2$ states of momentum, as there are $N/2$ unit cells in the lattice. Here, it is important to note the number of values of k are given by the number of unit cells and not by the number of atoms. Since each unit cell contains two atoms, one A atom and one B atom, the number of units defined by N atoms is just $N/2$ as noted. We will see later how this affects the manner in which the electrons are distributed in the bands.

While it is easy to consider this discussion of the one-dimensional linear chain mainly as a tutorial of limited interest, one-dimensional (or quasi-one-dimensional) materials have appeared in recent years. Foremost among these are the molecules bonded to metals or semiconductors for studies of their conduction properties. But one can also have pseudo-one-dimensional conduction in biological ion channels which translate potassium and sodium into, and out of, individual cells. But there have also been quasi-one-dimensional structures created by putting single phosphorous atoms in a chain embedded in un-doped silicon to produce an n^+ silicon nanowire. Then, one can make atomic chains which create nanowires in e.g., gold. But the primary approach of interest to us is the growth of nanowires on a silicon substrate. While these latter nanowires are not atomically thin, they remain quasi-one-dimensional in nature due to the strong quantization arising from their small lateral dimensions. The one-dimensional band structure provides a first estimate of the nanowire bands, while the transverse quantization provides more detail.

4.2 ANOTHER APPROACH

In the above, we considered that each atom in the array possessed a single electron which produced a single energy level, we we designated E_1. But that is only one assumption, key as it was. Suppose that, in Figure 4.1, the atoms sit down well into the quantum wells, and we really don't know how many bound states may exist in each of the potential wells. We can pursue another approach that is based upon the work in Chapter 2, using the wave function and the boundary conditions. For this, we will assumed that the nice barriers shown in Figure 4.1 are replaced by square potential barriers, which will simplify the mathematics.

We consider an array of quantum wells, which are spaced by barriers sufficiently thin that the wave functions can tunnel through in order to couple the possible bound states in the wells together (Kronig and Penney 1931). We still have a *periodic* potential. Although the rectangular potential model is only an approximation, it enables us to develop the essential features, which in turn will not depend crucially upon the details of the model. Just as in the previous section, this model has importance in the energy band structure of crystalline media, such as semiconductors in which the atoms are arranged in a periodic array, and the atomic potentials create a periodic potential in three dimensions in which the electrons must move. The important outcomes we will achieve from this model are the existence of ranges of allowed energies, called bands, and ranges of forbidden energies, called gaps, exactly as we have already found in Figures 4.2 and 4.4. Here, we shall see just how periodic potentials give rise to such bands and gaps in the energy spectrum.

Each potential in the array is represented by the simple model shown in Figure 4.5. Our interest is in the filtering effect such a periodic structure has on the energy spectrum of electron waves. The periodic potential has a basic lattice constant (periodicity) of $d = a + b$, where we take b to be the total thickness of each barrier and a is the spacing between the potentials. We are interested in states in which $\mathcal{E} \ll V_0$.The Schrödinger equation now becomes

$$-\frac{\hbar^2}{2m}\nabla^2\psi_1(x) - E\psi_1(x) = 0, \quad 0 < x < a, \tag{4.21a}$$

and

$$-\frac{\hbar^2}{2m}\nabla^2\psi_1(x) + V_0\psi_1(x) - E\psi_1(x) = 0, \quad -b < x < 0. \tag{4.21b}$$

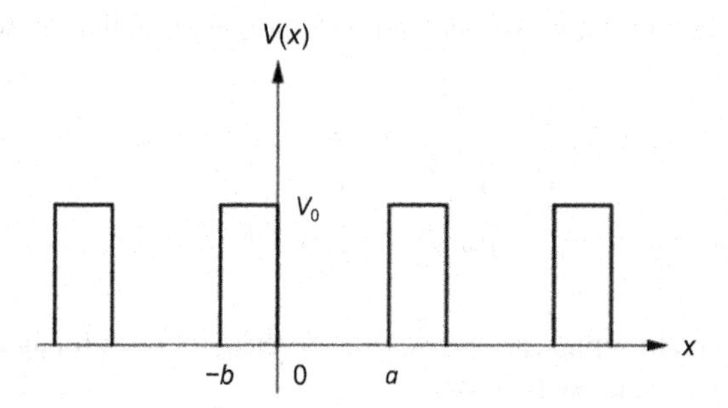

FIGURE 4.5 A simple periodic potential.

Here, we have taken the origin of our x-axis to sit at the right edge of one of the potential barriers for convenience. Of course, shifts of the x-axis by the amount d bring in other regions in which (4.21) is found to be the appropriate equation. Nevertheless, there will be a point at which we will *force* the periodicity onto the solutions. We also expect that the wave function will be periodic with the same periodicity as the potential, or

$$\psi_1(x) = e^{iKx}u(x), \tag{4.22}$$

where $u(x)$ has the periodicity of the lattice. Here, K is related to the wave numbers given in (4.8), but for now it is just a variable with the dimensions of inverse distance. A wave of the form (4.22) is termed a Bloch function. If we insert (4.22) into (4.21), the resulting equations are

$$\begin{aligned} &\frac{d^2u_1}{dx^2} + 2iK\frac{du_1}{dx} + \left(k^2 - K^2\right)u_1 = 0, \quad 0 < x < a, \\ &\frac{d^2u_2}{dx^2} + 2iK\frac{du_2}{dx} + \left(\gamma^2 + K^2\right)u_2 = 0, \quad -b < x < 0. \end{aligned} \tag{4.23}$$

In these equations the parameters γ and k have their normal meanings from Chapter 2 as

$$\begin{aligned} k &= \sqrt{\frac{2mE}{\hbar^2}} \\ \gamma &= \sqrt{\frac{2m}{\hbar^2}\left(V_0 - E\right)} \end{aligned}. \tag{4.24}$$

We notice that, in (4.23), we have introduced two different functions for the cell periodic part of the Bloch function, one in the area between the barriers and one within the barrier. To move forward, we may make the assumption that these two functions can be defined as

$$\begin{aligned} u_1 &= Ae^{-i(K-k)x} + Be^{-i(K+k)x} \\ u_2 &= Ce^{-(iK-\gamma)x} + De^{-(iK+\gamma)x} \end{aligned}. \tag{4.25}$$

These solutions again represent waves, in each case (either propagating or evanescent), one propagating in each direction.

There are now four unknowns, the coefficients that appear in (4.25). However, there are only two boundaries in effect. Hence, we require that both the wave function and its derivative be continuous at each boundary. However, it is at this point that we will force the periodicity onto the problem via the choice of matching points. This is achieved by choosing the boundary conditions to satisfy

$$\begin{aligned} u_1(0) &= u_2(0) \\ u_1(a) &= u_2(-b) \\ \frac{du_1(0)}{dx} &= \frac{du_2(0)}{dx} \\ \frac{du_1(a)}{dx} &= \frac{du_2(-b)}{dx} \end{aligned}. \tag{4.26}$$

The choice of the matching points, specifically the choice of $-b$ instead of a on u_2, causes the periodicity to be imposed upon the solutions. These four equations lead to four equations for the coefficients, and these form a homogeneous set of equations. There are *no forcing* terms in the equations. Thus, the coefficients can differ from zero only if the determinant of the coefficients vanishes. This leads to the determinantal equation

$$\begin{vmatrix} 1 & 1 & -1 & -1 \\ e^{-i(K-k)a} & e^{-i(K+k)a} & e^{(iK-\gamma)b} & e^{(iK+\gamma)b} \\ -i(K-k) & -i(K+k) & iK-\gamma & iK+\gamma \\ -i(K-k)e^{-i(K-k)a} & -i(K+k)e^{-i(K+k)a} & (iK-\gamma)e^{(iK-\gamma)b} & (iK+\gamma)e^{(iK+\gamma)b} \end{vmatrix} = 0. \tag{4.27}$$

Evaluating this determinant leads to

$$\frac{\gamma^2 - k^2}{2k\gamma} sinh(\gamma b) sin(ka) + cosh(\gamma b) cos(ka) = cos\left[K(a+b)\right]. \tag{4.28}$$

It is clear that all of the parameters are set by the system and the boundaries. The energy, which is yet to be determined, is in the parameter k, the barrier height is in γ,

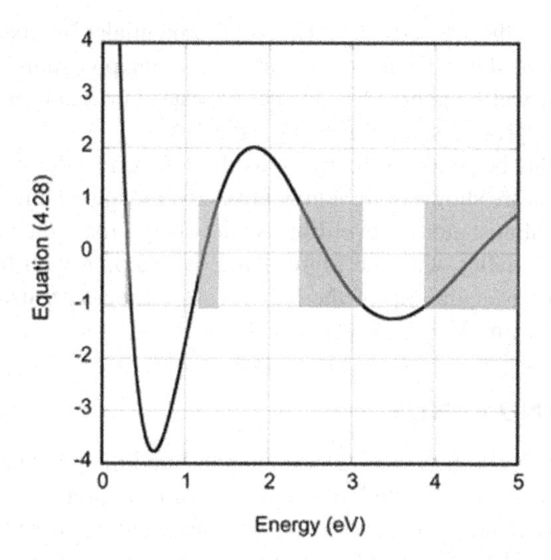

FIGURE 4.6 The allowed and forbidden values of *Ka*. The shaded areas represent the allowed range of energy values. The parameters used for the figure are discussed in the text.

and our wave (propagation) vector is K. Note that, in general, the magnitude of the left-hand side of (4.28) can be greater than unity. But the right-hand side always has a magnitude limited by unity. Hence, (4.28) *has solutions only when the magnitude of the left-hand side is less than unity*. For any other values of the parameters, the wave nature is forbidden. These are the energy gaps. The values of parameters that allow for a solution yield the energy bands.

The solution to (4.28) can be found graphically, as shown in Figure 4.6. The ranges of k, a for which (4.28) is satisfied are known as the allowed states. Other values are known as forbidden states. The allowed states group together in bands, given by the case for which the left-hand side traverses the range [−1, 1]. Each allowed band is separated from the next by a forbidden gap region, for which the left-hand side has a magnitude greater than unity.

In this model, k is a function of the energy of the single electron, so the limits on the range of this parameter are simply the limits on the range of allowed energies. If this is the case, then the results should agree with the results for free electrons and for bound electrons. In the former case, the prefactor of the first term in (4.28) vanishes, and we are left with $k = K$. Thus, the energy is just given by the wave vector in the normal manner. On the other hand, when the prefactor goes to infinity, we are left with

$$sin(ka) = 0, \quad k = \frac{n\pi}{a}. \tag{4.29}$$

which produces the bound-state energies of the last chapter. Thus, the approach used here does reproduce the limiting cases that we have already treated. The periodic potential breaks up the free electrons (for weak potentials) by opening gaps in the spectrum of allowed states. On the other hand, for strong potentials, the periodic tunneling couples the wells and broadens the bound states into bands of states. These are the two limiting approaches, but the result is the same.

One can see this behavior in the figure as the left-hand side of (4.28) oscillates. The allowed bands are shown in light gray boxes. At the lowest energy, the bands are very narrow, and almost atomic level-like. As the energy rises, the bands get broader and the gaps get smaller, with the fourth band overlapping with the continuum of states for energies over the top of the barriers. Here, we have used parameters as $a = 1.2$ nm, $b = 0.2$ nm, $V_0 = 5$ eV, and $m = 0.7m_0$.

4.3 BONDS AND BANDS

Let us summarize a bit on the above developments and think about what they mean to us. First, when we had a single electron/atom in the unit cell, we found that a single band formed and there were N states in this band, as there were N unit cells in the linear array. Now, the founders of quantum mechanics always seemed to find that the electron possessed an additional quantum number. That is, in one dimension, the electron seemed to have two quantum numbers, whereas one-dimensional confinement only involved one of these (the value of k and its corresponding energy). Similarly, in three dimensions, the electron seemed to have four quantum numbers, whereas three-dimensional confinement only accounted for three of these. What was this fourth quantum number?

At the end of the nineteenth century, Pieter Zeeman was studying the absorption spectrum of various atomic gasses and found that a magnetic field split the absorption line into a pair, whose energy difference was linearly dependent upon the magnetic field (Zeeman 1897a, 1897b, 1897c). Three-dimensional studies of the atomic structure of a simple atom like H could not produce this effect. Others repeated the experiments with the same results. For example, in 1921, Otto Stern and Walter Gerlach created an experiment with a very nonuniform magnetic field and sent a collimated beam of Ag atoms through it. In a normal (classical) system with a magnetic moment (as the electrons moving around the atom would have), a magnetic field will cause the magnetic moment to precess, much like a top spinning on its axis. In an inhomogeneous magnetic field, the force on the top and bottom of the "top" are different, and this should cause a deflection of the atom, which is detected by a sensitive screen. Because the magnetic moment is random in such an array of atoms, one expects to see a line on the screen, with the length being determined by the maximum range of this moment. What they found, however, was just two spots (Stern and Gerlach, 1921), just as Zeeman had found. Hence, the electron must have an extra magnetic moment which possessed only two values, taken to be "up" and "down." Later, Uhlenbeck and Goudsmit (1925) proposed that this extra magnetic moment was the electron spinning upon its own axis. Thus, this extra quantum number has come to be called the electron spin, as it is assumed to be spinning around its own axis. We will return to a fuller discussion of the spin and its properties in Chapter 10.

Here, however, we note that the angular momentum has dimensions $\sim mrv$, where m is the mass of the electron, r is the radius arm for rotation, and v is the angular velocity. These units combine to *kg•m²/s* or J-sec, the same units as $\hbar$. In fact, the two spots seen by Stern and Gerlach corresponded to angular momentum values of $\pm\hbar/2$, so that we say that the electron has spin of $\pm 1/2$. That is, the spin is a half-integer value, and particles with half-integer spins are known as *fermions*. We digress here to point out that there are only four quantum numbers for each state of the electron due to the *Pauli exclusion principle* which invoked this exact condition (Pauli 1926). Thus, Pauli stated that an electron can sit in a three-dimensional state described by three quantum numbers, but only two electrons could occupy this state and then only if they had opposite values of spin. Particles that satisfy this exclusion principle then satisfy Fermi–Dirac statistics, where the probability to occupy a state of energy E is given by

$$P(E) = \frac{1}{1 + exp\left[\left(E - E_F\right)/k_B T\right]}, \tag{4.30}$$

where E_F is termed the Fermi energy—the energy at which the probability is exactly ½. Because these particles satisfy these statistics, they are called fermions.

The uptake of this revelation is that our electron in the one-dimensional array discussed at the start of this chapter has a more complicated life. Now, each of the *N* values of the wave vector *k* can now hold two electrons, provided they have opposite spin (Pauli 1926). So the single band now can contain 2*N* electrons. But there are only *N* atoms and electrons. As a result, only the states in the lower half of the band will be filled. Since the electrons at the top of the distribution can easily be accelerated by an external electric field, this material will be classified as a metal. A metal has a large number of nearly-free electrons, who can easily be accelerated and possesses something like a half-filled energy band.

When we have two atoms, and two electrons, in each unit cell, as in the second example above, the situation is more complicated. Now, the two atoms combine to yield two energy bands. The lower band is termed the bonding band and arises from a cooperative interaction between the electrons on the two atoms per unit cell. The second, higher energy, band is termed the antibonding band. These two bands also are typically called the valence and conduction bands, respectively. For the diatomic lattice, we find that there are only *N*/2 unit cells, so that there are only *N*/2 values for *k*. Yet, each of these *k* states can hold two electrons, which have opposite spins. Thus, the lower, valence band can hold *N* electrons, and that is precisely how many electrons and atoms are present in the lattice. So, the lower, valence band will be completely full, and the upper, conduction band will be completely empty. Thus, this material will be an insulator or semiconductor, the difference being the size of the energy gap between the two bands. If electrons can be thermally excited over the gap, then it is a semiconductor. Otherwise, it is an insulator.

The last example we examined, in Section 4.2, produced several energy bands, which means that the single atom in the quantum well could lead to several distinct energy levels. Hence, it could correspond to a multi-electron atom. We can apply this to semiconductors. Typical semiconductors, such as our ubiquitous Si, come from column IVB of the periodic table. The atoms in this column have an outer shell that

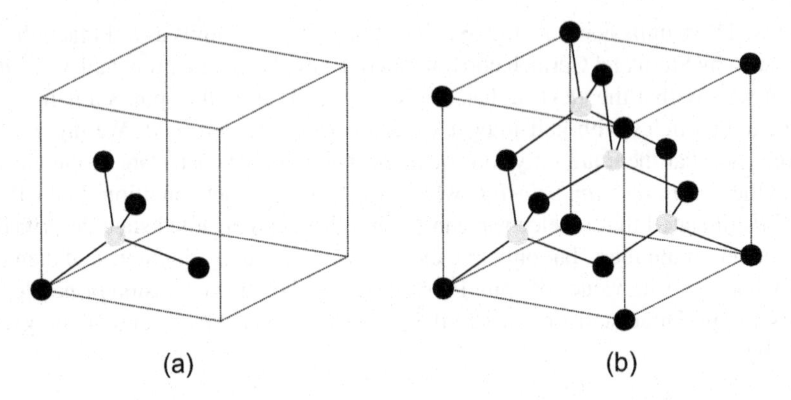

FIGURE 4.7 (a) The tetrahedral bond surrounding the central atom (light gray). (b) Groups of these atoms fit together for the diamond structure, which is a face-centered cubic lattice with a basis of two atoms (the lower left corner plus the light gray atom).

could contain a maximum of eight electrons but have only four electrons in this shell. Hence, they can come together with like atoms to form covalent bonds. Usually, in three dimensions, these outer shell electrons sit in an *s* state and 3*p* states (these states will be defined further in Chapter 9). These tend to form four hybrid wave functions which connect to the four nearest neighbors in the diamond lattice. These four nearest neighbors form a tetrahedron, so we refer to this as tetrahedral bonding. The tetrahedral bonding is depicted in Figure 4.7(a). The important aspect is that the crystal structure has two atoms per unit cell. The four tetrahedral bonds now will form four bands, so that there are $4(N/2)$ values for the k vectors. With spin, this means that there are 4N total states in the four bonding bands. Since we actually have N atoms, each with four electrons, we have just enough electrons to completely fill the bonding bands, leaving the conduction band empty. Si, and the semiconductors discussed here, crystallize in the face-centered cubic lattice, with two atoms per lattice site, as shown in Figure 4.7(b). Hence, Si is a semiconductor, as electrons can be thermally excited over its 1 eV band gap.

4.4 ELECTRON PAIRING—SUPERCONDUCTIVITY

Kammerlingh Onnes was the first person to be able to liquify helium at very low temperatures. Once he had done this, he spent most of his time examining the properties of various materials at this low temperature of 4.2 K. In studying Hg, he found a totally unexpected result, in that as he cooled Hg below the temperature of liquid helium, the resistance dropped to an unmeasurable value (Kammerlingh Onnes 1910). At 3 K, the resistance was less than 10^{-6} ohm. This phenomena is now known as *superconductivity*, and has been observed in a significant fraction of the known elements. The onset of superconductivity is considered to be a phase transition, and the critical temperature below which it occurs is called the *transition temperature* T_c. Our interest here, in the concept of superconductivity, is that it is an important quantum mechanical effect and one which modifies the simple energy

TABLE 4.1
Superconductor transition temperatures

Element	T_c (K)	Compund	T_c (K)
Pb	7.2	Nb_3Sn	18.05
V	3.72	V_3Ga	16.5
Ti	0.39	V_3Si	17.01
Nb	9.1	NbN	16.0
Ta	4.48	InSb	1.9
Hg	4.15	Nb_3Al	17.5
Zn	0.85	$YBa_2Cu_3O_7$	92
Sn	3.72	$HgBa_2Ca_2Cu_3O_8$	134

bands discussed in the previous sections of this chapter. This important modification was already mentioned in Section 3.4.5, and that is the pairing of electrons and their tunneling in metal-oxide-metal junctions when the two metals are superconducting—so-called Josephson tunneling.

4.4.1 Observable Properties

The most easily observed property of superconductors is the transition temperature, T_c. If one plots the resistance as a function of temperature, there is a clear drop in the resistance by orders of magnitude at this transition temperature. Just above the transition temperature, there is a gradual drop in resistance, that gives a "rounding" to the curve, and this is caused by thermal fluctuations at the transition temperature. These fluctuations are believed to be the fact that not all of the electrons are paired except at absolute zero temperature. Above T_c, on average none of the electrons are paired, so that the fraction that are paired is a function of temperature. In this region of enhanced conductivity above T_c, it is thought that small regions of the material are beginning to become superconducting, with the entire sample becoming so at further reductions of the temperature.

In general, metals which have very high conductivity do not become superconductors. On the other hand, metals which are poor conductors, such as Pb, Ta, Nb, Hg, and so on, become quite good superconductors. Some of the transition temperatures are given in Table 4.1. Also shown are a couple of members of a new class of materials known as high temperature superconductors, which have transition temperatures of tens of degrees Kelvin, up to ~200 K at high pressure. These latter materials tend to be cuprates and ceramics.

A magnetic field can be used to destroy the superconductivity. That is, there is a critical magnetic field H_c above which the superconductivity vanishes. Moreover, it is found that this critical field varies with temperature as

$$H_c = H_{c0}\left[1-\left(\frac{T}{T_c}\right)^2\right], \tag{4.31}$$

where H_{c0} is the critical field at absolute zero of temperature. Comparing different material, it is found that in general a higher critical temperature will lead to a higher critical magnetic field. This critical magnetic field will also limit the amount of current that the superconductor can carry, since the current gives rise to a magnetic field by Ampere's law. However, there are other materials, such as the compounds in Table 4.1, where the superconductivity does not end abruptly. In these materials, a lower critical magnetic field signals the beginning of the process and an upper critical field signals when superconductivity has finally gone away. These materials are sometimes called hard superconductors.

4.4.2 Pairing and Gaps

When we first discussed Josephson effects in Section 3.4.5, it was remarked that the electrons would pair up as Cooper pairs (Cooper 1956). Normally, electrons repel each other due to the Coulomb interaction between them. However, in metals there are typically something like a few times 10^{22} electrons/cm^3. Such a high electron concentration usually heavily screens the Coulomb interaction, so that it is a relatively weak interaction, especially if the electrons are not fairly close to one another. Cooper hypothesized that because of this weakness, electrons could actually interact with each other if there were another positive interaction. He suggested that one electron could interact with the atoms of the crystal in a manner that distorted the local atomic potential, creating a potential well into which a second electron could be drawn. He thought the process would work best if the two electrons had opposite spin. (We will deal with the dynamics of the lattice in Chapter 5, and the details of the interaction in Chapter 7.)

The combination of the two electrons, via the atomic interaction, creates the Cooper pair. Because the two electrons have opposite spin, the net spin of the Cooper pair is zero. This makes the Cooper pair into what is called a *composite boson*. Bosons are not required to satisfy the Pauli exclusion principle, so there is no limit to the number of bosons that can exist in any quantum state. Moreover, when we form the Cooper pair, it lies in a lower energy state—this lowering of the energy is the pair formation energy, which we called Δ in Section 3.4.5. The result of all of this process is observable by a gap that opens in the energy spectrum. We noted in the previous section that the energy band in Figure 4.2 was half-filled. Thus, the Fermi energy would sit at 3 eV in that figure. With the formation of the Cooper pairs, a gap will open around the value of 3 eV, as shown in Figure 4.8. The value of the gap is 2Δ, so that a Cooper pair must gain this energy to transition above the gap and become a pair of free electrons. The interesting point is that the two electrons in the Cooper pair need not be close to one another. Instead, it is generally thought that they can be some 100–400 nm apart, a length that is referred to as the *coherence length*. This means that there are a great number of other electrons near to the Cooper pair, or at very low temperature, the large number of Cooper pairs are all intertwined with one another. The energy gap that opens is also temperature dependent, varying as

$$E_G = E_{G0}\left(1-\frac{T}{T_C}\right)^{1/2}, \tag{4.32}$$

where this gap is the 2Δ mentioned above.

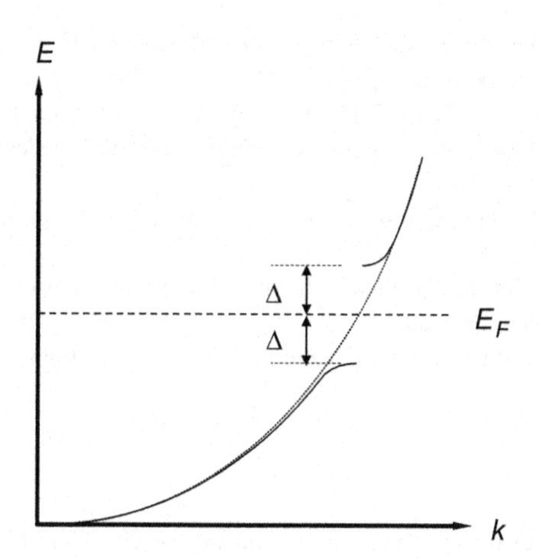

FIGURE 4.8 In superconductivity, the formation of Cooper pairs leads to the opening of a gap at the Fermi energy. This gap has a value of 2Δ, as shown.

It has been suggested that the Cooper pairs are the basic unit of superconductivity (Bardeen, Cooper, and Schrieffer 1957a, 1957b). In the BCS theory, named for these three authors, the critical temperature is found to depend upon the density of states at the Fermi energy and the strength of the interaction between the lattice and the electron that leads to forming the Cooper pairs. Because all the states are full below the energy gap, one would not expect conduction, but the nature of the bosonic Cooper pairs seems to indicate they can move through the lattice without dissipation. That is, they cannot be scattered, so move dissipation-free. And this seems to be the case. If one creates a coil of superconducting wire and induces a current through the wire, and then shorts the leads together, this current will continue to flow without dissipation. This current is known as a *persistent current* and is the heart of superconducting magnets.

REFERENCES

Bardeen, J., Cooper, L. N., and Schrieffer, J. R. 1957a. Microscopic theory of superconductivity. *Phys. Rev.* 106: 162–163

Bardeen, J., Cooper, L. N., and Schrieffer, J. R. 1957b. Theory of superconductivity. *Phys. Rev.* 108: 1175–1204

Cooper, L. N. 1956. Bound electron pairs in a degenerate Fermi gas. *Phys. Rev.* 104: 1189–1190

de L Kronig, R. and Penney, W. G. 1931. Quantum mechanics of electrons in crystal lattices. *Proc. Roy. Soc. A* 130: 499–513

Gerlach, W. and Stern, O. 1921. Der experimentelle Nachweiss des magnetischen Moments des Silberatoms. *Z. Phys.* 8: 110–111

Kammerlingh, Onnes K. 1910. Further experiments with liquid helium. C. On the change of electric resistance of pure metals at very low temperatures etc. IV. The resistance of pure mercury at helium temperatures. *Konig. Nederland. Akad. Wetensch. Proc.* 13: 1274–1276

Pauli, W. 1926. Über den Zusammenhang des Abschlusses der Elektronengruppen im Atom mit der Komplexstruktur der Spektren. *Z. Phys.* 31: 765–783

Uhlenbeck, G. and Goudsmit, S. 1925. Ersetzung der Hypothese vom unmechanischen Zwang durch eine Forderung bezüglich des inneren Verhaltens jedes einzelnen Elektrons. *Naturwiss.* 13: 953–954

Zeeman, P. 1897a. Phil. Mag., On the influence of magnetism on the light emitted by a substance. *Ser.* 5 43: 226–239

Zeeman, P. 1897b. Doublets and triplets in the spectrum produced by external magnetic forces. *Phil. Mag., Ser.* 5 44: 55–60

Zeeman, P. 1897c. The effect of magnetization on the nature of light emitted by a substance. *Nature* 55: 347

Problems

1 In (4.28), the values for which the right-hand side reach −1 must be satisfied by the left-hand side having $\cos(ka) = -1$, which leads to the energies being those of an infinite potential well. Show that this is the case. Why? The importance of this result is that the top of every energy band lies at an energy defined by the infinite potential well, and the bands form by spreading *downward* from these energies as the coupling between wells is increased.

2 Using the "Periodic potential—Kronig-Penney Lab" at NanoHub.org, repeat the calculation leading to Figure 4.6, using the parameters in the text. Vary the mass from 0.1 to 1.0 (in units of the free electron mass) and determine how many bands appear over this range.

5 The Harmonic Oscillator

One of the most commonly used quantum mechanical systems is that of the simple *harmonic oscillator*. The prototype system is the simple pendulum. Classically, the motion of the system exhibits harmonic, or sinusoidal oscillations. Quantum mechanically, the motion described by the Schrödinger equation is more complex. Although quite a simple system in principle, the harmonic oscillator assumes almost overwhelming importance due to the fact that it is one of the few systems that can be solved exactly in quantum mechanics. Almost any interaction potential may be expanded in a Tayor series with the low-order terms (up to quadratic order) cast into a form described by the harmonic oscillator. The sinusoidal motion of classical mechanics is the simplest system that produces oscillatory behavior, and therefore is found in almost an infinity of physical systems. Thus, the properties of the harmonic oscillator become important for their use in describing quite diverse physical systems. In this chapter, we will develop the general mathematical solution for the wave functions of the harmonic oscillator, using a simple operator algebra that allows us to obtain these wave functions and properties in a much more usable and simple manner. Classically, the harmonic oscillator is described in terms of Hermite polynomials, as the differential equation is one of the variants of the Sturm-Liouville problem. We can obtain these functions without the normal complication of solving the detailed differential equation, as the operator approach makes life easier. We then turn to two classic examples of systems in which we use the results for the harmonic oscillator to explain the properties: the simple *LC*-circuit and vibrations of atoms in a crystalline lattice. We the discuss motion in a quantizing magnetic field, as it is another system that reduces to a harmonic oscillator.

The simplest example of sinusoidal behavior in classical physics is that of a mass on a linear spring, in which the mass responds to a force arising from the extension or compression of the spring (Figure 5.1). The differential equation describing this is just

$$m\frac{d^2x}{dt^2} = F = -Cx, \tag{5.1}$$

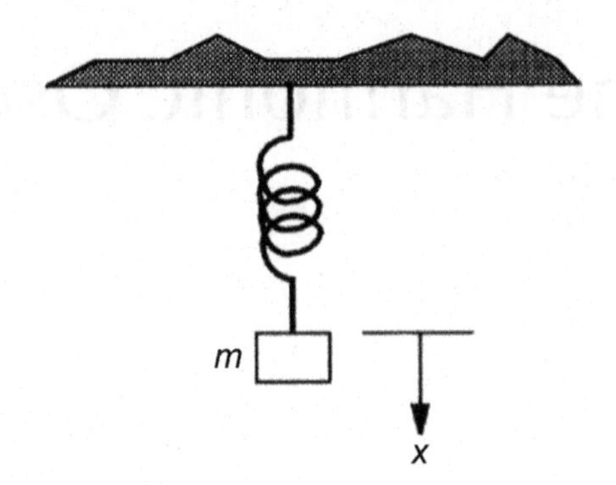

FIGURE 5.1 A simple mass on a spring.

where m is the mass and C is the spring constant. Instead of solving this equation classically, we instead write the total energy, using the square of the momentum $p = m(dx/dt)$ and the potential energy $V = Cx^2/2$ that arises from integrating the right-hand side, as

$$E = \frac{p^2}{2m} + \frac{m\omega^2 x^2}{2}, \tag{5.2}$$

where we have introduced the oscillator frequency ω through $C = m\omega^2$. This is the energy function that is used in the Schrödinger equation to find the quantum mechanical motion of the simple harmonic oscillator.

The simple pendulum is one of the first problems one attacks in introductory physics. Quite simply, a mass m is suspended by a rigid rod of length l from a fixed point, about which it can rotate. The angle that the rod makes with the vertical is given by θ. As the rod is pushed away from the vertical, gravity provides a restoring force of value mg, which is directed directly downward (Figure 5.2), and which leads to a force that tends to reduce the angular deflection of the rod, given by $-mg \sin \theta$. We may then write the differential equation for the angular deflection θ as

$$m\frac{d}{dt}\left(l\frac{d\theta}{dt}\right) = -mg\sin\theta. \tag{5.3}$$

The factor in the parentheses on the left-hand side is just the angular velocity of the pendulum, so the left-hand side is the time derivative of the angular momentum, and the right-hand side is the restoring force on the angle. We linearize this equation by expanding the sine function for small angles, so the resulting equation is just

$$m\frac{d}{dt}\left(l\frac{d\theta}{dt}\right) + mg\theta = 0. \tag{5.4}$$

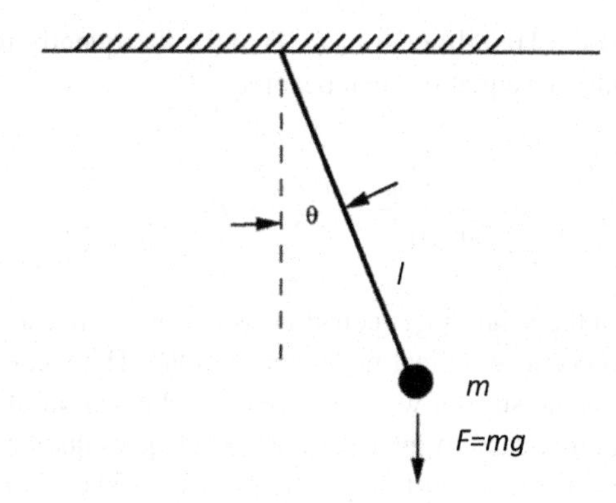

FIGURE 5.2 A simple pendulum.

We can easily develop the Hamiltonian for this, by remarking that this latter quantity is the sum of the kinetic and potential energies, just as for the mass on a spring discussed in the previous paragraph. The kinetic energy is obtained from the momentum as $p^2/2m$, and this is evaluated using the angular momentum in (5.3) and the mass of the pendulum (neglecting the mass of the rod itself).

The potential energy is obtained by noting that the right-hand side of (5.3) is the force, which is obtained from the spatial derivative of the potential energy. In the linearized version of (5.3), this leads to

$$F = -mg\theta = -\frac{1}{l}\frac{\partial V}{\partial \theta} \tag{5.5}$$

and

$$V(\theta) = \frac{1}{2}mgl\theta^2. \tag{5.6}$$

This leads to the Hamiltonian taking the form

$$H = \frac{1}{2m}\left(ml\frac{d\theta}{dt}\right)^2 + \frac{1}{2}mgl\theta^2. \tag{5.7}$$

To make (5.7) into a general Hamiltonian such as (5.2), which can be applied to many systems, we will make a few changes of variables. First, we take the 'position' angle into a position notation by making the change $l\theta \to x$. Next, we replace the gravity by a general frequency through making the change $g/l \to \omega^2$. Finally, we introduce quantization of the system by defining the momentum as a differential operator through

$$ml\frac{d\theta}{dt} \to -i\hbar\frac{d}{dx}. \tag{5.8}$$

This now leads us to the Hamiltonian of (5.2) and subsequently to the Schrödinger equation. The Schrödinger equation then becomes

$$-\frac{\hbar^2}{2m}\frac{d^2\psi}{dx^2}+\frac{m\omega^2x^2}{2}\psi=E\psi. \tag{5.9}$$

It is this formulation of the Schrödinger equation, as well as the linearized-angle Equation (5.7), that is found to occur in a great many applications. The recognition of this commonality is very important, since if we can recognize a known set of solutions to a new problem, it becomes quite easy to interpret the expected results for that problem. The harmonic oscillator is one of the most frequently studied problems for just this reason.

5.1 THE WAVE FUNCTIONS

The Hamiltonian developed for the harmonic oscillator is doubly quadratic; that is, it contains one term quadratic in the momentum and one term quadratic in the position. We want to rearrange this Hamiltonian, by introducing a set of operators that combine the position and momentum in a way that leads to a simpler approach to the problem. The rationale is based upon our familiarity with complex variables. We learned early in our career that we could expand

$$R^2+X^2=(R+iX)(R-iX). \tag{5.10}$$

If we separate the double quadratic Schrödinger equation in this manner, we will change it from a single second-order equation into a pair of first-order equations. The important difference with the Schrödinger equation from the simple complex variable approach of (5.10) is the problem of operator order; that is, the commutation relations that have to be satisfied with quantum mechanics.

Nevertheless, we suppose that a proper combination of the individual operators will provide a new set of operators that allow us to solve the Schrödinger equation in an easier manner that going through the details of Hermit polynomials and their generating functions. This will be the case, and as a result the entire approach becomes somewhat simpler. Because of the various coefficients in the Hamiltonian, the operators will be defined by combinations of these coefficients as well; the guiding principle will be that the Hamiltonian should be a simple product of the resulting operators.

With the above concepts in mind, we begin by defining the pair of adjoint operators, in order to rewrite (5.9) in the desired form, in terms of the position and momentum operators as

$$\begin{aligned} a &= \sqrt{\frac{m\omega}{2\hbar}}\left(x+i\frac{p}{m\omega}\right) \\ a^\dagger &= \sqrt{\frac{m\omega}{2\hbar}}\left(x-i\frac{p}{m\omega}\right) \end{aligned}. \tag{5.11}$$

Now, we have to address the fact that x and p do not commute. For this purpose, we first write the product

$$\begin{aligned} a^{\dagger}a &= \frac{m\omega}{2\hbar}\left(x - i\frac{p}{m\omega}\right)\left(x + i\frac{p}{m\omega}\right) \\ &= \frac{1}{\hbar\omega}\left(\frac{p^2}{2m} + \frac{m\omega^2}{2}x^2\right) - \frac{1}{2} \end{aligned}, \tag{5.12}$$

and, upon comparing this with (5.9), we can immediately rewrite the energy as

$$E = \hbar\omega\left(a^{\dagger}a + \frac{1}{2}\right). \tag{5.13}$$

Although we will show it later, we *assume* that the product of the two operators is the number of particles in a level, so this product is called the *number operator*. We will show below that this product of operators (unusually) produces a nonoperator, called a c-number (for constant number), that gives the quantum number of the energy level corresponding to a particular state, when operating on the Hamiltonian eigenstates.

The momentum and position operators satisfy a basic commutator relation since they are noncommuting operators. We expect that the above operators are also noncommuting since they arise from combinations of these noncommuting operators. The reverse product to (5.12) is given by

$$\begin{aligned} aa^{\dagger} &= \frac{m\omega}{2\hbar}\left(x + i\frac{p}{m\omega}\right)\left(x - i\frac{p}{m\omega}\right) \\ &= \frac{1}{\hbar\omega}\left(\frac{p^2}{2m} + \frac{m\omega^2}{2}x^2\right) + \frac{1}{2} \end{aligned}. \tag{5.14}$$

Now, if we subtract (5.12) from (5.14), we achieve the commutator

$$\left[a, a^{\dagger}\right] = aa^{\dagger} - a^{\dagger}a = 1. \tag{5.15}$$

This establishes the desired commutator relation. We note that both terms in (5.15) are c-numbers, from which the Hamiltonian can be defined.

Since the operator pair $a^{\dagger}a$ is a c-number, if we multiply a function by this quantity, when the function is one of the eigenfunctions of the harmonic oscillator Hamiltonian, we expect to obtain something proportional to the corresponding eigenvalue, that is

$$a^{\dagger}a\psi_n = \lambda_n\psi_n, \tag{5.16}$$

and similarly for the opposite choice of ordering, where λ_n is related to the eigenvalue still to be determined. We use the words "related to" because we have not yet assured ourselves that the wave function has been properly normalized. We shall address below. But, for now, we can certainly say that λ_n is positive, since from (2.30) we can write the expectation value as

$$\left\langle \psi_n \middle| a^\dagger a \middle| \psi_n \right\rangle = \left\langle a\psi_n \middle| a\psi_n \right\rangle \geq 0. \tag{5.17}$$

At this point, we have to devote some time to examine in detail the nature of these operators and the properties of the various eigen functions. To begin, let us examine the properties of the product $a^\dagger \psi_n$. We do this using (5.15) and (5.14), to demonstrate that

$$\begin{aligned} a^\dagger a\left(a^\dagger \psi_n\right) &= a^\dagger \left(aa^\dagger \psi_n\right) = a^\dagger \left(a^\dagger a + 1\right)\psi_n \\ &= a^\dagger \left(\lambda_n + 1\right)\psi_n = \left(\lambda_n + 1\right)a^\dagger \psi_n. \end{aligned} \tag{5.18}$$

Thus, if the wave function (and state) ψ_n has the eigenvalue λ_n, then operating on this state with the operator $a^\dagger$ produces a new state with the eigenvalue $(\lambda_n + 1)$. In essence, we have 'kicked' the energy of the initial state *upward* by one unit, which may be observed by reference to the Hamiltonian (5.13) and produced a new wave function corresponding to the level of higher energy. By the same token, if we operate with a, we find

$$a^\dagger a\left(a\psi_n\right) = \left(aa^\dagger - 1\right)a\psi_n = \left(\lambda_n - 1\right)a\psi_n. \tag{5.19}$$

Thus, operating with the operator a produces a new state with the eigenvalue $(\lambda_n - 1)$. In essence, we have 'kicked' the energy of the initial state *downward* by one unit and produced a new wave function corresponding to the level of lower energy. For these reasons, we normally refer to $a^\dagger$ as a *raising* operator (or *creation* operator since it creates one additional unit of energy in the system) and a as a *lowering* operator (or *annihilation* operator, since it destroys one unit of energy).

Now, there is always a lowest-energy state, the *ground state*. In the harmonic oscillator, it corresponds to $n = 0$, the state in which the wave function corresponds to the zero-order Hermite polynomial. Thus, if we try to lower the energy, and kick the state down by one unit, by operating with the lowering operator, the result must be zero, as

$$a\psi_0 = 0. \tag{5.20}$$

This means that $\lambda_0 = 0$, and repeated use of (5.18) leads to

$$\lambda_n = n. \tag{5.21}$$

We can also use the first of (5.10) to find the lowest wave function as

$$\sqrt{\frac{m\omega}{2\hbar}}\left(x+i\frac{p}{m\omega}\right)\psi_0=0. \tag{5.22}$$

Rearranging the terms leads to the first-order differential equation

$$-\frac{\hbar}{m\omega}\frac{d\psi_0}{dx}=x\psi_0, \tag{5.23}$$

and

$$\psi_0(x)=C_0 exp\left(-\frac{m\omega}{2\hbar}x^2\right). \tag{5.24}$$

This is the lowest (unnormalized) Hermite polynomial (called a polynomial because the prefactor, here 1, is usually a polynomial of order n). The exponential term is a weighting function that multiplies the Hermite polynomials. By direct normalization with (1.22), we find that

$$C_0=\left(\frac{m\omega}{\pi\hbar}\right)^{1/4}. \tag{5.25}$$

Now, by the use of (5.18), we can get the form of each eigen function by using the creation operator as

$$\psi_n\sim\left(a^\dagger\right)^n\psi_0\sim\left(\frac{m\omega}{2\hbar}\right)^{n/2}\left(x-i\frac{p}{m\omega}\right)^n\psi_0. \tag{5.26}$$

We have not put an equality in this last equation because there is another factor in the normalization of the higher basis functions that differs from the lowest wave function due to the zero value of λ_0. We have to probe a little bit deeper to resolve this last factor in the normalization.

To get the final parts of the normalization, we turn to the expectation values of the operators themselves. From the basic properties expressed in (5.18) and (5.19), we know that for unnormalized wave functions such that $(\psi_n,\psi_n)=C_n^2$, we can write (we take the coefficients as real)

$$\begin{aligned}\left\langle\psi_n\middle|a^\dagger\psi_k\right\rangle&=\left\langle\psi_n\middle|\psi_{k+1}\right\rangle=\alpha_n\delta_{n,k+1}C_n^2\\ \left\langle\psi_n\middle|a\psi_k\right\rangle&=\left\langle\psi_n\middle|\psi_{k-1}\right\rangle=\beta_n\delta_{n,k-1}C_n^2\end{aligned}. \tag{5.27}$$

We still have to determine the constants α_n and β_n, which in fact are related to the normalization of the eigenfunctions (the two eigenfunctions in the expectation value have different normalization constants). For this, we use

$$\langle a\psi_n | a\psi_n \rangle = \langle \psi_n | a^\dagger a\psi_n \rangle = nC_n^2 = \langle \psi_{n-1} | \psi_{k-1} \rangle = C_{n-1}^2. \tag{5.28}$$

Thus, we find that

$$C_{n-1} = \sqrt{n}C_n. \tag{5.29}$$

We could as easily have found this value by extending the arguments that led to (5.25) and the earlier discussion of λ_n, but it is instructive to repeat the work and reinforce this point of understanding. Thus, we find that $\alpha_n = C_{n+1} / C_n = \sqrt{n+1}$, and the normalized relationship becomes

$$\begin{aligned} \langle \psi_n | a^\dagger \psi_k \rangle &= \langle \psi_n | \psi_{k+1} \rangle = \sqrt{n+1}\delta_{n,k+1} \\ \langle \psi_n | a\psi_k \rangle &= \langle \psi_n | \psi_{k-1} \rangle = \sqrt{n}\delta_{n,k-1}. \end{aligned} \tag{5.30}$$

We can now rewrite (5.25) as

$$\psi_n = \frac{1}{\sqrt{(n+1)!}}\left(a^\dagger\right)^n \psi_0 = \frac{1}{\sqrt{(n+1)!}}\left(\frac{m\omega}{2\hbar}\right)^{n/2}\left(x - i\frac{p}{m\omega}\right)^n \psi_0. \tag{5.31}$$

Since we already have ψ_0, n begins with 1 in (5.31) and goes upward in unit steps. In Figure 5.3, we plot the lowest four Hermite polynomials. Like the bound states in a quantum well, each wave function has an additional peak in order to insure orthonormalization of the basis functions.

5.2 THE LC-CIRCUIT

Another version of the harmonic oscillator that is more familiar to electrical engineers is the resonant *LC*-circuit. The *LC*-circuit is one of the most pervasive systems found in electrical engineering. It represents not only a range of filters, but also the resonant cavity of microwave circuits. In this section, we want to show that the quantum mechanics of the *LC*-circuit is essentially that of the linear harmonic oscillator. We consider the circuit of Figure 5.4. The familiar equations for current and voltage in the inductor and capacitor (directions are indicated in the figure, so this will change some signs) are

$$I = -C\frac{dV}{dt}, \quad V = L\frac{dI}{dt}. \tag{5.32}$$

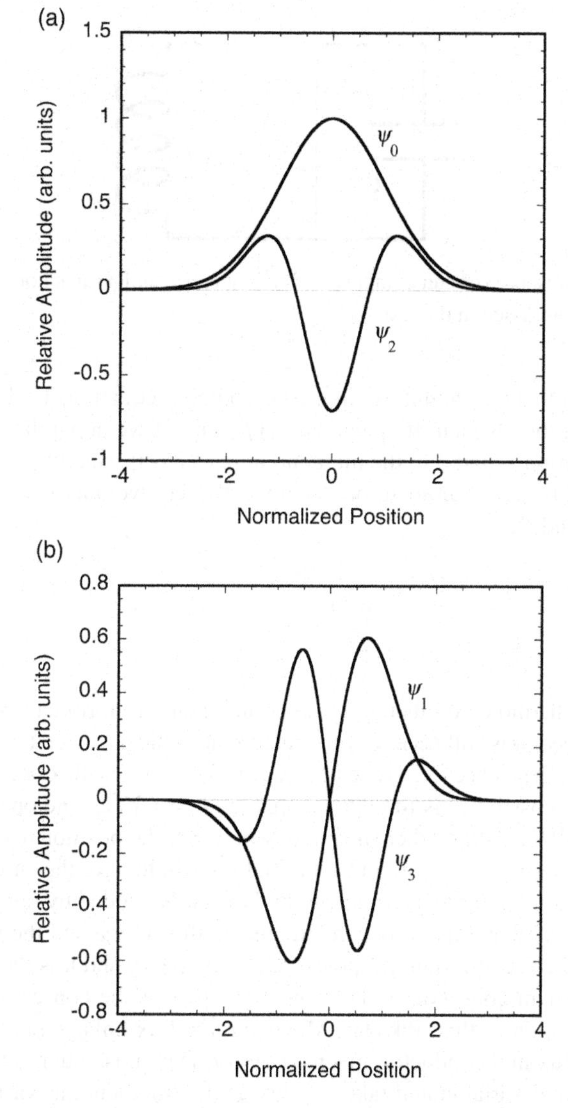

FIGURE 5.3 The lowest four basis functions for the harmonic oscillator.

On the other hand, the energy stored in the circuit is given by

$$E = \frac{1}{2}LI^2 + \frac{1}{2}CV^2 = \frac{\phi^2}{2L} + \frac{Q^2}{2C}. \tag{5.33}$$

where we have introduced the *flux linkages* $\phi = LI$, the charge $Q = CV$. The first term on the right-hand side is the energy stored in the capacitor, while the second term is that stored in the inductor. In fact, we can proceed immediately to proclaim that this is

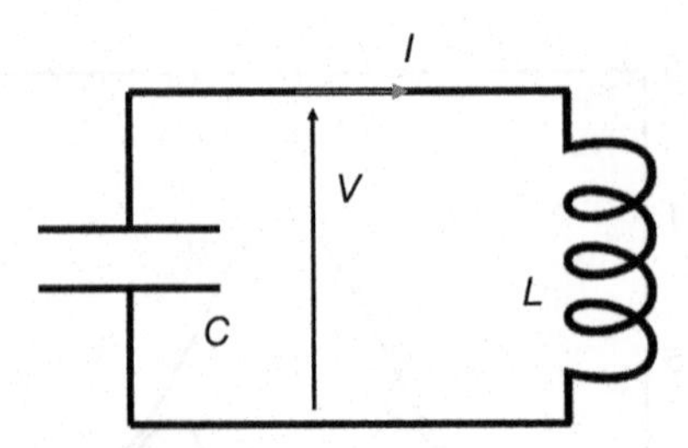

FIGURE 5.4 A resonant circuit composed of an inductor L and a capacitor C. The voltage V and current I are discussed in the text.

the Hamiltonian that we should use in the Schrödinger equation. To these quantities, we can also add the resonant frequency $\omega = 1/\sqrt{LC}$. If we now relate L to the mass m, Q to the position x, and ϕ to the momentum p, (5.33) is exactly (5.2). In this case, these relationships are *analogs* to one another. Indeed, we may now rewrite (5.32), for constant L and C, as

$$\frac{d^2\phi}{dt^2} + \omega^2\phi = 0. \tag{5.34}$$

This is just our harmonic oscillator equation, and a similar form can be found for Q. Before we proceed, it is still necessary to identify the conjugate operators and their relationship to each other, since (5.34) can be achieved for either the flux or the charge.

One obvious possibility is to take the charge Q (= CV) as one operator and then the current $I = dQ/dt$ as the other (in analogy between the position and the velocity). However, this doesn't quite work out. The analogy would have the current as the velocity (of the charge) and not the momentum. We can be guided by our understanding that the two fundamental quantities in circuits are the *charge* and the *flux*. It is these two quantities that are invariant in special relativity theory and it is these two that are subject to important conservation laws in circuit theory–the conservation of charge and the conservation of flux linkages. Moreover, the flux linkage in the present context is just the flux in the inductor, given by $\phi = LI$. Here it is better to take the charge as the 'coordinate' variable and take the flux as the 'momentum' variable. If this is done, we write the energy as the last form of (5.33). By introducing the resonant frequency, we can write this form as

$$E = \frac{\phi^2}{2L} + L\omega^2\frac{Q^2}{2}, \tag{5.35}$$

Thus, if we make the connection $L \rightarrow m$, this is the same equation as (5.2). Hence, we find that the resonant circuit is just a simple harmonic oscillator.

Finally, let us look at the commutator of charge and flux, when these are interpreted as operators. By making the analogy between the flux and the momentum coordinate, and between the charge and position coordinate, we have from (1.28) the relationship

$$\phi = -i\hbar \frac{\partial}{\partial Q}, \tag{5.36}$$

so that we can write the commutator relationship as

$$[Q, \phi] = i\hbar. \tag{5.37}$$

In this interpretation, the charge and flux are noncommuting variables, with the flux interpreted as the variation with respect to the charge. On the other hand, this relationship vanishes in the classical limit as $\hbar \to 0$. In this latter limit, charge and flux are independent quantities, but in the quantum case they are not independent and cannot be simultaneously measured. However, we should be very careful, as this limit cannot be interpreted as the going to the classical limit, as there are cases in which this limit cannot be taken as it introduces singularities.

The interpretation of the *LC*-circuit as a linear harmonic oscillator means that the energy in the resonator is a fixed quantity that is given in terms of the resonant frequency by (4.21). Energy storage can be increased or decreased only in units of the energy quantum $\hbar\omega = \hbar / \sqrt{LC}$. Thus, the photons that enter or leave an electromagnetic cavity (circuit) have well defined energies in terms of the resonant frequency of the cavity. This is obvious in the measurement of the frequency of the oscillating cavity, where we measure the energy of the photons emanating from the cavity, just as was done for the study of blackbody radiation discussed in Chapter 1. In many cases, such as was done by Planck, it is often useful to consider the resonant circuit quantum mechanically rather than classically. We have the full understanding of this in terms of the harmonic oscillator results of the previous sections.

With this discussion, it is now clear that electromagnetic waves satisfy the resonant *LC*-circuit, which itself is a linear harmonic oscillator. We previously established that the particles which satisfied the harmonic oscillator were *bosons*. Thus, it is clear that photons, the particles which are associated with electromagnetic waves, are bosons. The photons actually have spin, but it is quantized into values of ± 1, corresponding to the circular polarization of the plane-wave of electromagnetics. Thus, we may say that bosons have *integer* values of spin, while fermions have half-integer values of spin.

This approach also opens the door to a treatment of almost any oscillatory behavior to be treated by a quantum approach. As here, one only needs to identify the proper *generalized* coordinates, which are then subjected to an uncertainty principle. It is this step that introduces the quantum nature of the subsequent solutions in terms of harmonic oscillator coordinates. Generalized coordinates have long been treated in classical mechanics, and their use naturally flows over to quantum mechanics.

5.3 AN ATOMIC LATTICE AND PHONONS

In a solid, the motion of the atoms is much like the motion of free particles, with the important exception that the atoms are forced *on average* to retain positions within the solid that specify a particular crystal structure (we can also treat noncrystalline solids by similar methods, but our approach here will be limited to the crystalline solids, where the atoms are equally spaced on a *lattice*, such as in Figure 4.1). In any real solid, the lattice is a three-dimensional structure. However, when the motion is along one of the principal axes of the structure, it is usually possible to treat the atomic motion as a one-dimensional system. Although this is a simple model, its applicability can be extended to the real crystal if each atom represents the typical motion of an entire plane of atoms normal to the wave motion.

The total Hamiltonian for the atomic motion is given by the momentum of the individual atoms plus the potential that keeps the atoms in the lattice positions (on average). We can write this Hamiltonian as

$$H = \sum_i \frac{P_i^2}{2M_i} + \sum_{j \neq i} V\left(R_i - R_j\right). \tag{5.38}$$

(We use capitol letters here to denote that atoms and their properties.) In general, the atoms will vibrate around their equilibrium position, just like little harmonic oscillators. However, all of these oscillators are coupled through the potential term in (5.38). We have studied only the single oscillator. How will we treat the coupled oscillators? The answer to this is that we do a Fourier transform into the dominant Fourier modes of the overall vibration, and this will result in us obtaining a set of uncoupled oscillators, one for each mode of vibration of the entire set of coupled oscillators. Each of these modes will then be quantized, and the quantum unit of amplitude of each of the modes is termed a *phonon*.

The Hamiltonian in (5.38) describes the motion of the atoms about the equilibrium (or rest) positions. We expect this motion to be small, so shall use a Taylor series expansion of the potential about these equilibrium positions, and we can then rewrite (5.38) as

$$H = \sum_i \frac{P_i^2}{2M_i} + \sum_{j \neq i} r_i r_j \frac{\partial^2 V\left(R_i - R_j\right)}{\partial R_i \partial R_j} + \ldots. \tag{5.39}$$

The zero-order term in the potential is just an offset in energy and will be ignored, while the first partial derivative term must be zero if we are expanding about the equilibrium position. The lowercase letters in the second term are the *deviations* of the position from the equilibrium value. We also recognize that

$$P_i = M_i \frac{\partial r_i}{\partial t}. \tag{5.40}$$

Now, we introduce the appropriate Fourier series for the displacement as

$$r_i = \frac{1}{\sqrt{N}} \sum_q u_q e^{iqR_i}. \tag{5.41}$$

Note that the Fourier transform is based upon the equilibrium positions of the atoms. We assume that there is a lattice constant a that is the equilibrium distance between atoms and that, from some arbitrary reference point, $R_{i0} = n_i a$. Here, n_i is an integer specifying just how far along the chain of atoms the ith atom resides in equilibrium. The last term in the Hamiltonian (5.39) is the most complicated one. This is given by the Fourier terms

$$
\begin{aligned}
V_{qq'} &= \frac{1}{N}\sum_{j\neq i} u_q u_{q'} e^{iqR_i} e^{iq'R_j} \frac{\partial^2 V(R_i - R_j)}{\partial R_i \partial R_j} \\
&= \frac{1}{N}\sum_{j\neq i} u_q u_{q'} e^{iq(R_i - R_j)} e^{i(q'+q)R_j} \frac{\partial^2 V(R_i - R_j)}{\partial R_i \partial R_j}.
\end{aligned}
\tag{5.42}
$$

We assume that the partial derivative can be treated as constant (but may vary with the particular mode). Each of the position vectors is expanded in terms of its equilibrium position and its absolute position. The first exponential depends only upon the relative positions, while the second exponential mainly treats the equilibrium positions (the deviations are small). Thus, we bring this latter summation out to make a sum only over the equilibrium positions:

$$\sum_j e^{i(q'+q)R_j} = N\delta_{qq'}, \tag{5.43}$$

which is the *closure* relationship for Fourier transforms. The number N, which also appears in (5.41), is the number of atoms in the chain (and hence the number of Fourier modes, just as it was the number of values for the wave momentum in Chapter 4). The remaining part of the potential is just (summing over q', with $V_{qq'} = V_q$)

$$V_q = \sum_{j\neq i} u_q u_{-q} e^{iq(r_i - r_j)} \frac{\partial^2 V(R_i - R_j)}{\partial R_i \partial R_j}. \tag{5.44}$$

The mode amplitudes can be brought out of the summation, and the remainder is defined to be the spring constant for the particular mode C_q. The same summation over the lattice modes can be made in the kinetic energy term as well, and this allows us to write the final result as

$$H = \sum_q \left[\frac{P_q P_{-q}}{2M} + \frac{C_q u_q u_q^\dagger}{2}\right] = \sum_q \left[\frac{P_q P_{-q}}{2M} + \frac{C_q u_q u_q^\dagger}{2}\right], \tag{5.45}$$

where we recognize in the last form that the adjoint operator is the operator at negative q. By writing $C_q = M\omega_q^2$, we see that we have now obtained a set of uncoupled harmonic oscillators. Each mode is characterized by a wave vector q and frequency ω_q, and we can compute the wave functions as if that mode were totally isolated with a mode energy described by (5.13). Thus, any particular mode may contain a great many phonons as described by the energy level number for the mode. The total energy of the lattice vibration is now just the sum over the energy (and hence the number of phonons) in each mode. Quantization of the vibrations occurs by having the position u_q and the momentum P_q be noncommuting operators subject to a commutator relationship $[P_q, u_q] = -i\hbar$. We can at this point also introduce the creation and annihilation operators for each mode, which serve to create or destroy one unit of amplitude for that particular mode. Thus, the number of units of amplitude (or energy) in the mode of wave vector q is said to the number of phonons in that particular mode.

Because each mode of the lattice vibrations satisfies a harmonic oscillator equation, these *phonon* modes are *bosons*. The atoms that constitute the vibrations are such that the phonon mode has zero spin, which counts as an integer. In terms of the parameters of (5.45), the creation and annihilation operators are written as

$$\begin{aligned} a_q^\dagger &= \sqrt{\frac{M\omega_q}{2\hbar}}\left(u_q - i\frac{P_q}{M\omega}\right) \\ a_q &= \sqrt{\frac{M\omega_q}{2\hbar}}\left(u_q + i\frac{P_q}{M\omega}\right) \end{aligned}. \tag{5.46}$$

The extension of (5.12) now gives the energy in the particular mode as

$$E_q = \hbar\omega_q\left(a_q^\dagger a_q + 1\right) = \hbar\omega_q\left(N_q + 1\right). \tag{5.47}$$

In equilibrium conditions, the number of phonons in the mode is given by the *Bose-Einstein distribution* as

$$N_q = \frac{1}{exp\left(\hbar\omega_q / k_B T\right) - 1}, \tag{5.48}$$

which differs from the Fermi–Dirac distribution by the sign in the denominator and the lack of a Fermi energy. All bosons satisfy this distribution but, as with the Fermi–Dirac distribution, the limiting case in which the factor of unity can be ignored gives the classical Maxwell-Boltzmann distribution.

Finally, we need to address the question of just how the wave vector q and the frequency ω_q are related to each other. For this, we return to real space, but will use the fact that the Fourier modes require the motion to exist as a set of waves with propagation according to $e^{i(qR-\omega_q t)}$. Thus, if we write the equation of motion of any particular atom at position R_i we obtain from (5.38)

$$M\frac{d^2r_i}{dt^2} = -\sum_{j\neq i}\frac{\partial^2 V(R_i - R_j)}{\partial R_i \partial R_j}$$
$$\cong C(R_{i+1} - R_i) - C(R_i - R_{i-1}). \tag{5.49}$$

Here, the approximation is that only the nearest neighbors are important, and we have treated the potential as a linear spring. Because of the wave behavior of the lattice, we can rewrite the last term as

$$M\omega_q^2 r_i = -Cr_i\left[\left(e^{iqa} - 1\right) + \left(e^{-iqa} - 1\right)\right], \tag{5.50}$$

where the two exponentials are displacement operators, just as used in Chapter 4. We can then rewrite this last equation as

$$\omega_q^2 = \frac{2C}{M}\left[1 - cos(qa)\right] = \frac{4C}{M}sin^2\left(\frac{qa}{2}\right). \tag{5.51}$$

Any range of q that spans a total value of $2\pi/a$ will give all allowed values of the frequency that can be expected (which is of course positive). We generally choose this range to be $-\pi/a < q < \pi/a$, in keeping with the band structure arguments concerning periodic potentials in the last chapter. We show these frequencies in Figure 5.5. Real

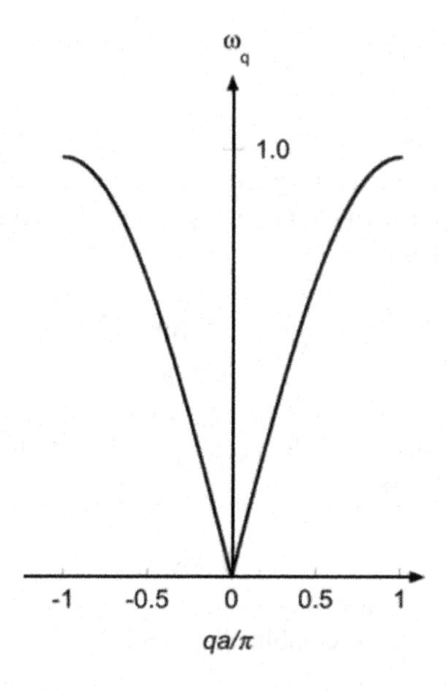

FIGURE 5.5 The relationship between the frequency and wave vector of a particular mode of lattice vibration.

lattices can be more complicated, containing for example, two different types of atom, or having the spacing between alternate atoms differing. In each case, the behavior is more complicated, and motion of the two atoms together (low frequency) plus relative motion of the two atoms (high frequency) can occur. This is beyond the level we treat here, where we want merely to introduce the periodic potential as an example of a harmonic oscillator example.

5.4 MOTION IN A QUANTIZING MAGNETIC FIELD

The last example of a harmonic oscillator we want to consider is the motion of an electron orbiting around magnetic field lines. Classically, the electron is pulled into a circular orbit such that the magnetic field is directed normal to the plane of the orbit, and the motion arises from the Lorentz force acting on the electron:

$$\mathbf{F} = -e\left(\mathbf{E} + \mathbf{v} \times \mathbf{B}\right). \tag{5.52}$$

The charge on the electron has been taken to be its proper negative value. Clearly, for motion in the $\boldsymbol{a}_\phi$ direction (in cylindrical coordinates) and a magnetic field in the $\boldsymbol{a}_z$ direction, the force is an inward centripetal force which causes the motion to be a closed circular orbit (ignoring the role of the electric field). Here, we want to examine the quantization of this orbit, and will deal only with the electric field that is induced by the magnetic field B, ignoring any explicitly applied field.

The approach we follow is exactly the one we used in Section 1.3. The accelerative force on the wave momentum of the electron may be expressed as in (1.6) as

$$\hbar \frac{d\mathbf{k}}{dt} = -e\left(\mathbf{v} \times \mathbf{B}\right). \tag{5.53}$$

In fact, we want to introduce the magnetic field through the vector potential. As we saw in Chapter 1, the magnetic field produces the electric field through Faraday's law, which may be expressed as

$$\nabla \times \mathbf{E} = -\frac{\partial \mathbf{B}}{\partial t}. \tag{5.54}$$

We now introduce the concept of the *vector potential* **A**, from which the magnetic field may be determined as

$$\mathbf{B} = \nabla \times \mathbf{A}. \tag{5.55}$$

These last two equations can be combined to yield

$$\mathbf{E} = -\frac{\partial \mathbf{A}}{\partial t}, \tag{5.56}$$

although any constant vector whose curl is zero can be added to the right-hand side. This would normally be the gradient of the scalar potential, here this term is not so important. Here, however, we want to define the electric field from the magnetic field alone, and hence from the vector potential that gives rise to the magnetic field. This suggests that if we now use (5.56) in place of the term on the right-hand side of (5.53), the proper total momentum to use in the Hamiltonian is just the combined quantity

$$\mathbf{p} \to \hbar\mathbf{k} + e\mathbf{A}. \tag{5.57}$$

While the substitution (5.57) has been derived and developed by a great many workers, it is usually referred to as the Peierls' substitution in solid-state physics.

We are now interested in using this value for the momentum in the Hamiltonian (5.2), except that we do not include any external potential. Nevertheless, the magnetic field will provide such a potential. To proceed, we must decide upon the vector potential $\mathbf{A}$. Our interest is in motion perpendicular to the magnetic field, which we take to be constant and oriented in the positive z-direction. Then, a vector potential defined as $\mathbf{A} = Bx\boldsymbol{a}_y$ will provide this magnetic field. We note that this definition is only *sufficient*, as many other definitions of the vector potential could as easily describe a uniform field in the z-direction. However, this particular definition, called the *Landau gauge*, will suit our purposes sufficiently well. Furthermore, we will only treat the motion in the plane normal to the magnetic field.

In the absence of any external potential, other than the vector potential giving rise to the magnetic field, Schrödinger's equation can be written as

$$-\frac{\hbar^2}{2m}\left(\nabla - \frac{eA}{i\hbar}\right)^2 \psi = E\psi. \tag{5.58}$$

where the wave momentum has been replaced by the normal gradient operator for the momentum. Expanding the momentum operator bracket, and using our form for the vector potential in the Landau gauge, leads to the differential equation

$$-\frac{\hbar^2}{2m}\left[\frac{\partial^2}{\partial x^2} - \frac{2eBx}{i\hbar}\frac{\partial}{\partial y} + \frac{\partial^2}{\partial y^2} - \left(\frac{eBx}{\hbar}\right)^2\right]\psi = E\psi. \tag{5.59}$$

The motion in the y-direction is essentially free motion, at least within this formulation of the Schrödinger equation, as the only deviation from the normal differential operators is the x-dependence arising from the magnetic field. Thus, we will take the wave function to have the general form

$$\psi(x, y) = e^{-iky}\phi(x). \tag{5.60}$$

With this substitution for the wave function, (5.59) becomes

$$-\frac{\hbar^2}{2m}\left[\frac{\partial^2}{\partial x^2}-\frac{2eBxk}{\hbar}+k^2-\left(\frac{eBx}{\hbar}\right)^2\right]\phi=E\phi. \tag{5.61}$$

Our goal is to put this into a form reminiscent of the harmonic oscillator. To this end, we introduce the cyclotron frequency $\omega_c = eB/m$, and rewrite (5.61) as

$$-\frac{\hbar^2}{2m}\frac{\partial^2\phi}{\partial x^2}+\frac{m}{2}\omega_c^2\left(x-x_0\right)^2\phi=E\phi, \tag{5.62}$$

where

$$x_0=\frac{\hbar k}{eB}. \tag{5.63}$$

we recall that k is the y-component of the wave motion. Comparison of this with (5.9) shows that the x-motion satisfies a harmonic oscillator equation, which yields Hermite polynomials as solutions. Thus, the energy levels are given as

$$E=\hbar\omega_c\left(n+\frac{1}{2}\right), \tag{5.64}$$

and the y-motion just shifts the center of the orbit. This circular motion is called cyclotron motion and the energy levels (5.64) are called Landau levels. The corresponding wave function is just

$$\phi_n\left(x\right)=\frac{1}{\sqrt{2^n n!}}\left(\frac{m\omega_c}{\pi\hbar}\right)^{1/4}e^{m\omega_c\left(x-x_0\right)^2/2\hbar}H_n\left(\sqrt{\frac{m\omega_c}{\hbar}}\left(x-x_0\right)\right). \tag{5.65}$$

These solutions are, of course, just the normal harmonic oscillator wave functions shifted by the offset (5.63). Here, both the scaling factor and the offset depend upon the magnetic field, just as the cyclotron frequency does. Noting the argument of the Hermite polynomial in (5.65) gives a natural scaling length described by $\sqrt{(\hbar/eB)}$, which is termed the *magnetic length*. To lowest order, the Hermite polynomial has its peak at the position $\sqrt{2n+1}$ times this basic length, so this is the natural cyclotron radius of the harmonic motion for a particular energy level.

5.4.1 Connection with the Classical Orbit

The form of the wave function (5.65) and the energy (5.64) do not give us a clear view of the orbiting nature of the electron in the magnetic field. True, it is a harmonic oscillator in the x-direction, but what about its other motion? As remarked earlier, the electron orbits around the magnetic field, and this motion is a solution of (5.53), which we can rewrite as

$$m\frac{d\mathbf{v}}{dt} = -e(\mathbf{v} \times \mathbf{B}). \tag{5.66}$$

The magnetic field is in the z-direction, according to the Landau gauge previously adopted, so that the motion in the (x, y)-plane is given by

$$\begin{aligned} m\frac{dv_x}{dt} &= -ev_y B \\ m\frac{dv_y}{dt} &= ev_x B. \end{aligned} \tag{5.67}$$

These two equations may then be solved for an arbitrary initial condition to give

$$\begin{aligned} v_x &= v_0 cos(\omega_c t) \\ v_y &= v_0 sin(\omega_c t) \end{aligned}. \tag{5.68}$$

Similarly, we can find the positions as

$$\begin{aligned} x &= \frac{v_0}{\omega_c} sin(\omega_c t) \\ y &= -\frac{v_0}{\omega_c} cos(\omega_c t) \end{aligned}. \tag{5.69}$$

We can now combine (5.68), (5.69), and (5.64) to give

$$\begin{aligned} E &= \frac{1}{2}mv_0^2 = \hbar\omega_c\left(n + \frac{1}{2}\right), \quad v_0 = \sqrt{(2n+1)\frac{\hbar eB}{m^2}}, \\ r^2 &= x^2 + y^2 = \left(\frac{v_0}{\omega_c}\right)^2 = \sqrt{(2n+1)\frac{\hbar}{eB}}. \end{aligned} \tag{5.70}$$

The last line gives us the result hypothesized at the end of the previous section, $\sqrt{\hbar / eB}$ is a magnetic length that gives us the radius of the lowest energy level and provides a natural scaling of the size of the cyclotron orbit around the magnetic field. The quantization of the orbit radius arises directly from the quantization of the energy levels of the harmonic oscillator. These quantized energy levels are termed *Landau levels*, as mentioned above.

One of the important aspects of the quantization in a magnetic field is the fact that the continuum of allowed momentum states is broken up by this quantization. The allowed states are coalesced into the Landau levels, and each spin-degenerate Landau level can hold

$$n_s = \frac{1}{\pi r_0^2} = \frac{eB}{\pi\hbar} \tag{5.71}$$

electrons. In some magneto-transport experiments, this quantization can be seen as oscillatory magnetoresistance, in which the periodicity in $1/B$ arises from (5.71) as

$$\Delta\left(\frac{1}{B}\right) = \frac{e}{\pi\hbar n_s}. \tag{5.72}$$

for spin-degenerate Landau levels. This oscillatory magnetoresistance is termed the Shubnikov–de Haas effect, which will be discussed further below.

5.4.2 Adding Lateral Confinement

Let us now consider what happens when the magnetic field is added to a simple harmonic oscillator potential in the x-direction—that is, we confine the electron in the lateral x-direction and add an additional magnetic field in the z-direction. Hence, we combine the Hamiltonians of (5.9) and (5.61) to seek a solution of

$$-\frac{\hbar^2}{2m}\left[\frac{\partial^2}{\partial x^2} - \frac{2eBxk}{\hbar} - k^2 - \left(\frac{eBx}{\hbar}\right)^2 + \frac{1}{2}m\omega^2 x^2\right]\phi = E\phi. \tag{5.73}$$

It is clear that this can now be rearranged to give

$$-\frac{\hbar^2}{2m}\frac{\partial^2\phi}{\partial x^2} + \frac{1}{2}m\Omega^2\left(x - x_0\right)^2\phi = \left[E - \frac{\hbar^2 k^2}{2m}\left(1 - \frac{4\omega_c^2}{\Omega^2}\right)\right]\phi, \tag{5.74}$$

with

$$\Omega^2 = \omega^2 + \omega_c^2, \quad x_0 = \frac{\hbar k\omega_c}{m\Omega^2}. \tag{5.75}$$

The motion remains that of a harmonic oscillator, but now the shift of the wave function and the energy levels are hybrids of the confinement harmonic oscillator and the magnetic harmonic oscillator. Each energy level of the magnetic field, the Landau levels, is raised by the confinement to a higher value. In essence, this coupling of the

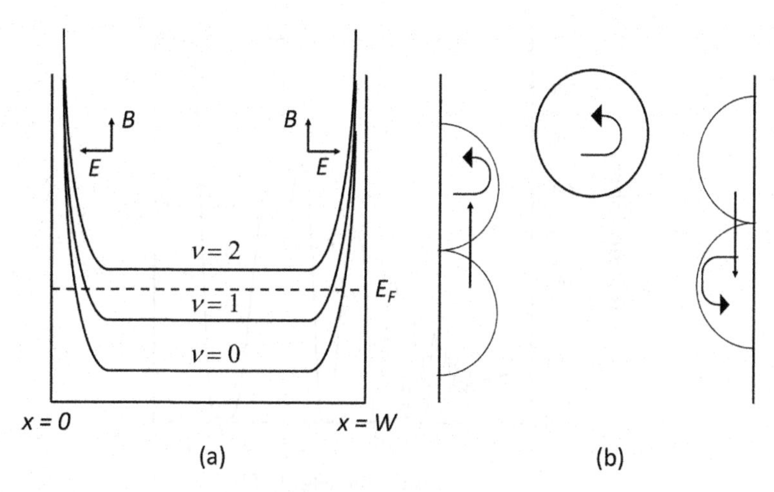

FIGURE 5.6 *(a)* The bending of the Landau levels at the edges of a confined sample. The electric field is shown for comparison at the two edges. *(b)* The confined and bouncing orbits for the situation of *(a)*. The magnetic field is out of the page.

two harmonic oscillators leads to enhanced confinement, and stronger confinement always costs energy—the result is that the energy levels lie at higher values.

This becomes more important if the harmonic oscillator is nonlinear. That is, if the value of the parameter ω varies with distance from the center of the harmonic oscillator, then the energy levels are further increased as they are found farther from the center. This reaches the extreme in hard-wall boundary conditions at certain points on the x-axis. This is shown in Figure 5.6(*a*). Here, the Landau levels rapidly increase in energy as they approach the hard walls due to the extra confinement, just as in (5.75). The importance of this is the fact that an electron whose orbit is within the orbit radius of the edge will strike the wall and create a bouncing orbit that moves along the wall, as shown in Figure 5.6(*b*). Instead of being trapped in a Landau orbit, these bouncing orbits can carry current along the walls of the confining region and are called *edge states*.

When the Fermi level lies in a Landau level, there are many states available for the electron to gain small amounts of energy from the applied field and therefore contribute to the conduction process. On the other hand, when the Fermi level is in the energy gap between two Landau levels, the upper Landau levels are empty and the lower Landau levels are full. Thus, there are no available states for the electron to be accelerated into, and the conductivity drops to zero in two dimensions. In three dimensions it can be scattered into the direction parallel to the field (the z-direction), and this conductivity provides a positive background on which the oscillations ride. Figure 5.7 shows a typical measurement of the longitudinal resistance. The 'zeros' of the longitudinal resistance correspond to the magnetic field for which there are full Landau levels. (R_{xx} has zeros as does the longitudinal conductance, since in the presence of the magnetic field, $R_{xx} = G_{xx} / G_{xx}^2 = G_{xy}^2$. Although the longitudinal conductance vanishes, the transverse conductance does not, and this means that the longitudinal resistance also vanishes.) However, the index that is shown (4, 6, 8, etc.)

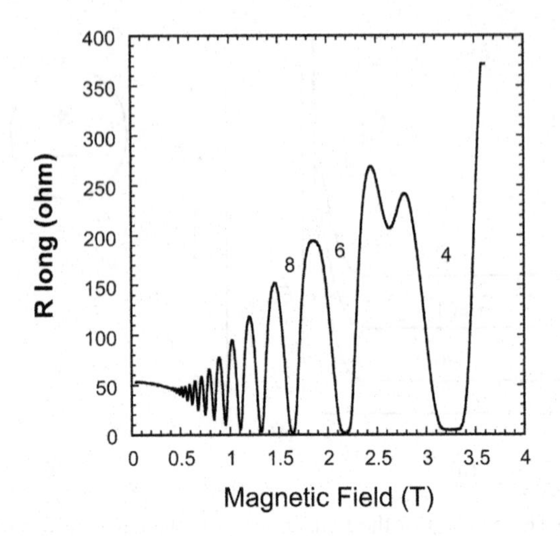

FIGURE 5.7 The longitudinal resistance for a quasi-two-dimensional electron gas in a high magnetic field. The oscillations are the Shubnikov–de Haas oscillations, and correspond to the sequential emptying of Landau levels.

is that for spin-resolved levels, rather than spin-degenerate Landau levels. Hence, for the case of the zero at index 4 ($B = 3.25$ T, termed the $\nu = 4$ level), the $n = 0$ and $n = 1$ Landau levels (both doubly spin-degenerate) are full. The zero corresponds to the transition of the Fermi energy between the 5/2 and 3/2 (in units of $\hbar\omega_c$) levels in Figure 5.6. These measurements are for a quasi-two-dimensional electron gas at the interface of a GaAlAs/GaAs heterostructure. From (5.72), we can determine the areal density to be approximately 3.3×10^{11} cm^{-2}.

5.4.3 The Quantum Hall Effect

The zeros of the conductivity that occur when the Fermi energy passes from one Landau level to the next-lowest level are quite enigmatic. They carry some interesting by-products. A full derivation of the quantum Hall effect is well beyond the level at which we are discussing the topic here. However, we can use a consistency argument to illustrate the quantization exactly, as well as to describe the effect we wish to observe. When the Fermi level lies between the Landau levels, the lower Landau levels are completely full. We may then say that

$$E_F \sim \nu\hbar\omega_c. \tag{5.76}$$

where ν is an integer giving the number of *spin-resolved* Landau levels. The presence of the edge states means that some carriers have moved to the edge of the sample, and this arises from including the electric field in the second of Equation (5.68) as

$$m\frac{dv_y}{dt} = -qE_y + ev_xB. \tag{5.77}$$

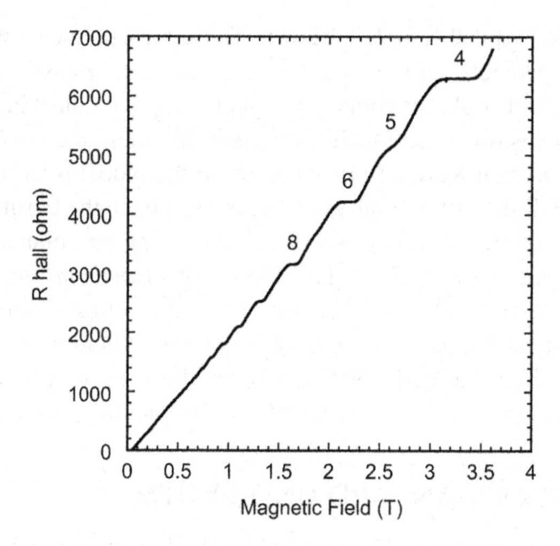

FIGURE 5.8 The Hall resistance of quantized Landau levels. The normalization is chosen so that the index value is of spin-resolved Landau levels. The first spin splitting is just being resolved at $\nu = 5$ (~2.6 T).

In a stable situation, the lateral acceleration must vanish at the edge, so that a transverse field must exist, given by

$$E_y = v_x B = -\frac{J_x}{ne} B. \tag{5.78}$$

Using (5.71) for the density in each Landau level, we have (inserting a factor of two to raise the spin degeneracy)

$$n_s = \nu \frac{eB}{2\pi\hbar}. \tag{5.79}$$

The density is constant in the material, so using (5.79) in (5.78) to define the *Hall resistance* gives

$$R_{xy} = -\frac{E_y}{J_x} = \frac{h}{\nu e^2}. \tag{5.80}$$

The quantity $h/e^2 = 25.81$ kΩ is a ratio of fundamental constants. Thus, the conductance (reciprocal of the resistance) increases stepwise as the Fermi level passes from one Landau level to the next-lower level. Between the Landau levels, when the Fermi energy is in the localized state region, the Hall resistance is constant (to better than 1 part in 10^7) at the quantized value given by (5.80) since the lower Landau levels are completely full. In Figure 5.8, the variation of the Hall resistance as a function of magnetic field for a typical sample is shown. These measurements were made in the same sample and geometry of Figure 5.7, so that they can be easily compared.

The magnetic field could, of course, be swept to higher values in both Figures 5.7 and 5.8. When this is done, new features appear, and these are not explained by the above theory. Klaus von Klitzing (1980) received the Nobel prize for the discovery of the quantum Hall. In fact, in high quality samples, once the Fermi energy is in the lowest Landau level, one begins to see fractional filling and plateaus, in which the resistance is a fraction of h/e^2 (Tsui et al. 1982). This *fractional quantum Hall effect* is theorized to arise from the condensation of the interacting electron system into a new many-body state characteristic of an incompressible fluid (Laughlin 1983). Tsui, Stormer, and Laughlin shared the Nobel prize for this discovery. However, the properties of this many-body ground state are clearly beyond the present level.

5.5 THE MODERN STANDARD UNIT SYSTEM

The MKS system of units was standardized in 1889 in order to provide some international agreement on what constituted a kilogram or a meter. Originally, the kilogram was a standardized hunk of Pt-Ir that the French called *Le Grande K*. Several similar pieces of this material are kept in the United States at the National Institutes of Standards and Technology. Similarly, the meter was defined as one ten-millionth of the distance from the north pole to the equator. Over the years, changes were gradually made in many of the various units that are part of the MKS hierarchy. What most of us knew and accepted in this system date from 1960 when an updated international set of units was published.

Recently, however, this system has been changed once again, with new standards being based upon revelations from quantum mechanics, such as Planck's constant and the quantum Hall effect. In 2019, the world standardized the value of Planck's constant to $6.62607015 \times 10^{-34}$ Joule-second. Similarly, the meter has been standardized as the distance light travels in $1/c$ seconds, where c is the speed of light, 2.99792458×10^8 meter/second. But this latter definition relies upon the second itself, which arises from the frequency separation of the Cesium-133 atom's outer electron spin resonance (transition from spin up to spin down), which is 9.1926317790×10^9 Hz.

Now, the quantum Hall effect gives us the resistance of the $\nu = 1$ plateau to be 25812.807 ohms, and we know from (5.80) that this is h/e^2. Hence, standardizing this value and Planck's constant means that we have standardized a value for the electronic charge e. Since Planck's constant can also be given its value in terms of kg-m^2/sec, standardization of this quantity plus the standard second and meter, now defines the new value for the kilogram as well. And this can be extended to other standards for current, voltage, and so on.

REFERENCES

Laughlin, R. B. 1983. Anomalous quantum Hall effect: An incompressible quantum fluid with fractionally charged excitations. *Phys. Rev. Lett.* **50** 1395–1398

Tsui, D., Stormer, H. L. and Gossard, A. C. 1982. Two-dimensional magnetoresistance in the extreme quantum limit. *Phys. Rev. Lett.* **48** 1559–1562

Von Klitzing, K., Dorda, G. and Pepper, M. 1980. New method for high accuracy determination of the fine-structure constant based on quantized Hall Resistance. *Phys. Rev. Lett.* **45** 494–497

Problems

1 Using the creation and annihilation operators, compute p^2 and x^2.
2 Determine the expectation values for the kinetic and potential energies in a harmonic oscillator.
3 Consider an electromagnetic resonator with a resonant frequency of 10^{10} Hz. What is the energy separation of the oscillator levels? How does this compare with the thermal energy fluctuation?
4 If the velocity of sound $d\omega/dq$ of a set of lattice vibrations is 10^5 cm s^{-1}, and the lattice constant is $a = 0.25$ nm, what is the phonon frequency at the zone edge $q = \pi/a$?
5 If one begins with the so-called symmetric gauge, where $\mathbf{A} = (eB/2)(-y\mathbf{a}_x + x\mathbf{a}_y)$, show that a more complicated harmonic oscillator solution results. Show that this still gives energy levels according to (5.64) and find the appropriate form of the wave functions that are the solutions.
6 Consider an electron moving in a z-directed magnetic field and constrained in a quadratic potential $\frac{1}{2}m\omega_0^2x^2$. In the simplest case, in which the system is homogeneous in the y-direction, determine the energy levels of the electron.
7 A semiconductor sample measures 1 cm by 0.5 cm and is 0.1 cm thick. For an applied electric field of 1 V cm^{-1}, 5 mA of current flows. If a 0.5 T magnetic field is applied normal to the broad surface, a Hall voltage of 5 mV is developed. Determine the Hall mobility and the carrier density.
8 Consider a free-electron 'gas' with an areal density of 2×10^{12} cm^{-2}. What is the periodicity (in units of $1/B$) expected for the Shubnikov–de Haas oscillation?

6 Operators and Bases

In the past few chapters, we have developed the Schrödinger equation and applied it to the solution of a number of quantum mechanical problems, mainly to develop experience with the results of simple and common systems. The results from these examples can be summarized relatively simply. In general, quantization enters the problem through the noncommuting nature of conjugate operators such as position and momentum. This has led to the Schrödinger equation itself as the primary equation of motion for the wave function solution to the problem. In essence, the system (e.g., the electron) is treated as a wave, rather than as a particle, and the wave equation of interest is the Schrödinger equation. When boundary conditions are applied, either through potential barriers, or through the form of the potential (as in the harmonic oscillator), the time-independent Schrödinger equation yields solutions that are often special functions. Examples of this are the sines and cosines in the rectangular-barrier case, the Airy functions in the triangular-well case, and the Hermite polynomials in the harmonic oscillator case.

It is generally true that any time we examine a bounded system (even a classical system), the allowed energy levels take on discrete values, each of which corresponds to a single one of the family of possible members of the special functions. Thus, in the rectangular-well case, each energy level corresponds to one of the sinusoidal Fourier harmonics; in the harmonic oscillator, each energy level corresponds to one of the Hermite polynomials. In the set of levels, there is always a lowest energy level, called the ground state. The higher energy levels are referred to as the excited states, even when some are occupied in thermal equilibrium.

It should not be surprising that the set of all possible solutions to a given problem formulation, for example, the set of all sines and cosines for the rectangular barriers, can be shown to form a complete set, and thus can serve as a *basis* for a many-dimensional *linear-vector-space* representation of the problem. In the preceding chapters, we have not employed this terminology, but rather treated the expansion in terms of these functions in a manner that had the time variation appearing directly within the function itself. This is the normal Schrödinger *picture*, in which the so-called basis functions arise from solving the Schrödinger equation subject to the boundary conditions and these basis functions contain any time variation. In this picture, the operators are not time-varying, but their projection onto any of the basis functions is time-varying. Here, the basis functions can be thought of as the unit vectors of a coordinate system, and the amplitudes such as those of (2.45) can be thought of as amplitudes along each axis of this coordinate system. In the *Schrödinger picture*, the coordinate system rotates around the direction of the vector operator as a function of time. But this is a relative view—it could as easily be that the coordinate system is fixed and the vector operator rotates in the coordinate frame in such a way that the projection on any axis still varies with time in the prescribed manner. This picture, or view, is referred to as the *Heisenberg picture* of quantum mechanics (Heisenberg 1925).

We can think about this connection between the Heisenberg picture and the Schrödinger picture in a relatively simple manner. We have found that the wave function $\psi(x,t)$ evolves under the action of the Hamiltonian operator via the time-dependent Schrödinger equation (2.3)

$$i\hbar\frac{\partial\psi}{\partial t} = H\psi = \left[-\frac{\hbar^2}{2m}\nabla^2 + V(\mathbf{r})\right]\psi. \tag{6.1}$$

In a sense, this wave function evolves (in one dimension) as

$$\psi(x,t) = e^{-iHt/\hbar}\psi(x,0). \tag{6.2}$$

If we consider any arbitrary complete set of functions $\{\phi_n(x)\}$, for example, the set of basis states for an infinite potential well lying in the range $0 < x < a$

$$\varphi_n(x) = \sqrt{\frac{2}{a}}sin\left(\frac{n\pi x}{a}\right), \tag{6.3}$$

we can expand the wave function as

$$\psi(x,0) = \sum_n c_n\varphi_n(x). \tag{6.4}$$

The basis functions ϕ_n create an infinite set of coordinates, and the coefficients $c_n(t)$ are the projections of the wave function onto each of these coordinates. Hence, the basis states create an infinite-dimensional coordinate system, which is the Hilbert space corresponding to these functions.

The rotation of the total wave function (6.2) in this new coordinate space, described by the basis states, can be found by inserting (6.4) into (6.2). This gives

$$\begin{aligned}\psi(x,t) &= e^{-iHt/\hbar}\sum_n c_n\varphi_n(x) = \sum_n c_n e^{-iHt/\hbar}\varphi_n(x) \\ &= \sum_n c_n e^{-i\omega_n t}\varphi_n(x)\end{aligned}. \tag{6.5}$$

In this last form, the frequency $\omega_n = E_n/\hbar$ has been introduced, and E_n is the energy corresponding to each of the time-independent basis functions in (6.4). The difference between the Heisenberg picture and the Schrödinger picture is one of just where to assign the exponential factor. If we take the view that the basis states are fixed and time-independent, as indicated in (5.4), then the coordinate axes are fixed and the wave function rotates in this fixed coordinate system with the time-varying projections

$$c_n(t) = e^{-i\omega_n t} c_n(0). \tag{6.6}$$

This is the Heisenberg picture. On the other hand, if we attach the time variation to the basis states, and allow them to vary with time as

$$\varphi_n(x,t) = e^{-i\omega_n t} \varphi_n(x,0). \tag{6.7}$$

In this last form, we take the various c_n as time-independent (and assign their projections at $t = 0$), then the coordinates 'rotate' around the wave function. This is the Schrödinger picture. In fact, we have used the latter in the previous chapters, although we will increasingly use the former in subsequent chapters. The difference is only one of detail.

In this chapter, we want to achieve a number of things. First, in the next section, we will examine how (6.2) leads us to be able to determine the time dependence of operators. For example, we introduced pseudo-position and pseudo-momentum operators in order to quantize some systems. In essence, this was the introduction of generalized conjugate operators, and we want to examine here the rules that allow us to do this. In fact, (6.4) creates a *linear vector space*, and we want to examine how operators function in this space. Then, we turn to using generalized operators, such as were encountered in Chapter 5 to actually solve cases with more complex potentials and study the symmetries that are introduced by these complex potentials (where this can be done). Then, we look at a more complex way of actually studying the probabilities rather than just the wave function and how this can lead to a phase space formulation of quantum mechanics.

6.1 TIME DEPENDENCE OF OPERATORS

In Chapter 1, Equation (1.23), we developed the expectation value of the position operator in its respective representation. We can do this quite generally for any operator, and the expectation value of such a general operator A is given by the generalization of (1.23) as

$$\langle A \rangle = \int \psi^\dagger(x,t) A \psi(x,t) dx, \tag{6.8}$$

where we have arbitrarily inserted a time dependence for the wave function. We do not write the dependence of the general operator A on position and momentum, but in general it is a function of these generalized conjugate variables. Thus, $A = A(p,x)$ is some function of the position and of the momentum. An example of this is the quadratic harmonic oscillator potential energy, in which we can write $A = m\omega^2 x^2/2$, so the expectation value (6.8) is the average of the potential energy.

How does the average defined in (6.8) change with time in a system in which the wave function $\psi(x,t)$ evolves in time? At this point, we want to develop a general approach for an operator. We will, however, assume that the operator A is *not a specific function of time itself* (time will not be one of the explicit variables upon which it depends). Thus, we can develop the time variation of (6.8) as

$$\begin{aligned}\frac{d\langle A\rangle}{dt} &= \int \frac{\partial \psi^\dagger(x,t)}{\partial t} A\psi(x,t)dx + \int \psi^\dagger(x,t) A \frac{\partial \psi(x,t)}{\partial t} dx \\ &= \frac{i}{\hbar}\int H^\dagger \psi^\dagger(x,t) A\psi(x,t)dx - \frac{i}{\hbar}\int \psi^\dagger(x,t) AH\psi(x,t)dx \\ &= \frac{i}{\hbar}\int \psi^\dagger(x,t)(HA - AH)\psi(x,t)dx = \frac{i}{\hbar}\langle [H,A]\rangle.\end{aligned} \tag{6.9}$$

This result is very important and is the basis of the Heisenberg approach to quantum mechanics, but it is applicable to all quantum mechanics. Previously, we needed to solve directly for the wave function and then develop the expectation value of specific operators. Here, however, *we may choose to construct the wave junction with any convenient basis set*. The time variation of the expectation values of the operators is entirely determined by the commutator relation between the specific operator and the total Hamiltonian. *Thus, the expectation value of any operator that commutes with the Hamiltonian remains constant in time*. This is because in the conservative (nondissipative systems) that we have treated so far, the Hamiltonian is not a function of time and the total energy is conserved. Therefore, any operator that commutes with the Hamiltonian is similarly conserved and can also be simultaneously measured.

In treating conjugate operators, such as position and momentum, it was not generally possible (with a potential term present) to write the Hamiltonian as a function of only position or of only momentum. In general, the Hamiltonian contains both position (in the potential energy term) and momentum (in the kinetic energy term). Thus, we cannot choose a basis in which both commute with the Hamiltonian, so we cannot measure both simultaneously. Hence, noncommuting operators generally appear in a manner such that they do not commute with the Hamiltonian.

Equation (6.9) leads us to the general view that we can choose a basis set of functions, in terms of which to expand the total wave function, that is *convenient*. Thus, we would choose a set that is the natural set of expansion functions for the coordinate system and boundary conditions present. The choice is only important in that we must be able to evaluate the integrals inherent in the averages expressed in (6.9), and as expansions of the wave functions in, for example (6.5), to which we return in the next section. So far, we have disregarded the time variation in the wave functions, although (6.9) begins to transfer this to the operator itself. Let us continue this process.

The time-varying Schrödinger equation allows us to write directly the time variation through a simple exponential term, as was done in (6.2). We know that the energy arises in the argument of the exponential as the eigenvalue for the Hamiltonian (which must be summed over in most cases), so when we introduce this into the averages in (6.9), we will retain use of the operator form; for example, we will introduce the formal solution of the time-varying Schrödinger equation. Following (6.9), this leads to

$$\begin{aligned}\frac{d\langle A\rangle}{dt} &= \frac{i}{\hbar}\int \psi^{\dagger}(x,t)(HA - AH)\psi(x,t)dx \\ &= \frac{i}{\hbar}\int \psi^{\dagger}(x)e^{iHt/\hbar}(HA - AH)e^{-iHt/\hbar}\psi(x)dx \\ &= \frac{i}{\hbar}\int \psi^{\dagger}(x)\left[H, e^{iHt/\hbar}Ae^{-iHt/\hbar}\right]\psi(x)dx.\end{aligned} \tag{6.10}$$

What we have done is to create a time-varying operator in the Heisenberg representation (where the wave functions form the nonrotating coordinates for the system), which is given as

$$A_H(t) = e^{iHt/\hbar}Ae^{-iHt/\hbar}, \tag{6.11}$$

when the operator does not have a specific time variation on its own. The subscript 'H' is often used to denote the Heisenberg representation of an operator, which varies with time only through the effects that arise from the noncommuting of this operator with the Hamiltonian. However, this subscript is just as often not explicitly given. From (6.11), we can work backwards toward our starting Equation (6.8) by dropping the time derivatives, so we can write the expectation of an operator in the Heisenberg picture as

$$\langle A\rangle = \int \psi^{\dagger}(x)A_H(t)\psi(x)dx. \tag{6.12}$$

This is the basis for the Heisenberg picture (or representation) of quantum mechanics. Here, the wave function is expanded in a convenient basis set of functions, which corresponds to a fixed coordinate system. Then, instead of the wave function evolving or varying with time, the operators themselves evolve in time (if they do not commute with the Hamiltonian). The averages are now computed using the arbitrary, fixed basis functions and the 'time-varying' Heisenberg representation of the operators. In essence, if we expand the wave function in terms of the basis set, the above average is a matrix multiplication.

But, this approach, as we said, can be applied in almost any situation. If fact, as a tutorial let us apply this to determine the time variation of the creation and annihilation operators of the harmonic oscillator itself. Here, the Hamiltonian is given as

$$E = \sum_n \hbar\omega_n \left(a_n^\dagger a_n + \frac{1}{2} \right). \tag{6.13}$$

We know that the operators with subscript n operate only on the wave function for that energy level. Hence, we can write for the creation operator for level n

$$\begin{aligned} \frac{da_n}{dt} &= \frac{i}{\hbar}\hbar\omega_n \left[a_n^\dagger a_n, a_n \right] = i\omega_n \left(a_n^\dagger a_n a_n - a_n a_n^\dagger a_n \right) \\ &= i\omega_n \left(a_n^\dagger a_n a_n - 1 - a_n^\dagger a_n a_n \right) = i\omega_n \end{aligned}. \tag{6.14}$$

In the last line, we have used the commutator relation (5.14). Thus, we can easily solve this equation to show that

$$a_n(t) = a_n(0) e^{i\omega_n t}. \tag{6.15}$$

In a similar manner, we can determine the time variation of the annihilation operator to be

$$a_n^\dagger(t) = a_n^\dagger(0) e^{-i\omega_n t}. \tag{6.16}$$

6.2 LINEAR VECTOR SPACES

When we talk about linear vector spaces, we are combining the concepts of state variables, used in linear-systems theory, and vector calculus, used in electromagnetic fields. In vector calculus, we choose a set of basis vectors to define a coordinate system; these can be for example, the unit vectors along the three Cartesian coordinate directions in a rectangular coordinate system. Then, any general vector can be written as the sum of components in each of these directions, as

$$\mathbf{B} = b_1 \mathbf{a}_x + b_2 \mathbf{a}_y + b_3 \mathbf{a}_z. \tag{6.17}$$

Here, the $\mathbf{a}_i$ are the unit vectors and the b_i are the components of the vector B in each of the various directions defined by the unit vectors. The same ideas carry over to linear vector spaces of operators. We choose a basis set of functions that define the coordinates, which are usually infinite in number. Then, we can expand a general function in terms of the components as (6.4), with any time dependence incorporated. Again, the $\varphi_i(x)$ are the unit vectors of this space and the c_i are the components of the 'vector' in each of the various directions defined by the unit vectors. The difference between (6.4) and (6.17) is that the Cartesian coordinate system has only three unit vectors (is only three dimensional) while the linear vector space generally has an infinite number of dimensions (von Neumann 1932).

In the rectangular coordinate system for vectors, the unit vectors satisfy the relationship $\mathbf{a}_i \cdot \mathbf{a}_j = \delta_{ij}$, that is they satisfy an orthonormality condition. Similarly, we require the basis vectors in the linear vector space to be orthonormal (normally, but there are cases where this is not required). This is expressed as

$$\int \varphi_i^\dagger(x)\varphi_j(x)dx = \delta_{ij}. \tag{6.18}$$

The basis set also satisfies a *closure* relation as

$$\sum_i \varphi_i^\dagger(x)\varphi_i(x') = \delta(x - x'). \tag{6.19}$$

The linear vector space has many standard properties associated with it being a *linear* vector space. Some of these are: (1) when two basis functions are added, the order is unimportant; (2) when a basis function is multiplied by two *c*-numbers (nonoperators), the order of these two numbers is unimportant; (3) when the sum of a number of basis functions is multiplied by a *c*-number, or an operator, each basis function is multiplied by this *c*-number or operator. Finally, the basis functions are linearly independent, so that a statement such as

$$\sum_i c_i\varphi_i(x) = 0 \tag{6.20}$$

implies that all $c_i = 0$. This leads us to a very important point. If we can write the wave function in terms of a single basis function as

$$\psi(x) = c_i\varphi_i(x), \tag{6.21}$$

we say that the wave function is in a *pure* state. Otherwise, it is often said to be in a *mixed* state. If a given wave function cannot be expanded into a sum over a set of basis functions, it is said to be in an *entangled* state.

6.2.1 Hermitian Operators

We recall from our earlier discussions that $A^\dagger$ is the Hermitian adjoint of the operator A. If A were represented by a matrix, then the Hermitian adjoint would correspond to the transpose complex conjugate of that matrix. Here, we want to emphasize that all *measurable* quantities correspond to operators in which the Hermitian conjugate is equal to the operator itself; for example, the operator is said to be a Hermitian operator. We first note that the average value of the operator A is given by (6.12). If the expectation value is a measurable quantity, then this average value must be a real quantity. With this in mind, we rewrite (6.12) as

$$\begin{aligned}\langle A\rangle = \langle A\rangle^{*} &= \left[\int \psi^{\dagger}(x,t) A\psi(x,t)dx\right]^{*} \\ &= \int \left[A\psi(x,t)\right]^{\dagger} \psi(x,t)dx \\ &= \int \psi^{\dagger}(x,t) A^{\dagger} \psi(x,t)dx.\end{aligned} \tag{6.22}$$

By comparison with (6.12), it is now obvious that for the expectation value to be a real quantity we must have $A = A^{\dagger}$. As mentioned, operators for which this relation holds are said to be Hermitian operators. We have used in (6.22) general properties of the complex conjugate and adjoint operations.

Let us now consider the situation in which the wave function is described in such a manner that the operator A produces the eigenvalue a, where a is a c-number (not an operator, but a scalar constant). Thus, we have the operator relation

$$A\psi(x,t) = a\psi(x,t). \tag{6.23}$$

Then, the expectation value of this operator is given again by (6.12) as

$$\langle A\rangle = \int \psi^{\dagger}(x,t) A\psi(x,t)dx = \int \psi^{\dagger}(x,t) a\psi(x,t)dx = a. \tag{6.24}$$

By the same token, we can compute the expectation value of the square of the operator

$$\langle A^{2}\rangle = \int \psi^{\dagger}(x,t) A^{2}\psi(x,t)dx = \int \psi^{\dagger}(x,t) a^{2}\psi(x,t)dx = a^{2}. \tag{6.25}$$

The uncertainty of an operator is given by $\sqrt{A^{2} - A^{2}}$ (the uncertainty relation is between the uncertainty values of two noncommuting operators), so that an operator with an eigenvalue, such as in (6.23), has no uncertainty! Its expectation value is the eigenvalue.

Suppose that we have two such wave functions for which the operator A produces eigenvalues, and so

$$\begin{aligned}A\psi_{1}(x,t) &= a_{1}\psi_{1}(x,t) \\ A\psi_{2}(x,t) &= a_{2}\psi_{2}(x,t)\end{aligned}. \tag{6.26}$$

Let us postmultiply the first of these equations by the complex conjugate of ψ_2, take the complex conjugate of the second and postmultiply by ψ_1, then subtract the two and integrate the result. This leads to

$$\left(a_1 - a_2^*\right)\int \psi_2^\dagger(x,t)\,\psi_1(x,t)dx =$$

$$= \int \psi_2^\dagger(x,t)A\psi_1(x,t)dx - \int\left[A\psi_2(x,t)\right]^\dagger \psi_1(x,t)dx$$
$$= \int \psi_2^\dagger(x,t)A\psi_1(x,t)dx - \int \psi_2^\dagger(x,t)A\psi_1(x,t)dx = 0. \quad (6.27)$$

We thus have two options for the integral on the left. The first option is that $\psi_2 = \psi_1$, which leads to $a_1 = a_2^* = a_1^*$. That is, if the two wave functions are the same, then the eigenvalue must be the same, and it must be real! The second option is that the eigenvalues actually are different, for which we must have the integral itself vanish, which means that the two wave functions are orthogonal to one another. Hence, they must be two different basis functions in an expansion of the wave function. It is this latter point that allows us to set up a 'coordinate' system with basis functions that are orthonormal (orthogonal and normalized). This is the purpose of the linear-vector-space approach. We can choose a set of orthogonalized (and normalized) wave functions $\varphi_i(x)$ (we write in now only the spatial variation, as we will pursue the Heisenberg picture). These can correspond to the infinite set of wave functions found in the potential well problems of Chapter 2, or the set of Hermite polynomials of Chapter 5, as examples. Then, we expect that any operator A will give a set of eigenvalues a_i corresponding to each member of the set of wave functions. To illustrate this, let us expand the total wave function in the set of basis functions, as

$$\psi(x) = \sum_n c_n \varphi_n(x). \quad (6.28)$$

With this expansion, we can find the expectation value of an operator as

$$\begin{aligned} A &= \sum_{n,m} c_n^* c_m \int \varphi_n^\dagger(x) A \varphi_m(x) dx \\ &= \sum_{n,m} c_n^* c_m \delta_{nm} a_m \\ &= \sum_n \left|c_n\right|^2 a_n. \end{aligned} \quad (6.29)$$

In going from the second line to the third, we have used the fact that the integral is zero (orthogonality) unless $n = m$. This provides the Kronecker delta function in the third line, which is then used in the summation over the index m. Thus, the coefficients provide the weight, or probability amplitude, for each of the individual eigenvalues of the basis set. The set of eigenvalues a_i is termed the *spectrum* of the operator A. The connection with a normal spectrum is quite straightforward: in the normal case, the strength of a component at frequency ω is given by the corresponding Fourier coefficient. Here, the eigenvalue a_i is the generalized Fourier coefficient giving the strength of the operator A in the particular basis coordinate φ_i. The coefficients c_i give a normalization to the excitation of the particular mode φ_i by the total wave function.

6.2.2 Some Matrix Properties

Once we have selected a basis set of functions, then it is no longer necessary to continue to restate these functions. Rather, we need only to keep track of the expansion coefficients for the wave function. This leads to matrix operations. Consider, for example, a wave function expanded in the series given in (6.28). If we operate on this wave function with the operator A, let us assume that it produces the new wave function given by $A\psi_a = \psi_b$. Thus, we write

$$A\psi_a(x) = \sum_n a_n A\varphi_n(x) = \sum_m b_m \varphi_m(x) = \psi_b(x). \tag{6.30}$$

Let us premultiply the last two expressions by $\varphi_j^*(x)$, and integrate this over all space. This gives the results

$$b_j = \sum_i a_i A_{ji}, \quad A_{ji} = \int \varphi_j^\dagger A\varphi_i dx. \tag{6.31}$$

The last expression defines the *matrix elements* of the operator A. This gives the matrix expression (we use the standard notation of only a single right bracket for a column matrix)

$$b] = [A]a]. \tag{6.32}$$

This leads to the expression for the expectation value of the operator A to be

$$\langle A\rangle = a^*]^T [A]a]. \tag{6.33}$$

where the superscript T implies the transpose operation (which here produces the row matrix). Since the operator is Hermitian, the matrix representation of the operator is a Hermitian matrix, which implies that $A_{ij}^* = A_{ji}$.

Suppose that after carefully choosing a set of basis functions, we discover that another choice would have been preferred. Just as one can do coordinate transformations in vector algebra, one can do coordinate transformations in the linear vector spaces. Initially, the wave function is defined in terms of the basis set $\varphi_{1i}(x)$ as

$$\psi(x) = \sum_i a_i \varphi_{1i}(x), \tag{6.34}$$

which we want to write in terms of a 'rotation' matrix with new coordinates defined by

$$\varphi_{2i}(x) = \sum_j S_{ij}\varphi_{1i}(x). \tag{6.35}$$

If the wave function is expanded in terms of the new coordinates as

$$\psi(x) = \sum_k b_k \varphi_{2k}(x), \tag{6.36}$$

it is easy to show that the coefficients are related by

$$a_i = \sum_k S_{ik} b_k. \tag{6.37}$$

We can now use our operator A of (6.31) and use the relationship between the different basis sets to show that

$$[A_a][a] = [S][A_b][b]. \tag{6.38}$$

In the last equation, we mean that the operator acts upon the individual basis functions, and hence will move through the matrix sum. Since the basis functions are changed on the right-hand side, the result is different and leads to there being different matrix elements, which have been denoted by the subscripts corresponding to the old and new sets of basis functions. By using (6.37), we can relate the two matrix representations (in the new and old basis sets) via

$$[A_a] = [S][A_b][S]^{-1}, \tag{6.39}$$

where the last term is the inverse of the transformation matrix. Equation (6.39) defines a *similarity* transformation. An important property of such transformations is that they satisfy

$$\sum_j S_{jk}^* S_{ji} = \delta_{ik}. \tag{6.40}$$

This means that the implied matrix product produces a unit matrix, and

$$[S]^\dagger = [S]^{-1}. \tag{6.41}$$

The last expression is the relationship defining a *unitary matrix*. The similarity transformations from one coordinate (orthonormal basis) set to another must be unitary transformations, and thus are generalized rotations.

6.2.3 The Eigenvalue Problem

As the final part of this section, we want to address the transformations that apply to the eigenvalue problem, as it is often called. In previous chapters, and sections, it was assumed that one could find a basis set in which an operator returned the same basis function multiplied by a c-number constant, as in (6.23). In general, we would have to be quite prescient to choose the basis set properly for this to occur for all, or even a significant fraction, of the operators of interest. How then do we find the eigenvalues of operators, which become importantly related to the expectation values of the operators? We know from the arguments of Section 6.2.1 that, if an operator produces two different values when operating upon a set of functions, these latter functions must be orthogonal to one another. This means, in general, that a selected basis set can be found that will lead to a diagonal matrix for any selected operator, but it is unlikely to be diagonal for other operators unless the latter commute with the selected operator. This is a very special situation, which is not found in general.

On the other hand, we have asserted above that the choice of the basis set is not particularly important to the physics but is generally made based upon other considerations such as the ease of solving the boundary condition problem. Thus, in general, we should expect that some linear combination of the chosen basis states, much like a Fourier series representation, will describe the basis states for any operator. The connection between the selected basis states and those that yield the diagonal representation for an operator is given by the similarity transformations described in the previous subsection. In this section, we show that this is the case, and that one can in fact diagonalize the basis set, or linear vector space, through a coordinate rotation defined by the similarity transformations.

In order to describe any physical problem, the first step is to choose a basis set of wave functions $\{\varphi_i(x)\}$ that describe the linear vector space. These functions are properly orthogonalized, so the inner product of any distinct pair is zero, but each is normalized. From this selection of a basis set, it is possible to determine the expectation value of a general operator in such a way that the operator lies in a space defined by these operators. It may be asserted that there exists a proper choice of basis functions, which describes a vector space rotated with respect to the initial choice, in which the operator is directed solely along one of the basis functions:

$$A\xi_i(x) = \lambda\xi_i(x), \tag{6.42}$$

where the various $\xi_i(x)$ are the proper basis functions in which the Hamiltonian is diagonal. The rotated basis function can be expressed in terms of the original set by the similarity transformation (6.38), or using the original basis set, we have

$$\sum_i c_i A\varphi_i(x) = \lambda \sum_i c_i \varphi_i(x), \tag{6.43}$$

where we have used the coefficients c_i in the actual expansion of the new basis functions in terms of the old. If we now multiply by any one of the original basis functions and integrate, it gives one row of the matrix equation

$$[A]c] = \lambda c]. \tag{6.44}$$

This latter is the *eigenvalue equation*. Since any choice of the c_i produces one of the ξ_j, we do not expect the vector $c]$ to be a null vector. Thus, the only solution of (6.44) is the choice

$$det\left([A]-\lambda[1]\right)=0. \tag{6.45}$$

In this last expression, [1] is the unit matrix which has the elements given by δ_{ij}. Solving this algebraic expression gives the allowed values of the eigenvalues of the operator A. If the space has n dimensions, there are then n values of these eigenvalues. Once these values are found, $n-1$ of these can be substituted, in turn, into (6.44) to find $n-1$ equations for the various coefficients, which relate the new basis states to the old (the last solution arises from normalization). These equations define the similarity transformation, since it is assumed that each value of an eigenvalue determined from (6.45) corresponds to defining one basis state of the rotated coordinate system.

The above also is dependent upon the values obtained for the eigenvalues from (6.45) all being distinct. If two or more values of λ are the same, then a problem arises. This case is called degeneracy, as two (or more) different basis functions in the rotated coordinate system yield the same value of the eigenvalue. The solution to this is to consider as a new subcoordinate system only those degenerate functions for which a linear combination (a second rotation) can be used to lift the degeneracy of the functions. That is, we form a linear combination of the functions that lifts the degeneracy, proceeding precisely as above in a second iteration. We will see an example of this in the next chapter.

The presence of eigenvalues raises an important issue with regard to the uncertainty discussed above. *An operator that satisfies an eigenvalue equation has zero uncertainty.* This is especially important for the time-independent Schrödinger equation. This equation is an eigenvalue equation for the total energy. On the left-hand side is the Hamiltonian, or energy operator, while the energy value is on the right-hand side. Since this is an eigenvalue equation, there is no uncertainty in the energy that results from its solutions. Each energy value is well described and subject to no variation that can be connected to an uncertainty principle.

6.3 SUPERSYMMETRY

In Chapter 5, where we treated the harmonic oscillator, the Hamiltonian was split by the use of a pair of operators, the creation and annihilation operators (5.10). The use of these operators greatly simplified the solution of the Schrödinger equation. In this special case, it was the quadratic potential in x that allowed the use of the operator reduction of the Hamiltonian. It turns out that this quadratic potential is one of a group of potentials that allow such a reduction of the Hamiltonian (Sukumar 1985, Cooper et al. 1995). Supersymmetry normally refers to quantum field theory and describes transformations between fermions and bosons. However, Witten (1981) described a quite simple supersymmetric system consisting of a single spin-1/2 particle moving in one dimension. This latter work had no connection to quantum field theory and provides the basis of what we now call supersymmetric quantum mechanics

(SUSYQM), and this involves the potentials mentioned just above. Just as the harmonic oscillator was simplified by operators, Hamiltonians containing these special potentials also benefit from reduction via the operators.

In general, we know that a second-order (in space) differential equation, such as the Schrödinger equation, has two types of solutions. In the harmonic oscillator, these are the even and odd solutions. In more complicated situations, they may be, for example, Bessel functions of the first and second kind. Suppose we have a one-dimensional Schrödinger equation of the form

$$H\psi(x) = -\frac{\hbar^2}{2m}\frac{d^2}{dx^2}\psi(x) + V_1(x)\psi(x) = E\psi(x). \tag{6.46}$$

We assert that we can find one of the solutions by defining the factorization of the Hamiltonian with two operators in the following form:

$$H_1 = A^+A^- + E, \tag{6.47}$$

where E is a constant (the energy, for example), and the two operators are defined by

$$A^{\pm} = W(x) \pm \frac{\hbar}{\sqrt{2m}}\frac{d}{dx}. \tag{6.48}$$

These operators have been given the name "super-charge operators." $W(x)$ is usually referred as the *super-potential*. The form (6.48) is often encountered in operator treatments of, for example, the harmonic oscillator, and the two super-charges are known as creation (A^+) and annihilation (A^-) operators, or raising and lowering operators which is more in keeping with their use in both the harmonic oscillator and SUSYQM. The relation between the potential $V_1(x)$ and the super-potential $W(x)$ may be found by direct substitution of (6.48) into (6.46), but is given as

$$V_1(x) = W^2(x) - \frac{\hbar}{\sqrt{2m}}\frac{dW(x)}{dx} + E. \tag{6.49}$$

If the ground state wave function is known, as in the case of the harmonic oscillater where the product of the annihilation operator and the wave function gave a first-order equation that was easily solved, then the super-potential may be found as

$$W(x) = -\frac{\hbar}{\sqrt{2m}}\frac{1}{\varphi_0(x)}\frac{d\varphi_0}{dx}. \tag{6.50}$$

For the second solution, we can define a partner Hamiltonian with a partner potential via the description

$$H_2 = A^- A^+ + E$$
$$V_2(x) = W^2(x) + \frac{\hbar}{\sqrt{2m}} \frac{dW(x)}{dx} + E. \tag{6.51}$$

Before proceeding, let us examine the two operators a little closer. For example, we may take the natural product of them as

$$\begin{aligned} A^+ A^- \psi &= \left(W(x) + \frac{\hbar}{\sqrt{2m}} \frac{d}{dx} \right) \left(W(x) - \frac{\hbar}{\sqrt{2m}} \frac{d}{dx} \right) \psi \\ &= W^2(x)\psi + \frac{\hbar}{\sqrt{2m}} \frac{d}{dx} (W\psi) - W \frac{\hbar}{\sqrt{2m}} \frac{d\psi}{dx} - \frac{\hbar^2}{2m} \frac{d^2}{dx^2} \psi \\ &= -\frac{\hbar^2}{2m} \frac{d^2}{dx^2} \psi + \left[W^2(x) + \frac{\hbar}{\sqrt{2m}} \frac{dW}{dx} \right] \psi. \end{aligned} \tag{6.52}$$

If we now use (6.49), we see that the bracketed term is just $V_1 - E$, and we recover the normal Schrödinger equation. Similarly

$$\begin{aligned} A^- A^+ \psi &= \left(W(x) - \frac{\hbar}{\sqrt{2m}} \frac{d}{dx} \right) \left(W(x) + \frac{\hbar}{\sqrt{2m}} \frac{d}{dx} \right) \psi \\ &= W^2(x)\psi - \frac{\hbar}{\sqrt{2m}} \frac{d}{dx} (W\psi) + W \frac{\hbar}{\sqrt{2m}} \frac{d\psi}{dx} - \frac{\hbar^2}{2m} \frac{d^2}{dx^2} \psi \\ &= -\frac{\hbar^2}{2m} \frac{d^2}{dx^2} \psi + \left[W^2(x) - \frac{\hbar}{\sqrt{2m}} \frac{dW}{dx} \right] \psi, \end{aligned} \tag{6.53}$$

and the bracketed term is $V_2 - E$. Quite often, the energy scale (and the potential) is uniformly shifted so that $E = 0$, and this simplifies the mathematics.

Now, in general, the commutator relationship is complicated. If we subtract (6.53) from (6.52), we have

$$\left[A^+, A^- \right] = A^+ A^- - A^- A^+ = \frac{2\hbar}{\sqrt{2m}} \frac{dW}{dx}. \tag{6.54}$$

That is, the commutator does not give a simple result. However, let us give an example using the harmonic oscillator. Here

$$V_1 - E = \frac{m\omega^2}{2} x^2, \tag{6.55}$$

where we have rescaled the energy scale. This leads us to consider that W is simply

$$W(x) = \sqrt{\frac{m}{2}}\omega x, \tag{6.56}$$

and

$$\left[A^+, A^-\right] = \hbar\omega. \tag{6.57}$$

Hence, if we renormalize the energy into units of $\hbar\omega$, the commutator relations would be the simple obtained in the harmonic oscillator.

Let us assume that the ground state of the system lies in the first solution, which arises from (6.46), with basis functions $\varphi_{1,i}$. Let us assert that there exists a ground state $\varphi_{1,0}$ such that the ground state energy is zero. Then, we should have

$$A^- \varphi_{1,0} = 0. \tag{6.58}$$

By the same token, if (6.58) is satisfied, then

$$\left\langle \varphi_{1,0} | A^+ A^- | \varphi_{1,0} \right\rangle = 0. \tag{6.59}$$

which assures that the ground state energy is zero. From (6.48), we then find that

$$\varphi_{1,0} = C_{1,0} exp\left(\frac{\sqrt{2m}}{\hbar}\int^x W(x')dx'\right). \tag{6.60}$$

When this occurs, H_2 will have no ground state with energy = 0. Now, the roles of H_1 and H_2 can be reversed, but that is simply a notation change. It is possible, however, that the two Hamiltonians are degenerate, but then the ground state energy must be greater than zero. Since the ground state energy for H_2 is greater than zero, we can state this as

$$A^- A^+ \psi_2 = E_2 \psi_2. \tag{6.61}$$

Now, let us premultiply (6.61) A^+ to give

$$A^+ A^- (A^+ \psi_2) = E_2 \left(A^+ \psi_2\right). \tag{6.62}$$

If $A^+\psi_2 \neq 0$, as must be the case if the eigenvalues are greater than zero, then this state must be an eigenfunction of H_1 as well and E_2 is also an eigenvalue of H_1. This has an amazing result: all eigenvalues of H_1, other than the ground state, are also eigenvalues of H_2, and all eigen values of H_2 are also eigenvalues of H_1. This leads to the normalized state

$$\psi_1 = \frac{1}{\sqrt{E_2}} A^+ \psi_2. \tag{6.63}$$

This means that the basis functions $\{\varphi_{1,i}\}$we use for ψ_1 is precisely the same set of basis functions, except for the normalizations, we use for ψ_2. These eigenvalues form a pair of ladders of energy that perfectly line up between the two solutions, with the exception of the ground state of the first set. The result is that the simplicity of the harmonic oscillator results can be extended directly to a more extensive set of potentials, which include the Coulomb, Morse, Rosen-Morse, extended Pöschl-Teller, Eckart, and Scarf potentials (Cooper et al. 1995).

6.4 THE DENSITY MATRIX

In many applications, people would prefer to have their quantum system defined in a manner that gives more relevance to the probability density (the magnitude squared of the wave function). This can be achieved in many different ways, but here we discuss only one—the density matrix. Here, we want to show it arises directly from the wave function. Generally, one can solve the Schrödinger equation by assuming an expansion of the wave function in a suitable basis set so that each basis function is an energy eigen function according to

$$H\varphi_i = E_i\varphi_i, \tag{6.64}$$

in which E_i is the energy level corresponding to the particular basis function. Then, the total wave function can be written as

$$\psi(x,t) = \sum_i c_i\varphi_i(x)e^{-iE_it/\hbar}. \tag{6.65}$$

For the general situation, we may think about this wave function and its adjoint, and we can then define the density matrix as

$$\rho(x,x',t) = \psi^\dagger(x,t)\psi(x',t) = \sum_{n,m} c_{nm}\varphi_n^\dagger(x,t)\varphi_m(x',t). \tag{6.66}$$

The diagonal terms are obviously the probability of that particular basis state. The off-diagonal terms are going to oscillate as

$$c_{nm}\varphi_n^\dagger(x)\varphi_m(x')e^{-i(E_m-E_n)t/\hbar}, \tag{6.67}$$

so that one could rightly assume that their time averages approached zero. But this is not always the case. Because the diagonal terms are the probability for the individual basis states, we can say that

$$Tr\{\rho\} = \sum_n c_{nn}|\varphi_n|^2 = 1. \tag{6.68}$$

Obviously, here $c_{nn} = c_n^* c_n$ is a real, positive number. This result follows precisely from the requirement that

$$\int \rho(x,x)dx = \int |\psi(x)|^2 dx = 1, \tag{6.69}$$

which is the standard normalization condition.

The above implies that we can use the density matrix quite generally to evaluate the expectation values of various operators. For example, we can write this expectation value as

$$\begin{aligned}\langle A\rangle &= \int \psi^\dagger(x)\psi(x)dx = \sum_{n,m} c_{nm}\int \varphi_n^\dagger(x)A\varphi_m(x)dx \\ &= \sum_{n,m} c_{nm}A_{nm} = \sum_{n,m} A_{nm}\rho_{mn} = Tr\{A\rho\} = Tr\{\rho A\}.\end{aligned} \tag{6.70}$$

The last two terms in this equation are important. Because the order of the two summations is not important, it means that one can move the operators in the trace by any cyclic permutation; in this case moving the ρ from the last position around to the first position. For multiple operators, this operation may be written as

$$Tr\{AB\rho\} = Tr\{\rho AB\} = Tr\{B\rho A\}. \tag{6.71}$$

As the density matrix satisfies other quantum equations, we can use (6.9) to write the time dependence of it as

$$\begin{aligned}i\hbar\frac{d\rho}{dt} &= [H,\rho] \\ &= \left[-\frac{\hbar^2}{2m}\left(\frac{\partial^2}{\partial x^2} - \frac{\partial^2}{\partial x'^2}\right) + V(x) - V(x')\right]\rho(x,x',t).\end{aligned} \tag{6.72}$$

This latter equation is known as the *Liouville equation*, or sometimes as the Liouville-von Neumann equation. Note that, this equation, like the Schrödinger equation has no dissipative terms included, so is reversible.

6.5 THE WIGNER FUNCTION

One problem that is always raised with respect to the density matrix is that it is a function only of time and the position variables. In classical physics, one has the Boltzmann equation which is a function of position, momentum, and time, so that one can generate a phase space formulation of the transport problem. None of the quantum functions we have discussed so far has this convenient phase space formulation. But it is often very convenient to describe the system with a phase space formulation. And, in fact, such a formulation exists (Wigner 1932). We first introduce the center-of-mass and difference coordinates through the transformations

$$X = \frac{x + x'}{2}, \quad s = x - x'. \tag{6.73}$$

We can now write the Wigner distribution function as a Fourier transform in the off-diagonal elements of the density matrix as

$$f_W(X,p,t) = \frac{1}{2\pi\hbar}\int_{-\infty}^{\infty} ds\rho\left(X + \frac{s}{2}, X - \frac{s}{2}, t\right)e^{ips/\hbar}. \tag{6.74}$$

This function is now similar to the Boltzmann distribution function, but this one is not necessarily positive definite. Normally, the ground state Wigner function (when we use a basis function expansion) is positive definite, because it represents the ground state. The ground state must be positive definite. But, excited states can have negative excursions, and these negative parts, as well as a non-Gaussian shape, are signals of the quantum nature of the system (Ferry and Nedjalkov 2018, Weinbub and Ferry 2018).

The Wigner function has certain normalizations, such as when we integrate it over the position, we obtain

$$\begin{aligned}\int f_W(X,p,t)dX &= \frac{1}{2\pi\hbar}\int_{-\infty}^{\infty} ds \int dX \psi^\dagger\left(X + \frac{s}{2}\right)\psi\left(X - \frac{s}{2}\right)e^{ips/\hbar} \\ &= \frac{1}{2\pi\hbar}\int_{-\infty}^{\infty} ds\delta(s)e^{ips/\hbar} \\ &= 1.\end{aligned} \tag{6.75}$$

Similarly, if we Fourier transform each wave function as in (1.25), and then integrate over the momentum, we will similarly achieve unity as the result. Hence, the Wigner function maintains the orthonormality of the wave function and its basis functions in the phase space.

Incorporating the coordinate transformations into the Liouville equation gives us a new equation of motion for the density matrix, which appears as

$$i\hbar\frac{\partial\rho}{\partial t}=\left[-\frac{\hbar^2}{2m}\frac{\partial^2}{\partial X\partial s}+V\left(X+\frac{s}{2}\right)-V\left(X-\frac{s}{2}\right)\right]\rho(X,s,t). \tag{6.76}$$

If we use the transformation (6.74) on this equation of motion, then we obtain the Wigner equation of motion for the Wigner function to be

$$\frac{\partial f_W}{\partial t}+\frac{p}{m}\frac{\partial f_W}{\partial X}-\frac{1}{i\hbar}\left[V\left(X+\frac{i\hbar}{2}\frac{\partial}{\partial p}\right)-V\left(X-\frac{i\hbar}{2}\frac{\partial}{\partial p}\right)\right]f_W(X,p,t)=0. \tag{6.77}$$

Using the propertiese of displacement operators, we can rewrite this equation as

$$\begin{aligned}&\frac{\partial f_W}{\partial t}+\frac{p}{m}\frac{\partial f_W}{\partial X}-\frac{1}{2\pi\hbar}\int dp' W(X,p')f_W(X,p+p',t)=0\\&W(X,p')=\int ds sin\left(\frac{p's}{\hbar}\right)\left[V\left(X+\frac{s}{2}\right)-V\left(X-\frac{s}{2}\right)\right]\end{aligned}. \tag{6.78}$$

Here, $W(X,p)$ is the Wigner potential. One can see immediately, that if the actual potential is of quadratic order or less, only the first-order derivatives appear, and the result is simply the classical Boltzmann equation. Moreover, only odd orders of X appear in the expansion of the square bracket in (6.76) or (6.78). The minus sign assures that all even orders cancel. Like the Boltzmann equation, one cannot simply solve (6.78), or even the Liouville equation. This is because the time derivative tells us we need an initial condition in order to solve these equations.

Now, suppose we have a wave function that is composed of two Gaussians, such as those in (2.68). Here, however, we will have one Gaussian moving in the positive x-direction (and the argument will be written as $x-vt$), while the second Gaussian is moving in the negative x-direction (and the argument will be written as $x+vt$). We are also going to write these wave functions with an exponent propagator so that

$$\psi_{1,2}=\frac{1}{\left(2\pi\sigma^2\right)^{1/4}}exp\left(\frac{(x\mp vt)^2}{4\sigma^2}+ik_0x\right), \tag{6.79}$$

where the velocity is related to the wave number by $v=\hbar k_0/m$. Now, we combine these two wave packets into a single wave function as

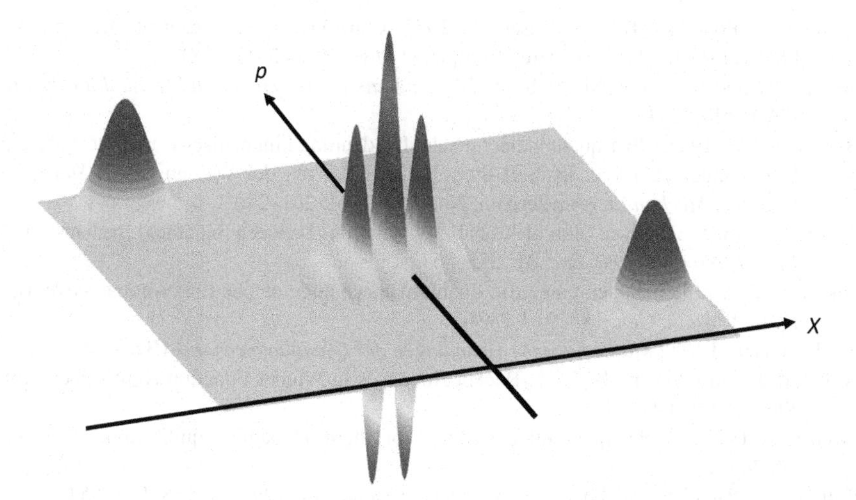

FIGURE 6.1 A composite wave function of two Gaussian packets produce a Wigner function with three components. The central one is the entanglement between the two original wave packets.

$$\Psi(x,t) = \frac{1}{\sqrt{2}}\left[\psi_1(x,t) + \psi_2(x,t)\right]. \tag{6.80}$$

Now, when we form the Wigner function of this composite wave function, we get three terms. Two of these terms are simply of the same form as (6.74), but the third term involves an off-diagonal term in the density matrix that has both X and X' in it. If we plot the Wigner function, it will appear as in Figure 6.1. If we had done to two beginning wave functions independently, we would have had only the two Gaussians of the form of (6.74). The third, mixed term represents the fact that we are really only using a single total wave function, and it represents the *entanglement* between the beginning wave functions.

The idea of entanglement surfaced originally in response to the Einstein-Podolsky-Rosen (1935) thought experiment, in which two classical particles interacted with one another and then went on their way. The term entanglement comes from Schrödinger (1935), who pointed out that, once the two particles had interacted, they were a single entangled quantum entity. Even when they separated and moved away, they remained a single quantum mechanical entity until some other process broke up the entanglement. In essence, this entanglement is a memory of the nature of the interaction that occurred between the two original particles. As we pointed out, the entanglement in Figure 6.1 arises from the off-diagonal terms in the density matrix, so it is not unusual that these terms are referred to as the entanglement.

REFERENCES

Cooper, F., Khareb, A., and Sukhatme, U. 1995. Supersymmetry and Quantum Mechanics. *Phys. Repts.* **251**: 267–385

Einstein, A, Podolsky, B., and Rosen, N. 1935. Can Quantum-Mechanical Description of Physical Reality be Considered Complete? *Phys. Rev.* **47**: 777–781.

Ferry, D. K. and Nedjalkov, M. 2018. *The Wigner Function in Science and Technology* (Bristol, IOP Publishing)

Heisenberg, W. 1925. Über quantentheoretische Umdeutung kinematischer und mechanischer Beziehunger. *Z. Phys.* **33**: 879–893; Tr. in B. L. van der Warden. 1967. *Sources of Quantum Mechanics*. (Amsterdam: North-Holland) 261–276

Schrödinger, E. 1935. Discussion of Probability Relations between Separated Systems. *Math. Proc. Cambridge Phil. Soc.* **31**: 555–563

Sukumar, C. V. 1985. Supersymmetric quantum mechanics of one-dimensional systems. *J. Phys. A: Math. Gen.* 18: 2917–2936

von Neumann, J 1932 *Mathematische Grundlagen der Quantummechanik* (Berlin: Springer)

Weinbub, J. and Ferry, D. K. 2018. Recent Advances in Wigner Function Approaches. *Appl. Phys. Rev.* 5: 041104

Wigner, E 1932 On the quantum correction for thermodynamic equilibrium. *Phys. Rev.* 40: 749–759

Witten, E. 1981. Dynamical breaking of supersymmetry. *Nucl. Phys. B* 188: 513–554

PROBLEMS

1 A dynamic system has $H = p^2 + Ax + Bpx$. Calculate the equations of motion for the operators x,p; that is, compute the time derivatives of the expectation values for these operators.

2 A particular operator A has been used to evaluate a dynamical property in an orthonormal set containing only three basis functions. When its expectation value is found, the resulting matrix for the operator is given by

$$[A] = \frac{1}{2}\begin{bmatrix} +\sqrt{2} & \sqrt{2} & 0 \\ -\sqrt{2} & \sqrt{2} & 0 \\ 0 & 0 & 2 \end{bmatrix}.$$

Determine the eigenvalues of this operator.

3 Compare the different dynamics that results from Hamiltonians that are written as $2x^2p^2$, $x^2p^2 + p^2x^2$ and $2(xp)^2$. What can you conclude about the proper symmetrization of operators?

4 Two operators A, B satisfy $[A,B] = C$. Compute the difference $A^3B^3 - (AB)^3$. Assume that C is a c-number and not an operator.

5 A particular projection operator $P_j = |j\rangle\langle j|$ is defined on the basis set

$$\psi_k(x) = \sqrt{\frac{2}{a}}\sin\left(\frac{k\pi x}{a}\right)$$

for $0 \leq x \leq a$ (k,i,j are integers). Expand the function $V(x)$ in the basis set $|j\rangle$ if $V(x) = 1$ for $0 \leq x \leq a$, and $V(x) = 0$ elsewhere; that is, determine the coefficients c_j in the expansion

$$V(x) = \sum_j c_j |j\rangle.$$

6 Consider a computational scheme in which the real axis (the position coordinate) is discretized to points $x = na$, where a is a 'lattice' constant. Instead of the continuous eigenvalue characteristics, each lattice site must be characterized by a function that allows the orthogonality to be retained. Determine a function localized at each lattice site that satisfies normalization and is orthogonal with all of its neighbors. (The functions will yield an integrated product that is one of the generalized distribution forms of the delta function.)

7 Consider a potential well which is described by the potential $V(x) = Fx$ for $0 < x < a$, and $V(x) \to \infty$ for $x < 0$ and $x > a$. Using the wave functions of the infinite potential well

$$\psi_n(x) = \sqrt{\frac{2}{a}} \sin\left(\frac{n\pi x}{a}\right)$$

compute the Hamiltonian matrix for the lowest five energy levels. Take $F = 0.02$ eV nm^{-1} and $a = 5$ nm. Then, diagonalize the matix to determine the energy eigenvalues and their wave functions. (While a proper approach would use an infinite set, approximate results can be found from these lowest five wave functions.) Compare the results with those of problem 15 of Chapter 2.

7 Stationary Perturbation Theory

It is generally nice to be able to solve a problem exactly. However, this is often not possible, and some approximation scheme must be applied. One example of this is the triangular potential well of Section 2.4 or the finite potential well of Section 2.3.2. In the former example, complicated special functions were required to solve Schrödinger's equation. In the latter example, the solution could not be obtained in closed form, and graphical solutions, or available software packages, were used to find the eigenvalues of the energy. These are examples of more common problems. In Chapter 6, however, it was pointed out that one could choose an arbitrary basis set of functions that formed a complete orthonormal set as a defining linear vector space. The only requirement on these functions (other than their properties as a set) was that they made the problem easy to solve. In the two previous examples mentioned, what choice would we have made for the set? One choice would have been to use the basis functions that arise from an infinitely deep potential well, but these are not really defined outside the range of the well itself. Nevertheless, it is usually found that it is difficult to find a proper basis set of functions for which the problem can be easily solved.

When the problem is not calculated easily with a direct method, one must then resort to approximate methods. One such method is *perturbation theory*. This approach is useful when the Hamiltonian can be split into two parts

$$H = H_0 + H_1, \tag{7.1}$$

in which the first part H_0 can be solved directly, and the second part H_1 is small. When this can be done, then the second part can be initially ignored, and the problem solved exactly in terms of a natural basis function set. After this is done, the second term in (7.1) is treated as a perturbation, and the eigenvalues and eigenfunctions are then developed in a perturbation series in terms of the natural basis function set.

In this chapter, we will develop the general methodology of the perturbation technique. We will then consider several examples of the technique. Finally, a totally different approach, the variational method, will be introduced. The latter is a useful approximation when the split of the Hamiltonian into two parts, as in (7.1), does not lead to a small perturbing term.

7.1 THE PERTURBATION SERIES

As was discussed above, the basic approach revolves around being able to solve the quantum mechanical problem with only a portion of the total Hamiltonian, with the remainder being a small term. For this, we rewrite (7.1) as

$$H = H_0 + \lambda V, \tag{7.2}$$

where λ is a small parameter (if V is small compared with H_0, we can eventually let λ go to unity without introducing any inconsistency). The introduction of this parameter is solely to help us to develop the proper terms in the perturbation series. The term V (= H_1) is the perturbing part of the Hamiltonian and contains all of the extra terms that have been removed from H to produce H_0. Now, it is asserted that the solutions to H_0 are easily obtained in terms of a set of basis functions for which

$$H_0 \left|k\right\rangle_0 = E_k^{(0)} \left|k\right\rangle_0 . \tag{7.3}$$

That is, we have a set of zero-order basis functions, that in this case are energy eigenfunctions where k is an integer denoting the term in the set. If the deviation of the actual basis functions and eigenvalues from these zero-order ones is small, then we can write the basis functions as

$$\left|k\right\rangle = \left|k\right\rangle_0 + \lambda \left|k\right\rangle_1 + \lambda^2 \left|k\right\rangle_2 + \ldots. \tag{7.4}$$

Similarly, the energies can be written as

$$E_k = E_k^{(0)} + \lambda E_k^{(1)} + \lambda^2 E_k^{(2)} + \ldots. \tag{7.5}$$

To solve the overall problem, we insert the two series (7.4) and (7.5) into the Schrödinger equation

$$H\left|k\right\rangle = E_k \left|k\right\rangle, \tag{7.6}$$

and collect all the terms with the same coefficient λ^n together. The idea is that if we move the right-hand side of (7.6) to the left, then all terms with the same power of λ must vanish independently of the other terms. The three lowest equalities give

$$\begin{aligned}
&\lambda^0 : H_0 \left|k\right\rangle_0 = E_k^{(0)} \left|k\right\rangle_0 , \\
&\lambda^1 : H_0 \left|k\right\rangle_1 + V\left|k\right\rangle_0 = E_k^{(1)} \left|k\right\rangle_0 + E_k^{(0)} \left|k\right\rangle_1 , \\
&\lambda^1 : H_0 \left|k\right\rangle_2 + V\left|k\right\rangle_1 = E_k^{(2)} \left|k\right\rangle_0 + E_k^{(1)} \left|k\right\rangle_1 + E_k^{(0)} \left|k\right\rangle_2 .
\end{aligned} \tag{7.7}$$

The first equation of (7.7) is obviously just the appropriate (7.3) for the unperturbed solutions. The second equation of (7.7) gives the basis for *first-order* perturbation theory.

The initial corrections to the basis functions and to the eigenvalues are found by working with the second equation of (7.7). This equation may be rearranged, to give a more direct approach to the solution, as

$$\left(H_0 - E_k^{(0)}\right)\left|k\right\rangle_1 = \left(E_k^{(1)} - V\right)\left|k\right\rangle_0. \tag{7.8}$$

The basic assumption is that the perturbed basis functions are only small deviations from the unperturbed ones, so that it is fruitful to expand the deviations in terms of the unperturbed functions as

$$\left|k\right\rangle_1 = \sum\nolimits_j c_j \left|j\right\rangle_0. \tag{7.9}$$

Introducing this expansion into (7.8) gives us

$$\sum\nolimits_j c_j \left(H_0 - E_k^{(0)}\right)\left|j\right\rangle_0 = \sum\nolimits_j c_j \left(E_j^{(0)} - E_k^{(0)}\right)\left|j\right\rangle_0 = \left(E_k^{(1)} - V\right)\left|k\right\rangle_0. \tag{7.10}$$

The left-hand side vanishes for the single term $j = k$, which will leave c_j undetermined. The right-hand side then gives us

$$E_k^{(1)} = V_{kk}, \tag{7.11}$$

where

$$V_{kk} = {}_0\left\langle k\middle|V\middle|k\right\rangle_0. \tag{7.12}$$

The first-order deviation of the wave function $|k\rangle_1$ is made up of small admixtures of the other members of the unperturbed basis set. Yet, we want the new wave function to remain normalized. To achieve this, we will require that $c_k = 0$ by definition. Then, using (7.3), Equation (7.10) can be rewritten as

$$\sum\nolimits_{j\neq k} c_j \left(E_j^{(0)} - E_k^{(0)}\right)\left|j\right\rangle_0 = \left(E_k^{(1)} - V\right)\left|k\right\rangle_0. \tag{7.13}$$

In order to evaluate the change in the wave function, let us multiply both sides of (7.13) by the adjoint function ${}_0\langle i|$, and perform the implied integral. The resulting Kronecker delta functions allow us to perform the summation, and rearrange (7.13) to give

$$\sum_{j\neq k} c_j \left(E_j^{(0)} - E_k^{(0)}\right)\delta_{ij} = -V_{ik}. \tag{7.14}$$

Using the δ-function, we now find

$$c_i = -\frac{V_{ik}}{E_i^{(0)} - E_k^{(0)}}. \tag{7.15}$$

The choice $c_k = 0$ is now seen also to avoid an inconvenient zero in the denominator of this equation. Thus, to first order, the eigenvalues are ($\lambda = 1$)

$$\begin{aligned} E_k &= E_k^{(0)} + V_{kk}, \\ |k\rangle_0 &= |k\rangle_0 + \sum_{j\neq k} \frac{V_{jk}}{E_k^{(0)} - E_j^{(0)}} |j\rangle_0. \end{aligned} \tag{7.16}$$

Let us now proceed to consider the second-order corrections to the eigenvalues and the basis functions. The third equation of (7.7) can be rearranged as

$$\left(H_0 - E_k^{(0)}\right)|k\rangle_2 = \left(V_{kk} - V\right)|k\rangle_1 + E_k^{(2)}|k\rangle_0. \tag{7.17}$$

Here, (7.12) has been introduced for the first-order perturbation of the energy levels. Again, the second-order perturbation of the wave function is expanded in terms of the unperturbed basis set as

$$|k\rangle_2 = \sum_{s\neq k} b_s |s\rangle_0. \tag{7.18}$$

Again, we omit the term b_k (= 0). Now, (7.18) and the last term of (7.16) are inserted into (7.17) to give

$$\sum_{s\neq k} b_s \left(H_0 - E_k^{(0)}\right)|s\rangle_0 = \sum_{j\neq k} \frac{V_{jk}\left(V_{kk}\delta_{jk} - V\right)}{E_k^{(0)} - E_j^{(0)}}|j\rangle_0 + E_k^{(2)}|k\rangle_0. \tag{7.19}$$

Again, in order to evaluate the change in the eigenvalue, let us multiply both sides of (7.19) by the adjoint function ${}_0\langle i|$, and perform the implied integral. The resulting Kronecker delta functions allow us to perform the summation, and rearrange (7.19) to give

$$b_i\left(E_i^{(0)}-E_k^{(0)}\right)=\sum_{j\neq k}\frac{V_{jk}\left(V_{kk}\delta_{jk}-V_{ij}\right)}{E_k^{(0)}-E_j^{(0)}}\delta_{ij}+E_k^{(2)}\left|k\right\rangle_0. \tag{7.20}$$

There are several possibilities in this equation, but again it will determine all that we need to know. First, consider the case for which $i = k$. For this case, the left-hand side vanishes (particularly as we have directly omitted such a term from the left-hand summation prior to the last step), but the term in V_{kk} in the numerator of the summation does not contribute $(j \neq k)$, and the second-order change in the energy is just

$$E_k^{(2)}=\sum_{j\neq k}\frac{V_{jk}V_{kj}}{E_k^{(0)}-E_j^{(0)}}. \tag{7.21}$$

When $i \neq k$, the last term on the right-hand side of (7.20) does not contribute, and it is straightforward now to determine the expansion coefficient for the wave functions. Some care must be taken with the diagonal terms (they are of the same order of magnitude and may cancel, especially when the first-order energy perturbation does not depend upon the index). This gives the final result as

$$b_i=\sum_{j\neq k}\left\{\frac{V_{jk}V_{ij}}{\left(E_k^{(0)}-E_j^{(0)}\right)\left(E_k^{(0)}-E_i^{(0)}\right)}-\frac{V_{jk}V_{kk}}{\left(E_k^{(0)}-E_j^{(0)}\right)^2}\delta_{ij}\right\},\quad i\neq k. \tag{7.22}$$

Thus, to second order, we may write the energy as

$$E_k=E_k^{(0)}+V_{kk}+\sum_{j\neq k}\frac{V_{jk}V_{kj}}{E_k^{(0)}-E_j^{(0)}}, \tag{7.23}$$

and the wave function as

$$\begin{aligned}\left|k\right\rangle=\left|k\right\rangle_0&+\sum_{j\neq k}\frac{V_{jk}}{E_k^{(0)}-E_j^{(0)}}\left|j\right\rangle_0\\&+\sum_{\substack{j\neq k\\ i\neq k}}\frac{V_{jk}\left(V_{ij}-V_{kk}\delta_{ij}\right)}{\left(E_k^{(0)}-E_j^{(0)}\right)\left(E_k^{(0)}-E_i^{(0)}\right)}\left|i\right\rangle_0.\end{aligned} \tag{7.24}$$

This procedure can obviously be extended to many higher orders of perturbation. In general, each higher order adds another summation over an intermediate state. For example, in (7.16), the first-order correction just couples two states, with a summation over all other states that couple to the kth one. At second order, however, this coupling generally goes through an intermediate state, with an additional ratio of a matrix element to an energy difference (in the denominator). Thus, at third order, it would be expected that an additional summation would appear as the connection moves through two intermediate states. In the world of perturbation theory, it is of course quite important to be sure that the correction terms are indeed quite small. If they are not, it is possible that the series does not converge, which means the entire approach violates the assumptions and is invalid. It is usual to check the results at an order above the desired one to make sure that the solution is stable. Otherwise other approaches must be pursued.

In the above discussion, it has been assumed that the eigenvalues $E_k^{(0)}$ are all discrete, or at least all different if the spectrum is continuous. This may not always be the case and there may be several members of the set of wave functions that have the same energy. Then, the approaches described above must be modified to 'lift' this degeneracy prior to proceeding with the perturbation series in order to avoid unwanted zeros in the denominators of the summations. Let us consider the case where there are several different wave functions that all have the same energy value $E_k^{(0)}$.The correct wave functions in the zero-order approximation are then some linear combinations of these wave functions

$$c_k \left|k\right\rangle_0 + c_{k'} \left|k'\right\rangle_0 + c_{k''} \left|k''\right\rangle_0 \ldots. \tag{7.25}$$

For the constants c_k, it suffices to take their zero-order values as the solutions of the process. We now will write out a set of equations with $i = k,k',k''$, etc. and substitute in them, as the first approximation, $\mathcal{E}_k = \mathcal{E}_k^{(0)} + \mathcal{E}_k^{(1)}$ obtained from the $H\left|k\right\rangle_0$ term in (7.8), which then gives

$$E_k^{(1)} c_k = \sum\nolimits_{k'} V_{kk'} c_{k'}. \tag{7.26}$$

Here, k and k' take on all the possible values that span the set of degenerate wave functions. This leads to a system of homogeneous equations

$$\sum\nolimits_{k'} \left(V_{kk'} - E_k^{(1)} \delta_{kk'}\right) c_{k'} = 0. \tag{7.27}$$

This system of homogeneous equations has solutions for the various values of the coefficients c_k if the determinant vanishes:

$$det\left|V_{kk'} - E_k^{(1)} \delta_{kk'}\right| = 0. \tag{7.28}$$

If there are n degenerate eigenvalues, then Equation (7.28) is of order n. The n roots of the equation then give the values of $E_k^{(1)}$. Equation (7.28) is called the *secular equation*. We note that the sum of the various new eigenvalue corrections is given by the sum of the diagonal matrix elements, a general result in degenerate perturbation theory. This process can be continued to second order if the degeneracy is not fully lifted by the first-order treatment.

7.2 SOME EXAMPLES OF PERTURBATION THEORY

In this section, we now want to consider a series of examples that will serve to illustrate the stationary perturbation technique. We will first consider a trapezoidal potential well, one with a linear potential term. Then we turn to a shifted harmonic oscillator, again obtained as a result of adding a linear potential to the harmonic oscillator.

7.2.1 The Stark Effect in a Potential Well

The system that we consider is that shown in Figure 7.1, where we have an infinitely deep potential well (one with infinitely high barriers, corresponding to Section 2.3.1), but with a linear potential existing within the well. Here, we take the unperturbed Hamiltonian as the quantity

$$H_0 = -\frac{\hbar^2}{2m}\frac{d^2}{dx^2} + V_0, \tag{7.29}$$

where

$$V_0 = \begin{cases} 0, & |x| \le a, \\ \infty, & elsewhere. \end{cases} \tag{7.30}$$

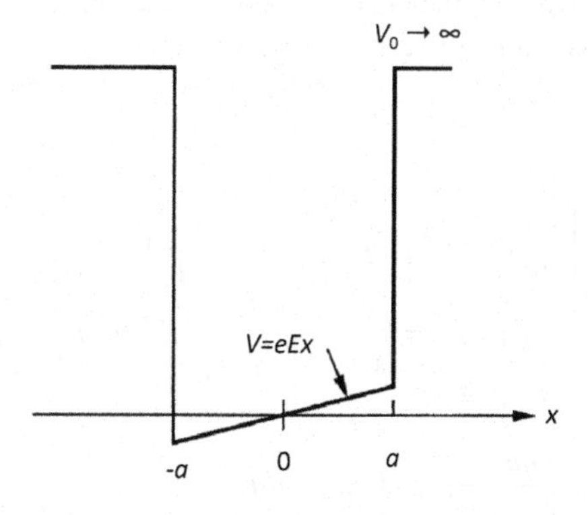

FIGURE 7.1 When an electric field is added to the infinite potential well, a trapezoidal well arises.

From Section 2.3.1, the eigenvalues and eigenfunctions are found to be

$$\psi_n(x)=\begin{cases}\dfrac{1}{\sqrt{a}}sin\left[\dfrac{n\pi}{2a}(x+a)\right], & |x|\le a,\\ 0, & elsewhere,\end{cases}$$
$$E_n=\frac{n^2\pi^2\hbar^2}{8ma^2}. \tag{7.31}$$

For the perturbing potential, we take a linear potential corresponding to adding an electric field as (hopefully, no confusion will arise from the double use of E)

$$V=eEx, \tag{7.32}$$

The entire problem is reduced to determining the matrix elements of this potential between the various basis states. There are two sets to be determined, the diagonal ones and the off-diagonal ones. First, the diagonal matrix elements are simply

$$\begin{aligned}V_{nn}&=\langle n|eEx|n\rangle\\ &=\frac{eE}{a}\int_{-a}^{a}xsin^2\left[\frac{n\pi}{2a}(x+a)\right]dx\\ &=\frac{eE}{2a}\int_{-a}^{a}x\left\{1-cos\left[\frac{n\pi}{2a}(x+a)\right]\right\}dx=0.\end{aligned} \tag{7.33}$$

This result is a particular result of choosing the potential to be symmetrical about the center of the well, so that the average under the perturbing potential is zero. Thus, there is no first-order shift in the energy levels themselves. In other words, the particular choice of the potential results in the center of the well not moving, which means that, on average, the energy levels are not shifted to first order. The off-diagonal terms, however, are nonzero, and we find that

$$\begin{aligned}V_{nm}&=\langle n|eEx|m\rangle\\ &=\frac{eE}{a}\int_{-a}^{a}xsin\left[\frac{n\pi}{2a}(x+a)\right]sin\left[\frac{m\pi}{2a}(x+a)\right]dx\\ &=\frac{eE}{2a}\int_{-a}^{a}x\left\{cos\left[\frac{(n-m)\pi}{2a}(x+a)\right]-cos\left[\frac{(n+m)\pi}{2a}(x+a)\right]\right\}dx\\ &=\begin{cases}-\dfrac{16eEa}{\pi^2}\dfrac{nm}{(n^2-m^2)^2}, & (i\pm j)odd,\\ 0, & (i\pm j)even.\end{cases}\end{aligned} \tag{7.34}$$

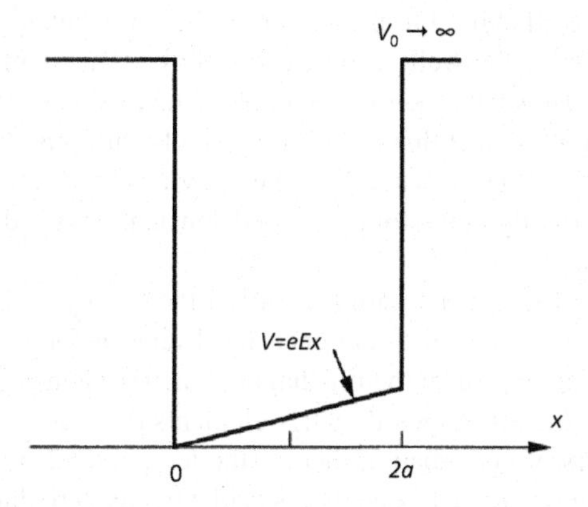

FIGURE 7.2 An asymmetrical quantum well with an applied electric field.

The last result tells us that the odd potential must mix only those states that have different parity. Thus, the odd potential produces matrix elements only between an even-parity state and an odd-parity state. (The plus/minus sign just reminds us that the if the difference between two integers is odd, then the sum of these same two integers is also odd, and thus the two terms that result from the two integrals in (7.34) have been combined in the final result.)

Normally, the addition of an electric field produces a shift of the energy levels, an effect called the Stark effect. Here, because the potential has been taken to be an antisymmetric one, the linear shift in the energy levels vanishes, and there is no first-order Stark effect. However, this is merely because of the choice of the reference level for the potential. If we had taken the zero of energy to lie at the point $x = -a$, then all the energy levels would have been pushed upward by the value of the potential at the center of the well. To see this, let us shift the origin to $-a$, so that the well goes from $-a$ to 0, 2a, as shown in Figure 2.3. Now, the basis functions are given by

$$\psi_n(x) = \frac{1}{\sqrt{a}} sin\left(\frac{n\pi x}{2a}\right), \quad 0 \le x \le a, \tag{7.35}$$

although the unperturbed energy levels do not change (we are only moving the reference in the axis, and the unperturbed well does not change). However, the symmetry of the perturbing potential has been changed. The diagonal elements become

$$\begin{aligned} V_{nn} &= \langle n | eEx | n \rangle \\ &= \frac{eE}{a} \int_0^{2a} x sin^2\left[\frac{n\pi x}{2a}\right] dx \\ &= \frac{eE}{2a} \int_0^{2a} x \left\{1 - cos\left[\frac{n\pi x}{2a}\right]\right\} dx = eEa. \end{aligned} \tag{7.36}$$

The energy levels are all shifted to first order by the same amount eEa, which is the potential at the center of the well, as suggested above. This is the linear Stark shift of the energy levels. However, the off-diagonal elements do not change from (7.34), which is a reflection of the fact that both energy levels shift, but their difference does not (the linear Stark shift is the same for all energy levels). It may be noted that the denominator of (7.34), other than for associated constants, is the difference (squared) of two energy levels.

Thus, whether or not we see a uniform shift of the energy levels in a perturbation approach often depends upon the reference level chosen for the overall potential. However, this does not appear in the off-diagonal matrix elements. This latter is significant, as it is only the off-diagonal matrix elements that appear in the summations for the various orders of perturbation theory (the term in (7.24) involving the difference between two diagonal matrix elements is identically zero due to the fact that all of these diagonal elements are equal).

7.2.2 The Shifted Harmonic Oscillator

As a second example of the application of perturbation theory, we want to consider a harmonic oscillator that is subjected to a linear potential that can arise from an applied electric field. Then, the Hamiltonian can be written as

$$\begin{aligned} H &= \frac{p^2}{2m} + \frac{1}{2}m\omega^2 x^2 + eEx \\ &= \hbar\omega\left(a^\dagger a + \frac{1}{2}\right) + eEx. \end{aligned} \tag{7.37}$$

In the last line, we have used the operator results of Chapter 5. The application of the linear potential still produces a harmonic oscillator, but one that is shifted in both position and energy. This can be seen, as the first line of (7.37) can be rewritten as

$$H = \frac{p^2}{2m} + \frac{1}{2}m\omega^2 \left(x + x_0\right)^2 - \frac{e^2E^2}{2m\omega^2}, \tag{7.38}$$

$$x_0 = \frac{eE}{m\omega^2}. \tag{7.39}$$

The last term in (7.38) is the downward shift of the quadratic potential in energy, while (7.39) is the shift of the center of the potential well in position. This is shown in Figure 7.3. There are now two ways to approach the solution of the new Hamiltonian in (7.38). In the first approach, the linear potential is taken to be the perturbing potential, and the wave functions of the harmonic oscillator centered at $x = 0$ are used. In

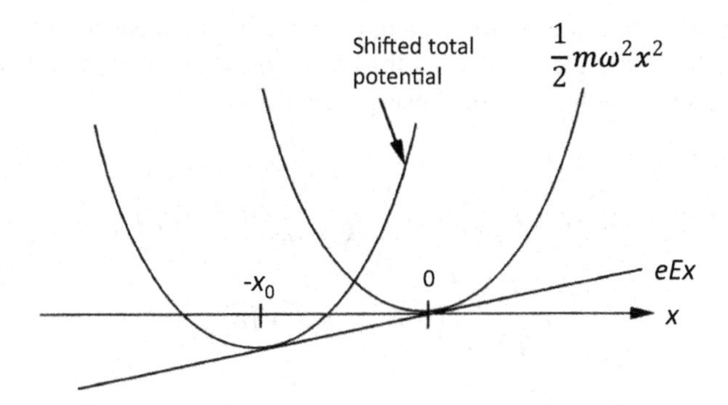

FIGURE 7.3 The combination of a quadratic potential and a linear potential produces a shifted quadratic potential.

the second approach, the perturbing potential is the uniform shift of the harmonic oscillator downward by the constant energy term, which is the last term on the right-hand side of (7.38). The wave functions are those corresponding to the harmonic oscillator centered at $x = -x_0$. We take the last approach first.

If we treat the harmonic oscillator directly as a shifted harmonic oscillator with the minimum centered at $x = -x_0$, then the wave functions are merely shifted Hermite polynomials of (5.25)

$$\psi_n(x) = \frac{1}{\sqrt{n!}}\left(\frac{m\omega}{2\hbar}\right)^{n/2}\left(x + x_0 - i\frac{p}{m\omega}\right)^n \psi_0(x), \tag{7.40}$$

where

$$a^\dagger = \left(\frac{m\omega}{2\hbar}\right)^{1/2}\left(x + x_0 - i\frac{p}{m\omega}\right). \tag{7.41}$$

The matrix elements are now

$$V_{ij} = \left\langle i\left|\frac{e^2E^2}{2m\omega^2}\right|j\right\rangle = \frac{e^2E^2}{2m\omega^2}\delta_{ij}. \tag{7.42}$$

Thus, only the diagonal elements are nonzero. A constant shift of the Hamiltonian provides only a uniform and constant shift of all energy levels. It provides no mixing of the states. Thus, when the linear potential is applied to the harmonic oscillator, the only result is a shift downward of all energy levels by the shift in the minimum, and a shift of the centroid of the wave functions to the new center position of the harmonic oscillator.

The first approach to the solution should yield a similar answer. For this, we can now compute the matrix elements using the linear potential as the perturbation. Then, the matrix elements can be computed using (5.10) as

$$V_{ij} = \langle i|eEx|j\rangle = eE\left\langle i\left|\sqrt{\frac{\hbar}{2m\omega}}\left(a^\dagger + a\right)\right|j\right\rangle$$
$$= eE\sqrt{\frac{\hbar}{2m\omega}}\left(\sqrt{i}\delta_{j+1,i} + \sqrt{i+1}\delta_{j-1,i}\right). \tag{7.43}$$

In this approach, there is no first-order shift of the energy, but the second-order shift involves only two terms in the summation. In each term, the product of matrix elements connects a given level only with its neighbours just above and just below, so only one of the two delta functions in (7.43) appears in each matrix element. We compute the second-order shifted energy to be (the signs arise from the energy denominators)

$$E_n = E_n^{(0)} - \frac{e^2E^2}{m\omega^2}. \tag{7.44}$$

This produces the full shift of the energy levels, and it also is independent of the value of the energy level under consideration. Thus, higher-order corrections to the energy levels should vanish, and this can be checked by examining the fourth-order term, which is the next one expected. However, it cannot always be ensured that the higher-order terms vanish, but rational thinking for this problem suggests that they should sum to zero, when all orders are included. The corrections to the wave functions are just those that try to move the centroid to the new center of the harmonic potential. This is a perfect example of a case where low-order perturbation theory does not produce a result that is close to the exact answer (at least for the wave functions), which we know from being able to solve the problem exactly. The first-order shift in the wave functions can also be obtained by using the matrix elements above. For the lowest energy level

$$|0\rangle^{(1)} = -\frac{eE}{\hbar}\sqrt{\frac{\hbar}{2m\omega}}|1\rangle^{(0)}, \tag{7.45}$$

and for the higher levels

$$|n\rangle^{(1)} = -\frac{eE}{\hbar}\sqrt{\frac{\hbar}{2m\omega}}\left(\sqrt{i}\,|i-1\rangle^{(0)} + \sqrt{i+1}\,|i+1\rangle^{(0)}\right). \tag{7.46}$$

To obtain the exact wave functions, we will need to perturbation theory to much higher orders.

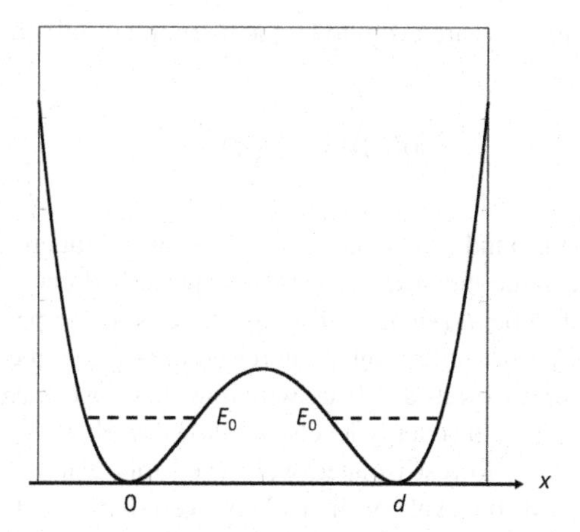

FIGURE 7.4

7.2.3 Multiple Quantum Wells

We now want to turn to a consideration of coupled quantum wells, which we will take each to be quadratic in nature. This treatment provides us with a discussion of *degenerate perturbation theory*, a process in which we decide how two levels that otherwise would have the same energy interact to cause a small splitting. This was discussed at the end of Section 7.1. Consider, for example, two harmonic wells that are displaced from one another by a distance d, as shown in Figure 7.4. Here, the unperturbed wave functions in each well are at the same energy, so the degeneracy is between wave functions centered in each of the two wells. The interaction between the two causes only a small barrier between the two wells, so the wave function of an electron localized in one of the wells actually extends somewhat into the other well. Thus, the two wave functions overlap with each other and there is an interaction between them. It is this interaction that causes a modification of the two wave functions, lifts the degeneracy and leads to two new levels, which are then separated slightly in energy. One of the levels has a wave function composed of the constructive interference of the two individual wave functions, while the second (the higher energy state) has a wave function that is antisymmetric in the combined wells and thus arises from the destructive interference of the two individual wave functions.

In each of the two wells, the unperturbed problem is just that of a harmonic oscillator, centered at the particular well, so the Hamiltonian and lowest wave function of the well at $x = 0$ are given by

$$\begin{aligned} H_{0,1} &= \frac{p^2}{2m} + \frac{1}{2}m\omega^2 x^2 \\ H_{0,2} &= \frac{p^2}{2m} + \frac{1}{2}m\omega^2 (x-d)^2, \end{aligned} \tag{7.47}$$

and the two wave functions are taken to be the harmonic oscillator ground states

$$|0(0)\rangle^{(0)},\quad |0(d)\rangle^{(0)}, \tag{7.48}$$

which both have the ground state energy $E_0^{(0)}$. Both wave functions, one centered in each well, have the same eigenvalues for the unperturbed energy. Thus, these two eigenvalues are said to be *degenerate*. The task here is to use perturbation theory to lift this degeneracy. We are not so much interested in how the interaction with higher levels changes the energy levels as we are with how these two wave functions interact with each other to change the energy levels. While we work with the energy level that we have indicated is the lowest isolated well level, this is not a limitation, and we could work with any of the levels of the isolated well that lie below the central peak between the two wells (for those levels above this peak, perturbation theory is not appropriate). Our interest then is in the matrix element

$$V_{00} = {}^{(0)}\langle 0(0)|V|0(d)\rangle^{(0)} = \int_{-\infty}^{\infty} \psi_0^{\dagger(0)}(x)V(x)\psi_0^{(0)}(x-d)dx. \tag{7.49}$$

Here, the potential is the actual total potential in the problem, but the main part is that part of the potential that lies between the two wells. Equation (7.49) is the *overlap* integral defining the *mixing* of the two wave functions.

The composite wave functions should be able to be written as a sum of the two individual wave functions, as

$$|0\rangle = A|0(0)\rangle^{(0)} + B|0(d)\rangle^{(0)}. \tag{7.50}$$

When we operate first with the conjugate of the wave function of the left-hand well, and then with the conjugate of the wave function of the right-hand well, we produce two equations. This is done by neglecting the variation of the potential difference from the single well for the diagonal terms (wave functions on the same sites). The process produces the resulting matrix equation

$$\begin{bmatrix} E_0 & V_{00} \\ V_{00} & E_0 \end{bmatrix}\begin{bmatrix} A \\ B \end{bmatrix} = E\begin{bmatrix} A \\ B \end{bmatrix}. \tag{7.51}$$

If this is to have a solution, then we must arrange for the determinant

$$det\begin{vmatrix} (E_0 - E) & V_{00} \\ V_{00} & (E_0 - E) \end{vmatrix} = 0. \tag{7.52}$$

This leads to the new energy levels

$$E = E_0 \pm V_{00}. \tag{7.53}$$

Thus, one eigenstate moves up slightly in energy while the other moves down in energy. The lowest of the two states is the bonding state with $A = B$, while the upper state is the antibonding state with $B = -A$. The total energy splitting between these two states is $2V_{00}$.

7.3 AN ALTERNATIVE APPROACH—THE VARIATIONAL METHOD

There are often times when the perturbation approach just does not work, because the basis set cannot be determined consistently with (7.3). An example is the triangular potential well of Section 2.4. The problem with this particular barrier is that the solution of the Schrödinger equation is complicated, and the wave functions are special functions that, in general, are evaluated numerically. Often, however, the exact eigenvalues and eigenfunctions are not required; rather, only good approximations to them are required. To accomplish this, another approximation technique is generally used—the Rayleigh–Ritz method, or variational method. Here, it may be assumed that we cannot begin with (7.3) or any equivalent form. Nevertheless, we may take an approximate wave function as an expansion in a convenient basis set:

$$\psi(x) = \sum_i a_i \varphi_i(x). \tag{7.54}$$

It may be assumed then that, if this is an orthonormal basis set, the expectation value of the Hamiltonian is given by

$$\langle H \rangle = \sum_i E_i |a_i|^2, \tag{7.55}$$

where we have constructed the expectation value by premultiplying by the conjugate of (7.54), postmultiplied by (7.54), integrated and used the orthogonality properties of the basis set. An inequality can be created by assuming that only the lowest energy level is used in the expansion (using the lowest energy level for each of the eigenstates), and

$$\langle H \rangle \ge E_0 \sum_i |a_i|^2 = E_0. \tag{7.56}$$

The sum over all the basis states is unity to ensure the normalization of the wave function in (7.54).

The principle of the Rayleigh–Ritz method is that we can adopt a trial wave function, which may have some parameters in it, and then minimize the energy calculated with this wave function. This energy will still be above the actual lowest energy level E_0. This gives a good approximation to the actual energy level. The next higher state can be developed by using a second parametrized basis function, which is made orthogonal to the first. The energy level of this wave function is then found by the same adjustment technique. Let us illustrate this with the actual triangular-potential-well problem of Section 2.4.

To begin, we note that the wave function vanishes at the origin, as the potential is typically defined by

$$V(x) = \begin{cases} \to \infty, & x \le 0, \\ eEx, & x > 0. \end{cases} \tag{7.57}$$

Now, we seek a trial function that vanishes at $x = 0$, and also vanishes for large x, where the energy lies at a lower level than the potential. For this, we take

$$|0\rangle = Axe^{-bx}, \tag{7.58}$$

which satisfies both limiting requirements. The prefactor coefficient A is determined by normalization to be

$$\langle 0|0\rangle = 1 = A^2 \int_0^\infty x^2 e^{-2bx} dx = \frac{A^2}{4b^3}. \tag{7.59}$$

Then, the expectation value of the Hamiltonian is

$$\begin{aligned} \langle H\rangle &= 4b^3 \int_0^\infty xe^{-bx} \left(-\frac{\hbar^2}{2m}\frac{d^2}{dx^2} + eEx \right) xe^{-bx} dx \\ &= \frac{\hbar^2 b^2}{2m} + \frac{3eE}{2b}. \end{aligned} \tag{7.60}$$

This value is now above the actual lowest energy level according to (7.56), and we can minimize this value by setting the derivative with respect to b to zero. This then leads to (we take the first level as the lowest level in keeping with the general results of Section 2.4)

$$b = \left(\frac{3eEm}{2\hbar^2} \right)^{1/3}, \tag{7.61}$$

and the actual lowest variational energy level is

$$E_0 = \frac{3\hbar^2}{2m}\left(\frac{3eEm}{2\hbar^2}\right)^{2/3} \tag{7.62}$$

This result should be compared with (2.65). The present result differs from (2.65) only by a factor of $3(2/3\pi)^2/3 \simeq 1.07$, or about a 7% error, which is quite small considering the approximations. As expected, the right-hand side of (7.62) lies above the actual value, given by (2.65).

To compute the next level, we must assume a wave function that is orthogonal to (7.58). The wave function of (7.58) has a single maximum, decaying away to zero at the origin and for large values of x. The next level should have a wave function with two maxima, but also decaying away for large values of x and at the origin. For this purpose, we assume the first-excited-state wave function may be approximated as

$$|0\rangle = C\left(x - Dx^2\right)e^{-fx}. \tag{7.63}$$

Our task is to determine the constants by: (1) making this wave function orthogonal to that of (7.58), (2) normalizing the wave function, and (3) minimizing its expectation energy. To begin, we approach the first of these tasks as

$$\begin{aligned}\langle 0|1\rangle = 0 &= 2b^{3/2}C\int_0^\infty xe^{-bx}\left(x - Dx^2\right)e^{-fx}dx \\ &= 2b^{3/2}C\left[\frac{2}{(b+f)^3} - \frac{6D}{(b+f)^4}\right].\end{aligned} \tag{7.63}$$

We may now write D in terms of the coefficients in the exponentials, as

$$D = \frac{1}{3}(b+f). \tag{7.64}$$

The second task is to normalize the wave function that leads to

$$\begin{aligned}\langle 1|1\rangle = 1 &= \frac{C^2}{4f^5}\left(f^2 - 3Df + 3D^2\right) \\ C^2 &= \frac{12f^2}{f^2 - bf + b^2}.\end{aligned} \tag{7.65}$$

In the last line, we have used (7.64) to replace the factor D. Now, the last factor is to minimize the expectation value of the energy for this level. The expectation value of the energy is

$$\langle H\rangle = \frac{\hbar^2 f^2}{6m}\left(1+\frac{15f^2}{f^2-bf+b^2}\right) + \frac{eE}{2f}\left(1+\frac{3b^2}{f^2-bf+b^2}\right). \tag{7.66}$$

This can now be minimized, to yield the value of f as above. While the tendency might be to ignore the second terms in each of the large parentheses, this would be a mistake, as it would lead to an energy level below the lowest level. These terms are actually the major terms. Unfortunately, this leads to a complicated, high-order algebraic equation for f in terms of b. This can be solved most easily by numerical calculations, which lead to the result $f \simeq 0.73b$, which in turn leads to $E_1 \approx 2.4E_0$.

A similar procedure can now be used to generate a variational wave function for the next higher eigenstate. A wave function, probably with a cubic term in the prefactor, is assumed and the coefficients are adjusted first to make it orthogonal to each of the two lower eigenfunctions. Then, the energy is minimized to provide an estimate of the energy level. Of course, each successive higher level becomes more complicated to determine, and most efforts using this method are addressed only to the lowest energy level.

Problems

1 In an infinite square well, in which $V(x) \to \infty$ for $x<0$ and $x>a$, $V(x)=0$ for $0 \le x \le a$, the basis states are given by $|n\rangle = \sqrt{2/a}\sin(n\pi x/a)$, and $E_n = \hbar^2\pi^2 n^2/(2ma^2)$. For the introduction of a perturbing potential of $V_1 = (x-\frac{a}{2})^2/2$, calculate the changes in the wave functions up to first order and in the energy levels up to second order.

2 Determine the third-order correction to the energy in perturbation theory.

3 Determine the corrections to second order necessary to lift the degeneracy of a set of states in an arbitrary perturbing potential.

4 Except for certain accidents, degeneracy usually does not arise in a one-dimensional problem. Degeneracy is usually found when the dimension is raised. Consider the two-dimensional infinite square well

$$V(x,y) = \begin{cases} 0 & \text{for} |x| \le a/2 \text{ and } |y| \le a/2 \\ \to \infty & \text{for} |x| \ge a/2 \text{ and } |y| \ge a/2. \end{cases}$$

Using this potential, separate the Schrödinger equation into two one-dimensional equations and solve for the allowed energy values using the fact that the potential can be written as the sum of two one-dimensional potentials. The solutions can then be written down using the results of Chapter 2. Consider the application of the simple perturbation $V_1(x,y) = V_0$ for $|x| \le a/4$ and $|y| \le a/4$, and is zero elsewhere (the perturbing potential exists only in the center of the well). Compute the splitting of the lowest two unperturbed energy levels.

5 Using the harmonic oscillator raising and lowering operators, compute directly the matrix element for $\langle n|x^4|n\rangle$. Then, introducing a complete set of states (as a resolution of the delta function) break this into two products x^2x^2, and compute the matrix element from

$$\left\langle n\middle|x^4\middle|n\right\rangle=\sum_j\left\langle n\middle|x^2\middle|j\right\rangle\left\langle j\middle|x^2\middle|n\right\rangle.$$

6 A one-dimensional harmonic oscillator is perturbed by a term αx^3. Calculate the change in each energy level to second order.

7 Calculate the lowest energy level of a harmonic oscillator using the Gaussian wave function as a variational wave function. Let the width of the Gaussian be the parameter that is varied.

8 Time-Dependent Perturbation Theory

In many physical systems, we are interested in small time-dependent changes in the state. The approach used in the last chapter cannot deal with this situation, and we must develop perturbation theory further. This will be done in the present chapter. In fact, we will develop the approach twice, first for the Schrödinger representation and then for the Heisenberg representation in a linear vector space. This double approach is followed for two reasons. First, it is important to grasp the concept of the time-dependent perturbation approach in a direct method, without the encumbrance of the mathematics that accompanies the latter approach–the interaction representation. Second, it is important to learn the mathematics of the interaction representation, but in doing so it is quite helpful to understand just where we are heading. The double approach achieves both objectives.

In time-dependent perturbation theory, we continue to assume that the Hamiltonian can be split into two parts

$$H = H_0 + H_1, \tag{8.1}$$

in which the first part H_0 can be solved directly and the second part H_1 is small. When this can be done, then the second part can be initially ignored, and the problem solved directly to give the wave function, or its expansion in a natural basis function set, typically the energy eigenstates. After this is done, the second term in (8.1) is treated as a perturbation, and the eigenvalues and eigenfunctions are then developed in a time-dependent perturbation approach in terms of the natural basis function set. But we need to remark that the formulation (8.1) remains a closed system, even though we may consider that the perturbation arises from an interaction outside the system, such as from phonons when we are considering the properties of the electrons.

In this chapter, we will first develop the approach in the Schrödinger representation and consider an example due to a harmonic potential. Then the interaction representation will be developed, and the perturbation theory redeveloped in the Heisenberg representation. Our attention will then be turned to an extended approach in which the initial state of the system has an exponential decay. It is this latter approach that allows us to return once again to the idea of an energy-time relationship as a result of the lifetime of the initial state. Finally, the scattering matrix approach (the T-matrix) is developed. One note of importance is that it will be assumed that the term H_0 is a time-independent quantity, while the perturbation term H_1 will be allowed to have an oscillatory nature.

8.1 THE PERTURBATION SERIES

If the Schrödinger equation is used with a time-independent and unperturbed Hamiltonian H_0, for which it may readily be solved, the solution may be written from (6.5) as (the position coordinate is suppressed)

$$\psi^{(0)}(t) = \sum_n c_n e^{-i\omega_n t} \varphi_n, \tag{8.2}$$

where

$$H_0 \varphi_n = E_n \varphi_n, \quad \omega_n = E_n / \hbar, \tag{8.3}$$

and

$$c_n = \int_{-\infty}^{\infty} \varphi_n^{\dagger} \psi^{(0)}(0) dx. \tag{8.4}$$

In the presence of the perturbation H_1, we seek a solution that has the same general form as (8.2)–(8.4), and this can be achieved by making the assumption that the perturbation $H_1 = V$ is small and produces a slow variation in the coefficients c_n. Then, we may write the new solution as

$$\psi(t) = \sum_n c_n(t) e^{-i\omega_n t} \varphi_n, \tag{8.5}$$

in which the time variation of the coefficient is very slow compared to any of the frequencies arising from the various energy levels. We now turn to determining the time variation of these coefficients. To begin, we put the solution (8.5) into the Schrödinger equation

$$i\hbar \frac{\partial \psi}{\partial t} = (H_0 + V)\psi. \tag{8.6}$$

The problem is now to determine the time variation of the coefficients. To proceed, the assumed solution (8.5) is introduced into the Schrödinger equation

$$\sum_n \left(c_n E_n + \frac{\partial c_n}{\partial t} \right) e^{-i\omega_n t} \varphi_n = \sum_n (E_n + V) c_n e^{-i\omega_n t} \varphi_n. \tag{8.7}$$

In this expansion, we have used (8.3) to insert the energies of the eigenfunctions. The first term on the left and the first term on the right thus cancel. We now multiply through by an arbitrary basis function $\varphi_k^\dagger$, and integrate over all space, using the orthonormality of these functions. We will further assume that $V = V_0 e^{-i\omega_0 t}$. This leads to

$$i\hbar \frac{\partial c_k}{\partial t} e^{-i\omega_k t} = \sum_n V_{kn} c_n e^{-i(\omega_n - \omega_0)t}, \tag{8.8}$$

where

$$V_{kn} = \langle k | V_0 | n \rangle \tag{8.9}$$

is the matrix element connecting the two states. To make life easy, we introduce the reduced frequency

$$\omega_{kn} = \frac{E_k - E_n}{\hbar} = \omega_k - \omega_n, \tag{8.10}$$

and (8.8) can be rewritten as

$$i\hbar \frac{\partial c_k}{\partial t} = \sum_n V_{kn} c_n e^{i(\omega_{kn} - \omega_0)t}. \tag{8.11}$$

What we now have is a complicated set of equations for the coefficients and their time dependence. In general, this can be solved by matrix techniques when the matrix elements are known.

The approach that we want to follow, however, assumes that the perturbation is very small. Moreover, we will also assume that the system is in one single eigenstate at the initial time, which will be taken as $t = 0$:

$$c_s(0) = 1, \quad c_k(0) = 0 \quad for\, all\, k \neq s. \tag{8.12}$$

To be consistent with this assumption, we also will assume that the perturbation V is zero for $t < 0$. With the above approximations, (8.12) now becomes a simple equation

$$i\hbar \frac{\partial c_k}{\partial t} = V_{ks} e^{i(\omega_{ks} - \omega_0)t}, \tag{8.13}$$

and

$$c_k(t) = -\frac{V_{ks}}{\hbar(\omega_{ks} - \omega_0)}\left[e^{i(\omega_{ks}-\omega_0)t} - 1\right]. \tag{8.14}$$

The change in the occupancy of the state is mainly determined by the change in the magnitude squared of the wave function. Thus, the quantity of interest is the magnitude squared of (8.14), which becomes

$$|c_k(t)|^2 = \frac{4|V_{ks}|^2}{\hbar^2(\omega_{ks} - \omega_0)^2} sin^2\left(\frac{(\omega_{ks} - \omega_0)t}{2}\right). \tag{8.15}$$

This is now the probability that density moves from the initial state s to the state k.

Our interest is in the rate at which this probability transfers, or the transition rate, which is the probability per unit time, or

$$\Gamma_{ks} = \frac{|c_k(t)|^2}{t} = \frac{4|V_{ks}|^2}{\hbar^2(\omega_{ks} - \omega_0)^2 t} sin^2\left(\frac{(\omega_{ks} - \omega_0)t}{2}\right). \tag{8.16}$$

For large values of time, we note that (Landau and Lifshitz, 1958)

$$\lim_{t\to\infty} \frac{sin^2(at)}{a^2 t} \to 2\pi\delta(a). \tag{8.17}$$

When $a \neq 0$, the limit is zero and the transition rate vanishes (we associate a with $(\omega_{ks} + \omega_0)/2$). Thus, we note that the delta function serves to ensure that the transition rate is nonvanishing only for $a = 0$. Thus, the transition rate can be written as

$$\Gamma_{ks} = \frac{2\pi}{\hbar^2}|V_{ks}|^2\,\delta(\hbar\omega_{ks} - \hbar\omega_0). \tag{8.18}$$

The last form in (8.18) is often referred to as the Fermi golden rule. The transition rate is given by a numerical factor times the squared magnitude of the matrix element and a delta function that conserves the energy in the transition. The latter is important, as it ensures that time-dependent variations arise only from those states in which the transition can conserve energy.

8.2 THE INTERACTION REPRESENTATION

There is another method of getting at the perturbation expression through a more formal approach, which will give rise to all orders of the perturbation (although no one really goes past first order without moving to more extensive fomulations, such as Green's functions). This approach is based upon the Heisenberg representation introduced in Chapter 6. To review this approach, the wave function is expanded in a linear vector space of basis functions, just as in the approach that leads to (8.3), but now it is assumed that the basis set is not time-varying, at least in the unperturbed state. Rather, the time variation is attached to the operators themselves, which rotate in the fixed linear vector space (which forms our coordinate system) as a function of time according to (6.11) as

$$\frac{\partial A(t)}{\partial t} = \frac{i}{\hbar}[H_0, A],$$
$$A(t) = e^{iH_0 t/\hbar} A e^{-iH_0 t/\hbar}. \tag{8.19}$$

This relationship is the basis of the Heisenberg representation. It is quite possible that the operator A actually varies with time, and, if this is the case, the final form of (8.19) leaves that time variation intact (as one can show by adding the partial derivative of A with respect to time to the right-hand side of the first equation, and then integrating the system).

In the presence of a time-varying perturbation, however, it will now be assumed that the perturbation introduces a slow variation in the basis set itself. Just as we earlier assumed that the coefficients of the expansion were now time-varying, we will assume that this slow variation is entirely due to the perturbing potential $V(t)$ defined according to (8.19). we will also assume below that there is an explicit time variation to this perturbation (such as might arise if the perturbation were due to an interaction with a harmonic oscillator). Hence, there is a slow variation of the wave function in the linear vector space that is introduced by V, while all other operators still respond to the unperturbed Hamiltonian according to (8.19). This mixed representation is termed the interaction representation.

Now, let us diverge a bit and talk about time variation in general. Earlier, we have written the time variation of a wave function as (6.2):

$$\psi(x,t) = e^{-iHt/\hbar}\psi(x,0). \tag{8.20}$$

This assumed that the time of interest began at $t = 0$ when we integrated (6.1). But suppose it began at t_0. Then, we would rewrite (8.20) as

$$\psi(x,t) = e^{-iH(t-t_0)/\hbar}\psi(x,t_0). \tag{8.21}$$

This leads us to assume that we can define a time-translation operator as

$$T(t,t_0) = exp\left[-\frac{i}{\hbar}H_0(t-t_0)\right]. \tag{8.22}$$

Then, we use (8.21) and (8.22) to write

$$\psi(x,t) = T(t,t_0)\psi(x,t_0). \tag{8.23}$$

Putting this form into the Schrödinger equation, we find that the time-translation operator has its own derivative given as

$$i\hbar\frac{\partial T}{\partial t} = V(t)T(t,t_0). \tag{8.24}$$

Now, let us investigate the properties of this time-translation operator. First, it must be such that the operator is unity when the two times are the same

$$T(t,t) = 1. \tag{8.25}$$

Then, it must also have the property, which is obvious from its definition (8.22), that

$$T(t,t_0) = T(t,t_2)T(t_2,t_0). \tag{8.26}$$

Let us now integrate (8.24) to get

$$T(t,t_0) = 1 - \frac{i}{\hbar}\int_{t_0}^{t} V(t')T(t',t_0)dt'. \tag{8.27}$$

Now, we can solve this iteratively, using the next lowest iteration in the integral to give

$$\begin{aligned}
&T_0(t,t_0) = 1,\\
&T_1(t,t_0) = 1 - \frac{i}{\hbar}\int_{t_0}^{t} V(t')T(t',t_0)dt' \cong 1 - \frac{i}{\hbar}\int_{t_0}^{t} V(t')dt',\\
&T_2(t,t_0) = 1 - \frac{i}{\hbar}\int_{t_0}^{t} V(t')dt' + \left(\frac{i}{\hbar}\right)^2 \int_{t_0}^{t} V(t')\int_{t_0}^{t'} V(t'')dt''dt',
\end{aligned} \tag{8.28}$$

and so on. This iteration can be continued, and proves that the result is

$$T(t,t_0) = exp\left[-\frac{i}{\hbar}\int_{t_0}^{t} V(t')dt'\right]. \tag{8.29}$$

This form is obtained by noting that the multiple integrations can all be changed to integrate over the full temporal range and this adds the 1/numerical factor that gives the exponential series.

Using this factor, we can then find the matrix element and the higher-order terms in the perturbation series between two states as

$$\begin{aligned}\langle k|T(t,t_0)|s\rangle &= \langle k|s\rangle - \frac{i}{\hbar}\int_{t_0}^{t} V_{ks}(t')dt' \\ &+ \left(\frac{i}{\hbar}\right)^2 \int_{t_0}^{t} \langle k|V(t')V(t'')|s\rangle dt''dt' + \dots\end{aligned} \tag{8.30}$$

Obviously, the first term on the right-hand side vanishes by the principle of orthogonality, while the second term is essentially the integral of (8.13) if we recognize that the exponential is buried in the time dependence of the perturbing potential. We note that the left-hand side is the coefficient c_k defined in Section 8.5. The result here tells us that the result obtained earlier is just the first term in a perturbation series expansion, which may be terminated only so long as the terms decrease sufficiently rapidly in order. The rest of the development follows Section 8.1 through to the Fermi golden rule.

8.3 EXPONENTIAL DECAY

In the preceding sections, it has been assumed that the initial state remains at unit amplitude. In fact, this is really unlikely, because it does not maintain a constant probability for the wave functions—the initial state cannot remain constant at unity if probability is being transferred to the other states via scattering by the Fermi golden rule. It is now desirable to see how this transfer of probability leads to the decay of the initial state, which in turn affects the transition probability. Each approach yields the same form for the transition rate, so we will use the direct form of (8.13)–(8.14). We can use these results to determine the decay of the initial state by summing over all the other states as

$$i\hbar\frac{\partial c_s}{\partial t} = \sum_{k\neq s} V_{ks}c_k(t)e^{i\omega_{ks}t}. \tag{8.31}$$

Now, we put in (8.14) for the coefficient within the summation to yield

$$\frac{\partial c_s}{\partial t} = -\frac{1}{\hbar^2}\sum_{k\neq s}\int_0^t |V_{ks}|^2 c_s(t')e^{i(\omega_{ks}-\omega_0)t'}dt', \tag{8.32}$$

where we have inserted (8.14) and reinserting $c_s(t)$ since it will no longer remain at unity. This last term is a convolution integral of this initial state amplitude and the exponential function. While this complicates the behavior, it is easy to handle with a Laplace transformation whose variable we take as z. Transforming (8.32) now leads us to

$$zC_s(z) - 1 = -\frac{1}{\hbar^2}\sum_{k\neq s}|V_{ks}|^2 C_s(z)\frac{1}{z - i(\omega_{ks}-\omega_0)}, \tag{8.33}$$

which leads to

$$C_s(z)=\left\{z+\frac{1}{\hbar^2}\sum_{k\neq s}\left|V_{ks}\right|^2\frac{1}{z-i\left(\omega_{ks}-\omega_0\right)}\right\}^{-1}. \quad (8.34)$$

If we now rationalize the last term in the curly brackets, we can rewrite it as

$$C_s(z)=\left\{z+\frac{1}{\hbar^2}\sum_{k\neq s}\left|V_{ks}\right|^2\frac{z}{z^2+\left(\omega_{ks}-\omega_0\right)^2}\right.$$
$$\left.+\frac{i}{\hbar^2}\sum_{k\neq s}\left|V_{ks}\right|^2\frac{\left(\omega_{ks}-\omega_0\right)}{z^2+\left(\omega_{ks}-\omega_0\right)^2}\right\}^{-1} \quad (8.35)$$

In the end, we will be taking the long time limit, which may be found by taking the $z\rightarrow 0$ limit of the expression. The same result will be found if we take the situation that

$$\int\frac{1}{z-i\omega}\rightarrow P+i\pi\delta(\omega), \quad (8.36)$$

so that the last term in the curly brackets becomes one-half of (8.18) for the scattering rate, although this time it is the total scattering rate out of the state s. It would be easy to set the second term to zero, this this is the principal part of the integral (or summation in this case) which leads to an energy shift of the state, and the two terms together furnish the self-energy of the state, so that (8.2) becomes

$$\psi(x,t)=\sum_n c_n e^{i(E_n-\Sigma_n)t/\hbar}\varphi_n(x), \quad (8.37)$$

where

$$\Sigma_n=\Delta E_n+i\frac{\Gamma_n}{2}. \quad (8.38)$$

We can use these results to return to (8.13) and see how it affects the scattering into the state k. We can rewrite (8.13) as

$$c_k=-\frac{i}{\hbar}\int_0^t V_{ks}e^{-\frac{\Gamma_s t'}{2}+i\left(\Omega_{ks}-\omega_0\right)t'}dt'. \quad (8.39)$$

Here, $\Omega_{ks}=\omega_{ks}-i\Delta E_s/\hbar$ is the energy shift due to the self-energy of the initial state. (The fact that there is no energy shift of the initial state is a result of assuming this state is zero at $t=0$. A more proper and advance treatment would give each state its own self-energy as implied above.) The integral may be easily done to give

$$c_k = -\frac{V_{ks}}{\hbar}\frac{e^{-\frac{\Gamma_s t}{2}+i(\Omega_{ks}-\omega_0)t}-1}{\Omega_{ks}+i\Gamma_s/2}. \tag{8.40}$$

Now

$$\begin{aligned} |c_k|^2 &= \frac{|V_{ks}|^2}{\hbar^2}\frac{1-2e^{-\frac{\Gamma_s t}{2}}\cos\left((\Omega_{ks}-\omega_0)t\right)+e^{-\Gamma_s t}}{(\Omega_{ks}-\omega_0)^2+(\Gamma_s/2)^2} \\ &\to \frac{|V_{ks}|^2}{\hbar^2\left[(\Omega_{ks}-\omega_0)^2+(\Gamma_s/2)^2\right]} \end{aligned}. \tag{8.41}$$

In the last line, we have taken the long time limit. Rather than dividing by t as was done in Section 8.1, we multiply by Γ_s, so that

$$\Gamma_{ks} = \frac{|V_{ks}|^2}{\hbar^2}\frac{\Gamma_s}{(\Omega_{ks}-\omega_0)^2+(\Gamma_s/2)^2}, \tag{8.42}$$

so that the delta function of (8.18) has been replaced by a Lorentzian line (one of the $\hbar$'s goes to the Lorentzian so that both the numerator and the denominator are energies). The integral over this Lorentzian gives the same value as that over the delta function, so that we may interpret the Lorentzian as a broadening of the delta function, just as the resistance in an L, R, C circuit broadens the resonance to a width determined by the Q of the circuit.

8.4 THE LIPPMANN-SCHWINGER EQUATION

In treating the Fermi golden rule transition rate of the last few sections, it was assumed that the perturbing potential was turned on at $t = 0$. This was done simply to avoid facing the evaluation of some of the integrals at $t \to -\infty$. The latter limit is problematic when the excitation is taken to be a simple sinusoid or a constant function. This has allowed us to use the Laplace transforms where necessary. A somewhat different approach is often used, in which Fourier transforms are utilized as necessary. In order to avoid the problem of the limit at large negative time, a different strategy is adopted. In this approach, we generate what is called the T-matrix, that in turn is used in the Lippmann-Schwinger equation to describe the scattered states.

8.4.1 The Scattering State T-Matrix

In order to describe negative time, we will assume that the perturbation is slowly turned on over a very long time period, but in such a manner that the matrix element can be written as

$$V_{ks} = T_{ks}e^{\alpha t}, \tag{8.43}$$

with

$$\begin{aligned} &\alpha > 0, \quad when\, t < 0, \\ &\alpha = 0, \quad when\, t > 0. \end{aligned} \tag{8.44}$$

In general, this factor can be set to zero at that point in the calculation in which all integrals have been evaluated at the lower limit of large, negative time. With this approach, (8.13) and (8.14) become

$$\begin{aligned} c_k &= -\frac{iT_{ks}}{\hbar}\int_{-\infty}^{t} e^{i(\omega_{ks}-\omega_0)t'+\alpha t'} dt' \\ &= \frac{T_{ks}}{\hbar}\frac{e^{i(\omega_{ks}-\omega_0)t'+\alpha t'}}{i\alpha-(\omega_{ks}-\omega_0)} + \delta_{ks} \end{aligned} \tag{8.45}$$

The last form is only for times $t \ll 1/\alpha$, so the growing exponential is not to be considered as a major factor, but we don't want to set α to zero just yet. Now, (8.15) becomes

$$|c_k|^2 = \frac{|T_{ks}|^2}{\hbar^2}\frac{e^{2\alpha t}}{\alpha^2+(\omega_{ks}-\omega_0)^2}, \tag{8.46}$$

and

$$\begin{aligned} \Gamma_{ks} &= \lim_{\alpha\to 0}\frac{\partial |c_k|^2}{\partial t} = \lim_{\alpha\to 0}\frac{|T_{ks}|^2}{\hbar^2}\frac{2\alpha e^{2\alpha t}}{\alpha^2+(\omega_{ks}-\omega_0)^2}. \\ &\to \frac{2\pi}{\hbar^2}|T_{ks}|^2\,\delta(\omega_{ks}-\omega_0) \end{aligned} \tag{8.47}$$

Here, we have taken the derivation only of the exponential term, and the result is just (8.18), the Fermi golden rule (the extra factor of $\hbar$ goes with the delta function to produce a delta function in energy rather than radian frequency).

The matrix of quantities T_{ks} is usually called the T-matrix, and it was introduced in the above discussion in a rather *ad hoc* manner. Let us pursue this a little more deeply and find out how it is really related to the matrix elements V_{ks}. To proceed, let us insert (8.43) into the earlier (8.11) for the coefficients of the expansion series. This leads to

$$i\hbar\frac{\partial c_k}{\partial t} = \sum\nolimits_n V_{kn}T_{ns}e^{i(\omega_{kn}+\omega_{ns}-\omega_0)t+\alpha t}, \tag{8.48}$$

and

$$c_k = T_{ks}e^{i(\omega_{ks}-\omega_0)t} = \frac{1}{\hbar}\sum\nolimits_n V_{kn}T_{ns}\frac{e^{i(\omega_{kn}+\omega_{ns}-\omega_0)t+\alpha t}}{i\alpha-(\omega_{ns}-\omega_0)}. \tag{8.49}$$

We note that the frequency term in the exponential reduces to ω_{ks} in keeping with the left-hand side of the equation where it relates the initial and final states of the T-matrix. For all practical purposes, we can now set α to zero and the exponential terms drop out of the equation. This now can be used in (8.47) to again regain the Fermi golden rule.

8.4.2 Gaining the Lippmann-Schwinger Equation

We now want to use the T-matrix to define scattering states that arise from the scattering of an ingoing wave by our scattering center. This incoming wave gives rise to the scattered outgoing wave $\varphi_s^{(+)}$ as

$$T_{ks} = \int \varphi_k^\dagger V(x,t)\varphi_s^{(+)}dx = \sum\nolimits_n \int \varphi_k^\dagger V(x,t)\varphi_n dx\cdot\left\langle \varphi_n^\dagger \middle| \varphi_s^{(+)}\right\rangle, \tag{8.50}$$

where we have inserted the resolution of the delta function as

$$\sum\nolimits_n \left|\varphi_n^\dagger(x')\varphi_n(x)\right| = \delta(x-x'). \tag{8.51}$$

This substitution leads to the second integral on the right-hand side of (8.50). Using this definition, (8.49) can be rewritten as

$$\left\langle \varphi_k \middle| V \middle| \varphi_s^{(+)}\right\rangle = \sum\nolimits_n \frac{\left\langle \varphi_k \middle| V \middle| \varphi_n\right\rangle\left\langle \varphi_n \middle| V \middle| \varphi_s^{(+)}\right\rangle}{\hbar\left(i\alpha-(\omega_{kn}+\omega_{ns}-\omega_0)\right)} + V_{ks}. \tag{8.52}$$

If we now let $k = s$, this becomes

$$\left\langle \varphi_s \middle| V \middle| \varphi_s^{(+)} \right\rangle = \sum_n \frac{\left\langle \varphi_s \middle| V \middle| \varphi_n \quad \varphi_n \middle| V \middle| \varphi_s^{(+)} \right\rangle}{\hbar\left(i\alpha - \left(\omega_{nk} - \omega_0\right)\right)} + V_{ss}. \tag{8.53}$$

This last summation can be decomposed to show that

$$\varphi_s^{(+)} = \sum_n \varphi_n \frac{\left\langle \varphi_n \middle| V \middle| \varphi_s^{(+)} \right\rangle}{\hbar\left(i\alpha - \left(\omega_{nk} - \omega_0\right)\right)} + \varphi_s. \tag{8.54}$$

We can now use (8.3) to rewrite this last expression as

$$\varphi_s^{(+)} = \varphi_s + \sum_n \varphi_n \frac{1}{\left(i\hbar\alpha - \left(H_0 - E_s - \hbar\omega_0\right)\right)} \left\langle \varphi_n \middle| V \middle| \varphi_s^{(+)} \right\rangle. \tag{8.55}$$

We can now back out the resolution of the delta function (8.51) and find that

$$\varphi_s^{(+)} = \varphi_s + \frac{1}{\left(i\hbar\alpha - \left(H_0 - E_s - \hbar\omega_0\right)\right)} V \varphi_s^{(+)}. \tag{8.56}$$

This is the Lippmann-Schwinger equation, and provides a self-consistent evaluation of the outgoing scattered wave function at the same energy level (Merzbacher 1970). As in Section 8.3, the inverse energy operator can be expanded in a series that will give rise to the perturbation series, and this will give the exact solution for the scattered wave in many cases. (8.56) can be rewritten as (with $\alpha = 0$)

$$\left(H_0 - E_s\right) \varphi_s^{(+)} = \left(H_0 - E_s\right) \varphi_s - V \varphi_s^{(+)}. \tag{8.57}$$

Since the first term on the right-hand side vanishes by (8.3), this can be rewritten as

$$\left(H_0 + \mathrm{V}\right) \varphi_s^{(+)} = E_s \varphi_s^{(+)}, \tag{8.58}$$

which tells us that the outgoing waves are solutions to the total Hamiltonian.

8.4.3 Orthogonality of the Scattering States

The Lippmann-Schwinger equation can now be used to show that the scattering states remain normalized and orthogonal. Before doing this, though, we diverge to illustrate another representation of the equation. If the resolvent operator, or Green's function (in this case a retarded Green's function due to the sign on the imaginary part), which is the fraction in the center of (8.56), is defined by this fraction as

$$G(E_s) = \frac{1}{E_s - H_0 + i\hbar\alpha}, \tag{8.59}$$

then (8.56) can be rewritten as

$$\varphi_s^{(+)} = \frac{1}{1 - G(E_s)V\varphi_s}. \tag{8.60}$$

This latter form is termed Dyson's equation, and clearly illustrates how the perturbation series is obtained by expanding the denominator in a power series. In fact, the form (8.60) is often termed the resummed perturbation series.

Let us now rewrite (8.56) to obtain another useful result. We premultiply all the terms by the denominator of the last term to give

$$(E_s - H_0 + i\hbar\alpha)\varphi_s^{(+)} = (E_s - H_0 + i\hbar\alpha)\varphi_s + V\varphi_s^{(+)}, \tag{8.61}$$

which can be rearranged to give

$$(E_s - H_0 - V + i\hbar\alpha)\varphi_s^{(+)} = (E_s - H_0 + i\hbar\alpha)\varphi_s$$

or

$$(E_s - H_0 - V + i\hbar\alpha)\varphi_s^{(+)} = (E_s - H + V + i\hbar\alpha)\varphi_s, \tag{8.62}$$

where we have used (8.1) on the right-hand side. This can now be rearranged to give

$$\varphi_s^{(+)} = \varphi_s + \frac{V}{E_s - H + V + i\hbar\alpha}\varphi_s. \tag{8.63}$$

Finally, we want to show that the scattering states retain their orthonormality. To begin, we use (8.63) to describe just one of the scattering states according to

$$\begin{aligned}
\left\langle \varphi_r^{(+)} \middle| \varphi_s^{(+)} \right\rangle &= \left\langle \varphi_r + \frac{V}{E_r - H + i\hbar\alpha} \middle| \varphi_s^{(+)} \right\rangle \\
&= \left\langle \varphi_r \middle| \varphi_s^{(+)} + \frac{V}{E_r - H - i\hbar\alpha}\varphi_s^{(+)} \right\rangle \\
&= \left\langle \varphi_r \middle| \varphi_s^{(+)} + \frac{V}{E_r - E_s - i\hbar\alpha}\varphi_s^{(+)} \right\rangle.
\end{aligned} \tag{8.64}$$

Here, we have used (8.58) to replace the total Hamiltonian with the energy of the scattered state. We can now factor out the minus sign of the denominator, and since the fraction is now a c-number, we can reverse the order of the terms multiplying the last scattering state. We then take the total operator *back* to the adjoint term, where the E_r-term becomes the operator H_0. Then, returning this operator to the position shown in (8.64), this latter equation can be written as

$$\left\langle \varphi_r^{(+)} \middle| \varphi_s^{(+)} \right\rangle = \left\langle \varphi_r \middle| \varphi_s^{(+)} \right\rangle + \left\langle \varphi_r \middle| \frac{V}{E_r - H_0 - i\hbar\alpha} \varphi_s^{(+)} \right\rangle. \tag{8.65}$$

Finally, using (8.63), we arrive at

$$\left\langle \varphi_r^{(+)} \middle| \varphi_s^{(+)} \right\rangle = \left\langle \varphi_r \middle| \varphi_s \right\rangle = \delta_{rs}, \tag{8.66}$$

which establishes the orthonormality of the scattering states.

REFERENCES

Landau, L. D. and Lifshitz, E. M. 1958. *Quantum Mechanics.* Pergamon: London, UK..
Merzbacher, E. 1970. *Quantum Mechanics.* Wiley: New York, USA.

PROBLEMS

1 For the unperturbed wave functions and energies of Section 7.2.1, estimate the decay rate

$$\Gamma_s = \frac{d}{dt} \sum_{k \neq s} \left| c_k \right|^2.$$

2 The propagator $U(t. t_0)$ satisfies the integral equation (8.24). If $V(t) = V_0 \delta(t - t_1)$, show that

$$U(t, t_0) = \frac{1}{1 + (iV_o / \hbar)\vartheta(t - t_1)}$$

where $\vartheta(x)$ is the Heaviside step function; $\vartheta(x) = 1$ for $x > 0$, and $\vartheta(x) = 0$ for $x < 0$. (Hint: break up the integral into segments $0 < t < t_1 - \delta$, $t_1 - \delta < t < t_1 + \delta$, $t_1 + \delta < t < \infty$, where δ is a very small parameter that is assumed to vanish, and use the property (8.23).)

3 Verify (8.28a). (Hint: the multiple time integrals in each term must be transformed all to have the same limits of integration. This introduces a factorial in the pre-factor and also changes the multiple integrals from a nested set to a product of individual integrals. The series is then summed.)

4 At $t < 0$, an electron is assumed to be in the $n = 3$ eigenstate of an infinite square potential well, which extends from $-a/2 < x < a/2$. At $t = 0$, an electric field is applied, with the

potential $V = Ex$. The electric field is then removed at time τ. Determine the probability that the electron is in any other state at $t > \tau$. Do not make any assumptions about the relative size of $\omega_{k_3}\tau$. What are the differences for the cases in which this latter quantity is small or is large?

5 Assume that an incident electron in a solid, in a state characterized by the wave vector $\boldsymbol{k}$, is scattered by the motion of the atoms of the solid. This scattering from the acoustic waves in the solid is characterized by a perturbing potential that is independent of the scattering wave vector q. Convert the sum over the final states into a sum over the final-state energies, and show that the scattering rate (the total transition rate Γ_s) is directly proportional to the magnitude of the incident wave vector.

9 Motion in Centrally Symmetric Potentials

The problem of interacting particles can usually be reduced in quantum mechanics to that of one particle, as can be done in classical mechanics. Normally, this problem is one of an electron orbiting around, or being affected by, a positive atomic core or a scattering center with the interaction governed by the Coulomb potential. In general, this is a multi-dimensional problem, and not one of the simpler one-dimensional problems with which we have been concerned in the previous chapters. Once we begin to treat multiple dimensions, then state degeneracy begins to arise more frequently, and the most common problem treated is that of the hydrogen atom. These extra dimensions provide more degrees of freedom and more complexity. In this chapter, we want to discuss the motion of a charged particle in a centrally symmetric potential, and so will discuss the hydrogen atom. First, however, we want to begin to understand how the degeneracies arise and just what they mean. To facilitate this, we will treat first the two-dimensional harmonic oscillator motion for a central potential. We will then consider the manner in which the degeneracies are split by a magnetic field. Following this, we will be ready to discuss the hydrogen atom with its three-dimensional potential. Finally, we will briefly discuss the energy levels that arise in atoms more complex than the hydrogen atom with real, but noncoulombic potentials. Finally, we will discuss hydrogenic impurities in semiconductors.

9.1 THE TWO-DIMENSIONAL HARMONIC OSCILLATOR

The harmonic oscillator in one dimension was discussed in Chapter 5. This simple problem of a particle in a quadratic potential energy is one of the typical problems of quantization. In two (or more) dimensions, the problem is more interesting, but not particularly more complicated. We want to treat only two dimensions here in order to understand better just how the degeneracies arise. However, as we will see, the two-dimensional problem is of particular interest for electrons in semiconductor interfaces. Here, this typical two-dimensional problem, similar to the one that we will consider, can arise when electrons are confined in an inversion layer at the interface between, for example, silicon and silicon dioxide or at the interface of a heterostructure or in a potential well in the direction normal to that considered here (see Section 2.6 for the first example). In any case, it is assumed that there is no z-motion, due to confinement of the carriers in this dimension. Furthermore, it will be assumed for simplicity that centrally symmetric properties of the potential require that the 'spring' constants of the harmonic potential are the same in the two free coordinates.

9.1.1 Rectangular Coordinates

We begin by treating the two dimensions as existing along two perpendicular axes in a normal rectangular coordinate system. Thus, the Hamiltonian may be obtained by expanding (5.9) to two dimensions, which leads us to

$$-\frac{\hbar^2}{2m}\left(\frac{\partial^2}{\partial x^2}+\frac{\partial^2}{\partial y^2}\right)\psi+\frac{m\omega^2}{2}\left(x^2+y^2\right)\psi=E\psi. \tag{9.1}$$

In order to simplify the results, we will introduce the creation and annihilation operators from (5.10), but with one set for the x-coordinates (denoted with a subscript x) and one set for the y-coordinates (with a subscript y). With the introduction of these operators, the Hamiltonian corresponding to Equation (9.1) may be expressed simply as

$$H=\hbar\omega\left(a_x^\dagger a_x+\frac{1}{2}\right)+\hbar\omega\left(a_y^\dagger a_y+\frac{1}{2}\right)=\hbar\omega\left(a_x^\dagger a_x+a_y^\dagger a_y+1\right). \tag{9.2}$$

Introducing the number operators from (5.15) and (5.20), this last expression may be written as

$$H=\hbar\omega\left(n_x+n_y+1\right). \tag{9.3}$$

We note from (9.2), or (9.3), that the lowest energy level E_0 is just $\hbar\omega$, which arises for $n_x = n_y = 0$. This level arises from just the zero-point motion of the two harmonic oscillators, and there is just one possible state that contributes to this level (the state where both number operators are identically zero). Thus, there is no degeneracy in this lowest energy level. On the other hand, the next highest energy level E_1 ($= 2\hbar\omega$) is doubly degenerate, since it can arise from either $n_x = 1$ and $n_y = 0$, as well as $n_x = 0$ and $n_y = 1$. In a similar manner, the third energy level is triply degenerate, since it arises from the three ways in which we can have $n_x + n_y = 2$. This discussion can be continued to higher energy levels, from which it is found that the energy value of the nth level is

$$E_n=\left(n+1\right)\hbar\omega. \tag{9.4}$$

This level has an n-fold degeneracy in which the two number operators add up to $n_x + n_y = n$.

How do we determine the x- and y-axes? The problem has no central property that allows us to determine uniquely the orientation of either of these axes. Thus, the selection of some arbitrary direction for the x-axis is one of convenience, but not one of basic physics. This means that the basic properties of the harmonic oscillator are not those associated with these axes. Rather, we will find that the various degeneracies arise from the radial and angular motion of the particle, with the energy level basically set by the radial motion (in cylindrical coordinates). To illustrate this better, we will change variables and work the problem in cylindrical coordinates, and this is done in the next section.

9.1.2 Polar Coordinates and Angular Momentum

If we specify the oscillator merely by its energy Hamiltonian (9.4), we cannot specify a complete set of commuting observables; that is, there is some degeneracy in the specification. This is because there are a number of combinations of x- and y-oscillator states that can combine into any single value of n. If we expand the total Hamiltonian into polar coordinates (cylindrical coordinates with no z-variation), and separate the resulting Schrödinger equation into the radial and the angular parts, we will find that the radial part generally determines the energy level, and the degeneracy goes into the various angular variations that result. For example, if we consider the $n = 2$ level of (9.4), one of the solutions arises for $n_x = n_y = 1$. Thus, each of the two one-dimensional harmonic oscillators is in the first excited state. How do we adjust the phase difference between the two oscillations? This phase difference can make the total oscillation actually rotate in the (x, y) plane as a phasor. The rotation of the oscillation amplitude corresponds to angular momentum, and it is this momentum operator that arises from the angular parts of the Hamiltonian in polar coordinates. By treating this angular momentum, we can understand the overall motion of the two-dimensional harmonic oscillator in polar coordinates.

Angular momentum in classical mechanics arises from the motion of a body around some center of rotation. This is defined by the vector

$$\mathbf{L} = \mathbf{r} \times \mathbf{p}. \tag{9.5}$$

In the two-dimensional harmonic oscillator in quantum mechanics, these variables become operators, but the only component (there is no z-variation in either position or momentum in our current approach) that arises is L_z. This is given by

$$L_z = xp_y - yp_x. \tag{9.6}$$

We want to put this into the operator notation used above, and so we introduce the operators defined by inverting (5.60) for each coordinate as

$$x = \sqrt{\frac{\hbar}{2m\omega}}\left(a_x + a_x^\dagger\right), \qquad y = \sqrt{\frac{\hbar}{2m\omega}}\left(a_y + a_y^\dagger\right),$$
$$p_x = -i\sqrt{\frac{\hbar m\omega}{2}}\left(a_x - a_x^\dagger\right), \quad p_y = -i\sqrt{\frac{\hbar m\omega}{2}}\left(a_y - a_y^\dagger\right). \tag{9.7}$$

This now leads to

$$\begin{aligned} L_z &= -i\hbar\left[\left(a_x + a_x^\dagger\right)\left(a_y - a_y^\dagger\right) - \left(a_y + a_y^\dagger\right)\left(a_x - a_x^\dagger\right)\right] \\ &= -i\hbar\left(a_x^\dagger a_y - a_y^\dagger a_x\right). \end{aligned} \tag{9.8}$$

In arriving at (9.8), we have used the fact that operators with different subscripts always commute, while operators with the same subscript satisfy (5.14). Moreover, we find that

$$\begin{aligned} \left[a_x^\dagger a_y, a_x^\dagger a_x + a_y^\dagger a_y\right] &= \left[a_x^\dagger, a_x^\dagger a_x\right]a_y + a_x^\dagger\left[a_y, a_y^\dagger a_y\right] \\ &= \left(a_x^\dagger a_x^\dagger a_x - a_x^\dagger a_x a_x^\dagger\right)a_y + a_x^\dagger\left(a_y a_y^\dagger a_y - a_y^\dagger a_y a_y\right) \\ &= -a_x^\dagger a_y + a_x^\dagger a_y = 0, \end{aligned} \tag{9.9}$$

where we have used (5.14) in evaluating the various commutators. A similar result holds for the second term in the second line of (9.8) when it is put into a commutator with the energy terms as above. This establishes that the z-component of the angular momentum in this system commutes with the energy and therefore is a constant of the motion. Since the energy (9.4) has a set of degeneracies, these must correspond to different values of the angular momentum. In the lowest level ($n = 0$), there is only one state which must correspond to a single angular momentum value. Because there is automatically a degeneracy between positive and negative angles of rotation (e.g., for every state with positive angular momentum, there must be a state with the opposite negative value of angular momentum, as there is nothing in the problem to give a preferred rotation direction), this single level must have $L_z = 0$. In the next energy level ($n = 1$), there are two degenerate states, one for each of the x- and y-axes. These must also correspond, in polar coordinates, to a nonzero value of angular momentum, with one state for each direction of rotation. This can be continued to the higher energy levels. We shall therefore seek to find the basis of the eigenvalues for the total Hamiltonian H and the angular momentum L_z. We will basically follow the treatment of Cohen-Tannoudji et al. (1977).

In electromagnetic fields, it is useful to decompose linear polarized waves into left- and right-circularly polarized waves. These circularly polarized waves have angular momentum just as we are discussing here. This suggests that we introduce rotating creation and annihilation operators to correspond to the rotating coordinates and to remove the meaningless cartesian coordinates. These are done by the new definitions:

$$a = \frac{1}{\sqrt{2}}\left(a_x - ia_y\right),$$
$$a^\dagger = \frac{1}{\sqrt{2}}\left(a_x^\dagger + ia_y^\dagger\right), \tag{9.10}$$

and

$$b = \frac{1}{\sqrt{2}}\left(a_x + ia_y\right),$$
$$b^\dagger = \frac{1}{\sqrt{2}}\left(a_x^\dagger - ia_y^\dagger\right). \tag{9.11}$$

We note from these definitions that both a and b will produce the result that

$$a\left|n_x, n_y\right\rangle = c_1\left|n_x - 1, n_y\right\rangle + c_2\left|n_x, n_y - 1\right\rangle, \tag{9.12}$$

where the c_i are constants. This means that the operation of a or b produces a linear combination of states that are at the energy level one quantum ($\hbar\omega$) down, and so removes one quantum of energy from the system just as the isolated component of an annihilation operator in a single dimension does from that coordinate's harmonic oscillator. Similarly, the adjoint operators (creation operators) increase the energy of the state by one unit of energy $\hbar\omega$. We will see below that this energy reduction, or increase, is accompanied by a corresponding change in the angular momentum of the state.

The rotational operators satisfy a commutation relation that can be obtained from those of the independent ones for each axis, and

$$\left[a, a^\dagger\right] = \left[b, b^\dagger\right] = 1, \tag{9.13}$$

and all other commutators vanish as expected. The defininitions (9.10) and (9.11) can be used to find the connections

$$a^\dagger a = \frac{1}{2}\left(a_x^\dagger a_x + a_y^\dagger a_y - ia_y^\dagger a_x + ia_x^\dagger a_y\right)$$
$$b^\dagger b = \frac{1}{2}\left(a_x^\dagger a_x + a_y^\dagger a_y + ia_y^\dagger a_x - ia_x^\dagger a_y\right), \tag{9.14}$$

so that (9.2) and (9.3) can be replaced with

$$H = \hbar\omega\left(a^\dagger a + b^\dagger b + 1\right) = \hbar\omega\left(n_a + n_b + 1\right). \tag{9.15}$$

Similarly, the angular momentum (9.8) can be replaced with

$$L_z = \left(n_a - n_b\right)\hbar. \tag{9.16}$$

The Hamiltonian remains in a form as simple as that for rectangular coordinates, and the angular momentum has been considerably simplified over its representation above.

Using the operators a and b, we can go through the complete arguments of Section 5.1 to determine the wave functions. This will lead to the orthonormal states

$$\left|n_a, n_b\right\rangle = \frac{1}{\sqrt{n_a! n_b!}}\left(a^\dagger\right)^{n_a}\left(b^\dagger\right)^{n_b}\left|0,0\right\rangle, \tag{9.17}$$

and these states are eigenstates of *both* the Hamiltonian and the angular momentum. The normal parameters, such as the energy index in (9.4), are given by the integers

$$\begin{aligned} n &= n_a + n_b, \\ m &= n_a - n_b. \end{aligned} \tag{9.18}$$

Here, we have introduced the traditional integer m for the angular momentum. We note that if we act with the operator $a^\dagger$ on (9.17), we not only raise the energy by one unit (we increase n_a by one unit), we also raise the value of m by one unit, hence increasing the angular momentum by one unit of Planck's (reduced) constant. Operating with $b^\dagger$ raises the energy but reduces the angular momentum by decreasing m through increasing n_b. This tells us that this operator corresponds to the positive (anticlockwise) direction of rotation, while the other operator set refers to the negative (clockwise) direction of rotation. The eigenvalues of the angular momentum are then $m\hbar$. The eigenvalues with a given value of n are $(n + 1)$-fold degenerate, since we can have

$$\begin{aligned} n_a &= n, & n_b &= 0, \\ n_a &= n-1, & n_b &= 1, \\ n_a &= n-2, & n_b &= 1, \\ &\cdots \end{aligned} \tag{9.19}$$

This continues until we get to $n_a = 0$, $n_b = n$. Hence, m can be a positive or negative integer, since it can range from $n_{a,\max}$ to $-n_{b,\max}$, or from n to $-n$. For $n = 0$, there is only a single level that requires $m = 0$. For the next energy level, $n = 1$ and there are two levels. Thus, $m = \pm 1$. For the third level, $n = 2$ and there are three levels. Here, m takes on the values 0 and ± 2. Hence, for a given energy level, the allowed values of m are

$$-n, -n+2, -n+4, \ldots, n-4, n-2, n. \tag{9.20}$$

Hence, any particular state is specified by the integers n and m, and the wave function is defined through

$$\left| n_a = \frac{n+m}{2}, n_b = \frac{n-m}{2} \right\rangle. \tag{9.21}$$

We now need to turn our attention to finding the lowest wave function, as the higher wave functions can be obtained from this lowest level with (9.17).

In seeking the wave function for the ground state, we will now introduce the polar coordinates, through the quantities

$$\begin{aligned} x &= r\cos\varphi, \qquad r \geq 0, \\ y &= r\sin\varphi, \quad 0 \leq \varphi \leq 2\pi. \end{aligned} \tag{9.22}$$

Using (5.10), (9.11), and (9.12), we find that

$$a = \frac{e^{-i\varphi}}{2}\left[Br + \frac{1}{B}\frac{\partial}{\partial r} - \frac{i}{Br}\frac{\partial}{\partial \varphi} \right], \tag{9.23}$$

and

$$b = \frac{e^{+i\varphi}}{2}\left[Br + \frac{1}{B}\frac{\partial}{\partial r} + \frac{i}{Br}\frac{\partial}{\partial \varphi} \right]. \tag{9.24}$$

Similarly

$$a^{\dagger} = \frac{e^{+i\varphi}}{2}\left[Br - \frac{1}{B}\frac{\partial}{\partial r} - \frac{i}{Br}\frac{\partial}{\partial \varphi} \right], \tag{9.25}$$

and

$$b^{\dagger} = \frac{e^{-i\varphi}}{2}\left[Br - \frac{1}{B}\frac{\partial}{\partial r} + \frac{i}{Br}\frac{\partial}{\partial \varphi} \right]. \tag{9.26}$$

In these equations

$$B = \sqrt{m\omega / \hbar}. \tag{9.27}$$

Note, that because of the rotating coordinates, *the adjoint operators are not simply complex conjugates of the annihilation operators* but must be carefully calculated from the definitions themselves. Now, either a or b will attempt to lower the energy and change the angular momentum. The lowest eigenstate should satisfy both of these operators, as

$$a|0,0\rangle = b|0,0\rangle = 0. \tag{9.28}$$

The two operators a and b differ only by the sign of the imaginary part. Since (9.28) must satisfy both the real and imaginary parts simultaneously, it leads to

$$\begin{gathered}\frac{1}{Br}\frac{\partial}{\partial\varphi}|0,0\rangle = 0 \\ \frac{1}{B}\frac{\partial}{\partial r}|0,0\rangle = -Br|0,0\rangle.\end{gathered} \tag{9.29}$$

The first line just leads to a function that is independent of the angle φ. The second of these equations leads to $\exp(-B^2r^2/2)$ behavior, just as for the linear harmonic oscillator. After normalization, the ground-state wave function is just

$$|0,0\rangle = \frac{B}{\sqrt{\pi}}e^{-B^2r^2/2}. \tag{9.30}$$

The higher-lying levels are now found by using (9.17). We note that angular variation is introduced into the wave function by the operators $a^\dagger$ and $b^\dagger$ themselves through the exponential prefactors. For example, the wave function for the $n = 1$, $m = 1$ state is given by

$$a^\dagger|0,0\rangle = \frac{e^{+i\varphi}}{2}\left[Br - \frac{1}{B}\frac{\partial}{\partial r} - \frac{i}{Br}\frac{\partial}{\partial\varphi}\right]\frac{B}{\sqrt{\pi}}e^{-B^2r^2/2} = \frac{B^2r}{\sqrt{\pi}}e^{+i\varphi}e^{-B^2r^2/2}. \tag{9.31}$$

The state $|0,1\rangle$ ($n = 1$, $m = -1$) is given by the complex conjugate of this expression. Thus, the two angular momentum states have counterrotating properties, but both are made up of linear combinations of the linear harmonic oscillators along the two axes. In the next section, an additional potential due to a magnetic field will be used to raise the degeneracy of the two angular momentum states.

9.1.3 Splitting the Angular Momentum States with a Magnetic Field

The motion of an electron in a magnetic field was discussed earlier in Section 5.4, where it was shown that the magnetic motion also introduces a harmonic oscillator potential. Here, we want to examine the coupling of the magnetic harmonic oscillator potential with the two-dimensional harmonic oscillator potential of (9.1). The first people to study the two-dimensional harmonic oscillator in a magnetic field were apparently Fock (1928) and Darwin (1931), but the operator approach that we want to use follows the treatment of Rössler (1991). The total Hamiltonian can be found by coupling (9.1) with (5.56), and

$$H = \frac{1}{2m}(\boldsymbol{p} + e\boldsymbol{A})^2 + \frac{m\omega^2}{2}\left(x^2 + y^2\right). \tag{9.32}$$

Because of the two-dimensional nature of the electrostatic harmonic oscillator, it is more convenient to take the vector potential in the symmetric gauge, rather than the Landau gauge used in Section 5.4, with $\mathbf{A} = (-By, Bx, 0)/2$, so that the magnetic field is oriented in the z-direction, normal to the two-dimensional plane of the motion ($\mathbf{B} = \nabla \times \mathbf{A} = B\mathbf{a}_z$). The momentum term can now be expanded as

$$\begin{aligned}(\boldsymbol{p} + e\boldsymbol{A})^2 &= -\hbar^2\left(\frac{\partial^2}{\partial x^2} + \frac{\partial^2}{\partial y^2}\right) \\ &\quad - i\hbar eB\left(x\frac{\partial}{\partial y} - y\frac{\partial}{\partial x}\right) + \frac{e^2B^2}{4}\left(x^2 + y^2\right).\end{aligned} \tag{9.33}$$

The last term can be combined with the electrostatic harmonic oscillator if we define the new oscillator 'spring' frequency to be

$$\Omega^2 = \omega^2 + \left(\frac{\omega_c}{2}\right)^2, \quad \omega_c = \frac{eB}{m}. \tag{9.34}$$

This last frequency is the cyclotron frequency introduced below (5.60). The second term in (9.33), which is linear in the magnetic field, can be rewritten using (9.6) as

$$BL_z. \tag{9.35}$$

With these definitions, the Hamiltonian, with the change in frequency in the definitions of the operators as $\omega \rightarrow \Omega$, can then be rewritten as

$$H = \left(n_a + n_b + 1\right)\hbar\Omega + \frac{1}{2}\hbar\omega_c m. \tag{9.36}$$

where $m = n_a - n_b$ is the angular momentum quantum number, and not the mass that appeared in the some of the early equations in this section. The reader should check that this new Hamiltonian still commutes with the angular momentum, and hence that both are simultaneously measurable. However, the energy now has a contribution from the angular momentum directly. This term raises the degeneracy of the previous eigenvalues (energies).

For small values of the magnetic field, $\omega_c \ll \Omega$, each energy level is split into $n + 1$ levels as the degeneracy is raised by the magnetic field. Each pair of these levels is separated by the amount $\hbar\omega_c$ since δm was determined in the previous section to be 2. As the magnetic field value is increased, this spread in energies changes significantly.

For slightly larger values of the magnetic field, but still with $\omega_c \ll \Omega$, the uppermost energy level is described by the frequency

$$\omega_+ = (n+1)\left(\omega + \frac{\omega_c^2}{4\omega}\right) + \frac{m\omega_c}{2}, \tag{9.37}$$

where m is the angular momentum quantum number. This is only slightly above the degenerate energy level that exists in the absence of the magnetic field. The lowest energy level is now described by the frequency

$$\omega_- = (n+1)\left(\omega + \frac{\omega_c^2}{4\omega}\right) - \frac{m\omega_c}{2}, \tag{9.38}$$

so all the levels begin to show a weak quadratic upward motion away from the linear spreading of the levels for very small magnetic fields.

In the case of very large magnetic fields, $\omega_c \gg \Omega$, the energy levels are quite different. The energy levels are now given by the frequencies

$$\omega_{n,m} = (n+m+1)\frac{\omega_c}{2} + (n+1)\frac{\omega^2}{\omega_c}. \tag{9.39}$$

If we neglect the last term, we recover the Landau level energies, since $n + m = 2n_a$ is an even integer. Moreover, we note that for example for the lowest Landau level, where $n + m = 0$, one level from each of the electrostatic harmonic oscillator levels (when $B = 0$) merges into the Landau level (see Figure 9.1). Similarly, for the next higher Landau level, one level from each electrostatic harmonic oscillator level for which $n \geq 1$ converges into the Landau level. This continues upward throughout the spectrum, with the ith Landau level being formed from states arising from the levels for which $n \geq i$. Thus, as the magnetic field is increased in size, the energy levels move smoothly from those values associated with the electrostatic harmonic oscillator to those values associated with the Landau levels.

While a given Landau level has contributions from all equal-index and higher-index harmonic oscillator levels, a given harmonic oscillator level contributes to only a fixed number of Landau levels. The lowest harmonic oscillator level, for example, contributes only to the lowest Landau level, since it has only one nondegenerate state. This is the case for which $n + m = 0$. Similarly, each harmonic oscillator level

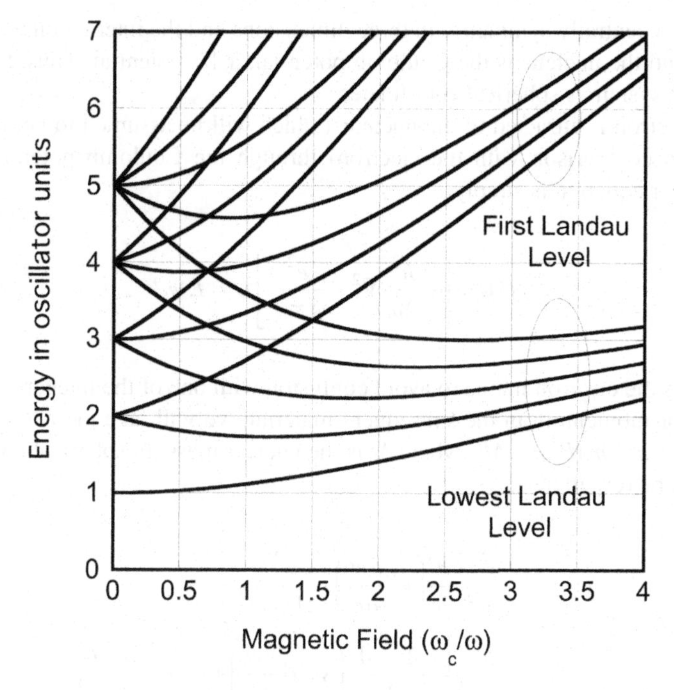

FIGURE 9.1 Lowest five harmonic oscillator levels in the presence of a magnetic field, illustrating the splitting of the angular momentum states and the formation of the Landau levels.

contributes its lowest level to the lowest Landau level. However, its highest level ($m = n$) goes into the nth Landau level. So the first harmonic oscillator level contributes to the lowest two Landau levels, the second to the lowest three Landau levels, and so on. The spacing of the levels that merge into a particular Landau level is given by the frequency so the electrostatic harmonic oscillator potential splits the degeneracy of the Landau levels. These are depicted in Figure 9.1.

The spectral positions of these 'quantum box' levels, shown in Figure 9.1, have been measured by far-infrared absorption measurements for InSb by Sikorski and Merkt (1989) and via the conductance arising from single-electron charging in GaAs by McEuen et al. (1991) and in GaAs tunneling structures by Tarucha et al. (1996).

9.2 THE HYDROGEN ATOM

In three dimensions, we could extend the above treatment to a three-dimensional harmonic oscillator, but a more important problem has spherical symmetry in the potential, and the latter is not the quadratic potential of the harmonic oscillator. It is of more interest to turn our attention to the motion of an electron about the nucleus of an atom. The case of interest is that of the hydrogen atom, really the only problem that can be solved exactly (the motion of two electrons about a nucleus includes the interaction between the two electrons and produces a three-body problem, which is well beyond the scope of this text). For simplicity, we will deal only with the *relative* motion of the electron about the nucleus, taking the latter as fixed in position.

The atom is centrally symmetric in three dimensions and the interaction between the electron and the nucleus is the Coulomb potential, a $1/r$ potential. Thus, the natural coordinate system is spherical coordinates.

The electron is attracted to the nucleus (which will be assumed to be immensely massive in comparison with the electron) through the Coulomb potential, so the Schrödinger equation is simply

$$H\psi = \left[-\frac{\hbar^2}{2m}\nabla^2 - \frac{e^2}{4\pi\varepsilon r} \right]\psi = E\psi, \tag{9.40}$$

where m is the electron mass. To avoid confusion with one of the integers describing the angular momentum in the subsequent material, we will take the mass as the reduced mass $\mu = mM/(m + M)$, where M is the nuclear mass. In spherical coordinates, (9.40) can be rewritten as

$$\begin{aligned} E\psi = &-\frac{\hbar^2}{2\mu r^2}\frac{\partial}{\partial r}\left(r^2\frac{\partial\psi}{\partial r}\right) - \frac{e^2}{4\pi\varepsilon r}\psi \\ &-\frac{\hbar^2}{2\mu r^2}\left[\frac{2}{\sin\vartheta}\frac{\partial}{\partial\vartheta}\left(\sin\vartheta\frac{\partial\psi}{\partial\vartheta}\right) + \frac{1}{\sin^2\vartheta}\frac{\partial^2\psi}{\partial\phi^2}\right] \end{aligned} \tag{9.41}$$

The traditional method of solving this is to assume that $\psi(r, \vartheta, \phi)$ can be split into two product terms as $R(r)\ \psi(\vartheta, \phi)$. By inserting this wave function, and then dividing by the wave function itself, (9.41) can be split into two terms, one a function of r alone and the other a function of ϑ and ϕ alone. If this is to be true for all values of the position and angles, each term must be equal to a constant. Thus, the term in the square brackets is a constant, say λ, times the wave function. While this constant is arbitrary, it is known from treatments of spherical harmonics that it is more appropriate to set this separation constant equal to $\ell(\ell + 1)$ (the separation constant for Legendre polynomials which are the know solutions in this type of problem). It is easy to show that there is a one-to-one correspondence to any value of λ and ℓ, so long as $\lambda > 0$, so no generality is lost by making this latter substitution. We will deal with these two solutions separately, and then consider the combined results.

9.2.1 The Radial Equation

By eliminating the angular variables with the above constant interpretation, which still must be shown to be valid (and we do this below), the radial equation now becomes

$$\frac{1}{r^2}\frac{\partial}{\partial r}\left(r^2\frac{\partial R}{\partial r}\right) - \frac{\ell(\ell+1)}{r^2}R + \frac{2\mu}{\hbar^2}\left[E + \frac{e^2}{4\pi\varepsilon r}\right]R = 0 \tag{9.42}$$

To simplify the equation, we introduce the reduced units for energy and length

$$\alpha^2 = \frac{8\mu E}{\hbar^2}, \tag{9.43}$$

and

$$\rho = \alpha r \tag{9.44}$$

respectively, and set

$$\lambda = \frac{\mu e^2}{2\pi\varepsilon \alpha\hbar^2} = \frac{e^2}{4\pi\varepsilon \alpha\hbar}\sqrt{\frac{\mu}{2|E|}} \tag{9.45}$$

With these substitutions, we can rewrite (9.42) as

$$\frac{1}{\rho^2}\frac{\partial}{\partial\rho}\left(\rho^2\frac{\partial R}{\partial\rho}\right) + \left[\frac{\lambda}{\rho} - \frac{1}{4} - \frac{\ell(\ell+1)}{\rho^2}\right]R = 0 \tag{9.46}$$

The choice of ¼ in the square bracket is arbitrary, but it actually puts this equation into a standard form that is well studied in the field of boundary value problems.

As we have done so far, the subsequent procedure is a well-traveled path, and we begin by noting that for sufficiently large values of the normalized radius, it is clear that the behavior of R is as $\rho^n e^{-\rho/2}$. This follows by retaining just the second-order derivative and the term with ¼ as the prefactor. The factor n will have to be determined, though, but this suggests that we seek solutions of the form $R(\rho) = F(\rho)e^{-\rho/2}$, where $F(\rho)$ will be a polynomial of finite order in ρ. Using this substitution leads to

$$\frac{\partial^2 F}{\partial\rho^2} + \left(\frac{2}{\rho} - 1\right)\frac{\partial F}{\partial\rho} + \left[\frac{\lambda - 1}{\rho} - \frac{\ell(\ell+1)}{\rho^2}\right]F = 0 \tag{9.47}$$

We now will assume that $F(\rho)$ varies as $\rho^s L(\rho)$, so (9.47) becomes

$$\begin{aligned}\rho^2\frac{\partial^2 L}{\partial\rho^2} &+ \rho\left[(2s+1) - \rho\right]\frac{\partial L}{\partial\rho} \\ &+ \left[\rho(\lambda - s - 1) + s(s+1) - \ell(\ell+1)\right]L = 0\end{aligned} \tag{9.48}$$

If we evaluate this at $\rho = 0$ (we assume the derivatives are well behaved and are finite at this point), we see clearly that $s = \ell$ or $-(\ell + 1)$. Since we require the functions to be finite at $\rho = 0$, only the former value is allowed. Equation (9.48) then becomes

$$\rho\frac{\partial^2 L}{\partial \rho^2}+\left[(2\ell+1)-\rho\right]\frac{\partial L}{\partial \rho}+(\lambda-\ell-1)L=0 \tag{9.49}$$

Finally, we will assume that $L(\rho)$ is a power series in ρ, but one that is finite and terminates at some order. This leads to the ratio of coefficients of successive orders v and $v + 1$:

$$a_{v+1}=a_v\frac{v+\ell+1-\lambda}{(v+1)(v+2\ell+2)} \tag{9.50}$$

For this to terminate, there must be a maximum value of v, which we call v', such that

$$\lambda = v'+\ell+1 \tag{9.51}$$

Here, we set this value of λ to n and this is the total quantum number (this and thereby λ determines the total energy of the level), n' is the radial quantum number and l is the angular momentum quantum number; all are integers. With this total quantum number, the energy levels are given by (9.45) as

$$E_n=\frac{\mu e^4}{32\pi^2\epsilon^2\hbar^2 n^2} \tag{9.52}$$

Thus, as in the two-dimensional harmonic oscillator, the energy levels are specified by the total quantum number, except that there is no $n = 0$ level here. This may be used in (9.43) to show that the effective radius of the ground state is given by $2/\alpha = a_0 = 4\pi\epsilon\,\hbar^2/\mu e^2 = 5.3 \times 10^{-9}$ cm. This quantity is called the *Bohr radius*. Note, however, that the actual normalization radius, and the factor α, depend upon the index n.

The coupling between the radial quantum number and the angular momentum quantum number means that as the angular momentum increases, the radial variations decrease (the order of the polynomial decreases). Furthermore, for the lowest energy level ($n = 1$), both the radial and angular momentum quantum numbers are zero, so the only variation is in the exponential term. In this lowest level, the wave function is spherically symmetric, decaying exponentially away from the center of the atom. For the second level ($n = 2$), ℓ can be either 0 ($n' = 1$) or 1 ($n' = 0$), so there is a spherically symmetric state and one that has preferential directions due to the angular momentum. This continues, with ℓ taking on values 0, 1, …, $n - 1$,so there are n values for this variable. We can write out a few of the lower-level wave functions in terms of the components written as $R_{n\lambda}(\rho)$, where n and λ are the two eigenvalue integers that we have discussed above. Thus, the lowest three energy levels have the unnormalized radial wave functions ($\rho = \alpha r$)

$$\begin{aligned}
R_{1,0}(\rho) &\sim e^{-\rho/2}, \\
R_{2,0}(\rho) &\sim \left(\rho^2 - 2\right)e^{-\rho/2}, \\
R_{2,1}(\rho) &\sim \rho e^{-\rho/2}, \\
R_{3,0}(\rho) &\sim \left(\rho^2 - 6\right)(\rho - 1)e^{-\rho/2}, \\
R_{3,1}(\rho) &\sim \rho(\rho - 4)e^{-\rho/2}, \\
R_{3,2}(\rho) &\sim \rho^2 e^{-\rho/2}.
\end{aligned} \tag{9.53}$$

We see that the highest-order term is determined by the radial quantum number n, but that the number of lower-order terms is related to the angular momentum quantum number. Note, again, that we use the normalized radius, and the normalization of each wave function depends upon the level number n.

9.2.2 Angular Solutions

The angular equation that arises from (9.41) represents the behavior of the wave function with the two spherical angles: ϑ, the polar angle measured from the z-axis (the polar axis), and ϕ, measured in the (x,y)-plane from the x-axis. This equation may be written as

$$\frac{1}{\sin\vartheta}\frac{\partial}{\partial\vartheta}\left(\sin\vartheta\frac{\partial\psi}{\partial\vartheta}\right) + \frac{1}{\sin^2\vartheta}\frac{\partial^2\psi}{\partial\phi^2} + \ell(\ell+1)\psi = 0 \tag{9.54}$$

This can be separated further by introducing $\psi = \Theta(\vartheta)\Phi(\phi)$, and follow the same separation procedure as was done above. This allows us to separate (9.54) into two equations, one for each of the two variables. These are given as

$$\begin{aligned}
&\frac{\partial^2\Phi}{\partial\phi^2} + m^2\Phi = 0 , \\
&\frac{1}{\sin\vartheta}\frac{\partial}{\partial\vartheta}\left(\sin\vartheta\frac{\partial\Theta}{\partial\vartheta}\right) + \left[\ell(\ell+1) - \frac{m^2}{\sin^2\vartheta}\right]\Theta = 0 .
\end{aligned} \tag{9.55}$$

Here, we let the separation constant be the square of an integer, m^2, and, in particular, we expect this value of m to be related to the angular motion of the azimuthal angle around the polar axis, just as in the two-dimensional case discussed previously. The fact that it is an integer is in recognition that Φ has to be periodic in 2π. Thus, we can write

$$\Phi(\phi)=\frac{1}{\sqrt{2\pi}}e^{im\phi},\quad |m|>0. \tag{9.56}$$

We note that m can take on both positive and negative values, but that Φ must be continuous and have a continuous derivative for any value of (ϕ), and, in addition, and Φ is a constant for $m = 0$.

In addressing the last angular equation, it will be convenient to let $w = \cos\theta$. With this substitution, the second equation of (9.55) becomes

$$\frac{d}{dw}\left[\left(1-w^2\right)\frac{d\Theta}{dw}\right]+\left[\ell(\ell+1)-\frac{m^2}{1-w^2}\right]\Theta=0. \tag{9.57}$$

Since the angle varies from 0 to π, the domain of w is 0 to 1. To begin, we let the wave function be assumed to vary as $(1 - w^2)^s G(w)$. With this substitution, we find that

$$\left(1-w^2\right)\frac{d^2G}{dw^2}-2w(2s+1)\frac{dG}{dw}+\left[\ell(\ell+1)-2s(2s+1)\right]G=0. \tag{9.58}$$

Here, $4s^2 = m^2$. This leads us to say that $2s = |m|$. G is now a polynomial of finite order, and for this to be the case, we must have the order of the polynomial equal to $\ell - 2s$. The parity of the polynomial (even or odd) is that of $\ell - 2s = \ell - |m|$. These polynomials are the associated Legendre polynomials, and, therefore, the $\Theta(w)$ are the associated Legendre polynomials.

9.2.3 Angular Momentum Again

In Section 9.1.2, we introduced the angular momentum, and particularly the z-component of this quantity. We now want to consider the total angular momentum of the particle orbiting the central potential in the hydrogen atom. By continuation of (9.6), we can immediately write the three components of angular momentum as

$$\begin{aligned}
L_x &= -i\hbar\left[y\frac{\partial}{\partial z}-z\frac{\partial}{\partial y}\right],\\
L_y &= -i\hbar\left[z\frac{\partial}{\partial x}-x\frac{\partial}{\partial z}\right],\\
L_z &= -i\hbar\left[x\frac{\partial}{\partial y}-y\frac{\partial}{\partial x}\right].
\end{aligned} \tag{9.59}$$

It is easily then shown using the normal commutation relations between position and momentum that

$$\left[L_x, L_y\right] = i\hbar L_z, \quad \left[L_y, L_z\right] = i\hbar L_x \qquad \left[L_z, L_x\right] = i\hbar L_x. \tag{9.60}$$

Finally, the square of the total angular momentum can be written as

$$L^2 = L_x^2 + L_y^2 + L_z^2 \tag{9.61}$$

and, moreover, this total angular momentum commutes with each of the components individually. Thus, we have four relatively independent quantities—the three components of the angular momentum and the total angular momentum—that can all be measured simultaneously. Normally, we can only specify two of these independently (we have only two parameters, l and m, that arise in the treatment of the orbital motion).

Let us first examine L_z, since the second of the parameters, m, is related to the motion about the polar axis, the z-axis, as it were. If we convert this term from (9.59) to spherical coordinates, with $x = \rho sin\vartheta cos\phi, y = \rho sin\vartheta sin\phi, z = cos\vartheta$, then we find that

$$L_z = -i\hbar \frac{\partial}{\partial\phi} \tag{9.62}$$

Obviously, this involves only the quantum number m, so L_z is a good component of the angular momentum to take as one of the independent quantities in the problem. The eigenvalues of the z-component of the angular momentum are then $\pm im\hbar$, with m an integer, as mentioned above. From the discussion of (9.58), it is clear that $|m| \le l$, so the values of m are

$$-\ell, -\ell+1, \ldots, -1, 0, 1, \ell-1, \ell. \tag{9.63}$$

Thus, there are $2\ell + 1$ values for the z-component of angular momentum, for a given value of ℓ.

We expect that the final independent quantity will be the total angular momentum. Indeed, this will be the case. To see this, we begin by taking combinations of the x- and y-components:

$$L_\pm = L_x \pm iL_y = \hbar\left[\pm\frac{\partial}{\partial\vartheta} + icot\vartheta\frac{\partial}{\partial\phi}\right] \tag{9.64}$$

Now, we can create the products of these circular momentum terms as

$$L_+L_- = (L_x + iL_y)(L_x - iL_y) = L_x^2 + L_y^2 + \hbar L_z,$$
$$L_-L_+ = (L_x - iL_y)(L_x + iL_y) = L_x^2 + L_y^2 - \hbar L_z. \tag{9.65}$$

Now, we can write the total angular momentum as

$$L^2 = L_+L_- - \hbar L_z + L_z^2$$
$$= \left[\frac{1}{\sin\vartheta}\frac{\partial}{\partial\vartheta}\left(\sin\vartheta\frac{\partial}{\partial\vartheta}\right) + \frac{1}{\sin^2\vartheta}\frac{\partial^2}{\partial\phi^2}\right] \tag{9.66}$$

The last line tells us that the total angular momentum is the angular part of the differential equation for the total wave function! We need only show now that the eigenvalue is indeed given by $\ell(\ell + 1)$.

To begin with, let us examine the first line of (9.66) in terms of the total angular momentum. We may rewrite this first line as

$$L^2 - L_z^2 = L_x^2 + L_y^2 \geq 0 \tag{9.67}$$

Now, this means that $L_z^2 \leq L^2$, or that

$$-\sqrt{L} \leq L_z \leq \sqrt{L} \tag{9.68}$$

From (9.63), one might jump immediately to a wrong conclusion ($L \sim \ell$), but the angular momentum is more subtle than this. However, (9.63) is important to us. To proceed, let us consider using the commutator relationships (9.60), which lead to

$$L_z(L_x \pm iL_y) = (L_x \pm iL_y)(L_z \pm 1) \tag{9.69}$$

Let us now use this result to operate on an arbitrary angular wave function in the manner of

$$L_z(L_x \pm iL_y)\Theta\Phi = (m \pm 1)(L_x \pm iL_y)\Theta\Phi \tag{9.70}$$

This tells us that the operator L_- reduces the angular momentum by one unit while L_+ raises it by one unit. If $m = \ell$, the raising operator must give a zero result, or

$$(L_x + iL_y)\Theta\Phi_l = 0 \tag{9.71}$$

Now, let us operate on this equation with the lowering operator (with the minus sign), to obtain

$$\left(L_x - iL_y\right)\left(L_x + iL_y\right)\Theta\Phi_l = \left(L^2 - L_z^2 - L_z\right)0. \tag{9.72}$$

Introducing the eigenvalues for the z-component of angular momentum, with $m = l$ in these results, we find

$$L^2 = \hbar^2\left(\ell^2 + \ell\right) \tag{9.73}$$

Thus, we achieve a value that accounts for the separation constant used in (9.42).

9.2.4 Atomic Energy Levels

In treating the hydrogen atom, the central potential was taken to be the Coulomb potential that existed between the atomic core and the single electron. It would be desirable to continue to do this for the general atom, but there are problems. Certainly, we can continue to use the central-field approximation. However, the simple Coulomb potential of (9.40) is only valid for the first. Then, each additional electron sees that the core is shielded by the electrons already bound to the core. In addition, for the innermost level, the charge is Ze, rather than e, where Z is the atomic number of the atom, or the number of protons in the core. This means that the potential is not simply a $1/r$ potential but rather one whose amplitude varies with the electron under consideration. The simple energy levels obtained in (9.52) are no longer valid, in that the angular momentum states are no longer degenerate with the states with no angular momentum—in essence, the energy levels now depend upon both n and ℓ. This change can be found by treating the difference between the actual potential (for a given electron) and the Coulomb potential as a perturbation and calculating the shift in energy for various values of n and ℓ. In fact, this is not particularly accurate, because the inner-shell electrons also interact with the outer electrons (and, in fact, each electron interacts with each other electron) through an additional Coulomb (repulsive) potential. Many schemes have been proposed for calculating the exact eigenvalues through various approximations, but usually some form of variational approach yields the best method. These approaches are, however, well beyond the level we want to treat here. The most important result is that, for a given level index n, the states of lowest ℓ lie at a lower energy (are more tightly bound to the nucleus). This is sufficient information to begin to construct the periodic table of the elements.

The lowest, and only exact, energy level is that for hydrogen, where there is a single electron and the atomic core of unit charge. The lowest energy level from (9.52) is the one for $n = 1$. The allowed values of radial and angular momentum indices must satisfy $n' + \ell = 0$ from (9.51). Thus, the only occupied state has no angular momentum, and the radial wave function varies (in the asymptotic limit of a simple Coulomb potential) simply as $e^{-\rho/2}$, in reduced units. Because of electron spin, this energy level can hold two electrons, so this behavior actually holds also for helium (although the energy level is shifted to lower energy due to the non-Coulomb nature of the potential in helium). Because these wave functions are spherically symmetric, they have come to be termed the 1*s* levels. The 1s ($n = 1, \ell = 0$) level can hold two electrons, which then fill the complete shell for $n = 1$. This *shell* comprises the first row of the periodic table.

The next energy level, according to (9.52) has $n = 2$. For this level, (9.51) would tell us that $n' + \ell = 1$, so the lowest angular momentum state has $(n', \ell) = (1,0)$. This lowest state gives rise to the pair of electrons in the 2*s* state. For this state, the wave function is again spherically symmetric and has an asymptotic variation (for a simple Coulomb potential) proportional to $(\rho - 2)e^{-\rho/2}$. This wave function has two nodes (in the amplitude-squared value), and this behavior continues as the order of the *s* levels increases. The two elements that are added via these two energy states are Li and Be. For more electrons, the angular momentum states begin to be filled. These levels arise for $(n',\ell) = (0,1)$. There are $2\ell + 1 = 3$ values of m for these levels, and each will hold two electrons because of the electron spin angular momentum. This means that there can be an additional six electrons accommodated by these levels. From the discussion of Section 9.2.2, we know that the angular part of the wave function for $m = 0$ varies as cos θ only. This gives amplitude peaks along the z-axis, so this state is the P_z state. Similarly, the $m = \pm 1$ states give rise to the $L_\pm$ states, which are combinations of the P_x and P_y states. These states have an angular variation as $(\sin\theta)e^{\pm i\phi}$. In Figure 9.2, we indicate the orientation of these three angular momentum wave functions, which have been taken along the three priniciple axes in accordance with their notation given just above. The filling of these six states, which are termed the 2*p* states (we use the symbol *p* to signify the $\ell = 1$ states), is carried out by progressing through the elements B, C, N, O, F, and Ne ($Z = 5 - 10$). Thus, the $n = 2$ level can hold eight electrons, and this shell is completely filled by the element of atomic number ten (Ne). The elements Li through to Ne comprise row two of the periodic table.

The states for $n = 3$ arise in a similar manner. Here, $n' + \ell = 2$, so ℓ can take the values 0, 1, and 2. These give rise to the two 3*s* states and the six 3*p* states, which do

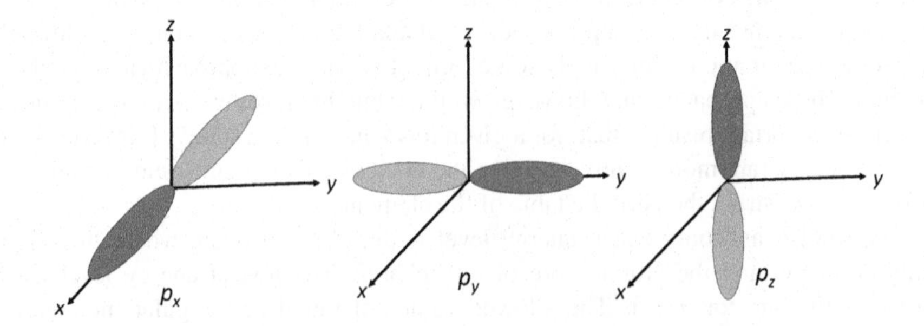

FIGURE 9.2 The orientation of the 2*p* orbitals is illustrated here. The darker blue represents the positive part of the wave function; the lighter blue represents the negative parts.

not differ in principle from those described above. However, the ten possible states for $\ell = 2$ raise additional complications and are described by quite complicated wave functions. These are the *d* levels, as we use the symbol *d* to describe the $\ell = 2$ levels. First, however, the third row of the periodic table is formed by filling the two 3*s* states and the six 3*p* states, in that order, by the elements Na, Mg, Al, Si, P, S, Cl, and Group A (elements number 11 through to 18). The 3*d* levels lie rather high in energy, and actually *lie above* the 4*s* levels! Thus, the next two elements, K and Ca, actually have their electrons in the 4*s* levels, with the 3*d* levels completely empty. This is where the complications begin, as the 3*d* levels are now filled by the next elements, and the ten elements required to fill these levels are the first set of *transition metals.* In the periodic table, these are the elements that fill the Group B columns, as the original series is designated the Group A columns. The transition metals formed from the 3*d* levels are known as the Pb series, those formed by the 4*d* levels are known as the Pd series, and so on.

After the first of the transition metals rows is completed, the 4*p* levels are filled by Ga, Ge, As, Se, Br, and Kr. Thus, in going across row 4 of the periodic table, there is a difference between columns 2A and 3A, in that the inner *d* levels are filled for the latter column and are empty for the former column, and it is between these two columns that the entire Group B set of columns is inserted. This same behavior continues in row 5, where the second transition metal series is encountered. In the outer shells, further complications arise from the filling of the $\ell = 3$ (designated the *f* levels) states, and this gives two new series of elements of elements. The first of these, the rare-earth series, involves elements 58–71, and fills their *f* levels in such a manner that they fit into the periodic table between the first two transition metals in the 5*d* shells. The second series of *f*-shell elements follows Ac and is known as the actinides. In Table 9.1, the energy levels of the s and p levels are given for a variety of atoms that find use in semiconductor technology (Herman and Skillman, 1963).

TABLE 9.1
Outermost atomic energy levels for selected atoms (eV)

Atom	n	$-E_s$	$-E_p$	Atom	n	$-E_s$	$-E_p$
H	1	13.6		Ga	4	11.37	4.9
B	2	12.54	6.64	Ge	4	14.38	6.36
C	2	17.52	8.97	As	4	17.33	7.91
N	2	23.04	11.47	Se	4	20.32	9.53
O	2	29.14	14.13	Cd	5	7.7	3.38
Al	3	10.11	4.86	In	5	10.12	4.69
Si	3	13.55	6.52	Sn	5	12.5	5.94
P	3	17.1	8.32	Sb	5	14.8	7.24
S	3	20.8	10.27	Te	5	17.11	8.59
Zn	4	8.4	3.38	Hg	6	7.68	3.48

9.3 THE COVALENT BOND IN SEMICONDUCTORS

The basic structure of the energy bands in a semiconductor can be inferred from a knowledge of the atomic lattice and its periodicity. There are basically two steps in this process: (1) forming the covalent bond in these materials, and (2) extending the bond into a set of bands using the periodicity of the lattice. The second of these steps was extensively studied in Chapter 4. In that chapter, the interaction between the atoms was simply expressed as a coupling energy. That energy is the covalent bond in semiconductors and all organic compounds. Here, it is the covalent bond that we want to discuss.

The semiconductors in which we are interested are *tetrahedrally* coordinated, by which we mean that there are four electrons (on average) from each of two atoms in the basis of the diamond structure. (This is the face-centered cubic structure with a basis of two atoms per lattice site.) These four electrons are in the s and p levels of the outer shell. For example, the Si bonds are composed of $3s$ and $3p$ levels, while GaAs has bonds composed of $4s$ and $4p$ levels, as does germanium. In each case, the inner shells are not expected to contribute anything at all to the bonding of the solid. This is not strictly true, as the inner d levels often lie quite close to the outer s and p levels when the former are occupied. This would imply that there is some modification of the energy levels in GaAs due to the filled $3d$ levels. This correction is small and will not be considered further. The p wave functions are quite directional in nature (see Figure 9.2), and this leads to a very directional nature of the bonding electrons. Thus, the electrons are not diffusely spread, as in a metal, but are quite localized into a set of hybrids that join nearest-neighbor atoms together. This hybrid orbital bonding is depicted in Figure 9.3.

The bonds are composed of *hybrids* that are formed by composition of the various possible arrangements of the s and p wave functions. There are, of course, four

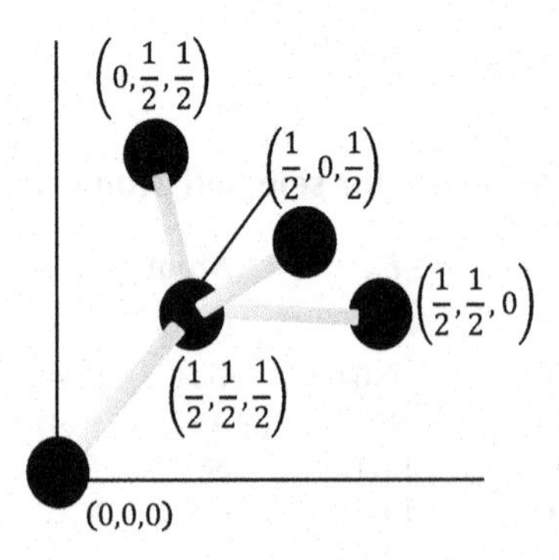

FIGURE 9.3 The bonding configuration of the tetrahedral coordination in the diamond and zinc-blende lattice common to most semiconductors. The inner atom of the two on the basis located at (0, 0, 0) is bonded to its four nearest neighbors (on the adjacent faces of the cube) by highly directional *orbitals*. These sp^3 orbitals give highly directional bonds.

hybrids for the tetrahedrally coordinated semiconductors. These four hybrids may be written as

$$\begin{aligned}
|h_1\rangle &= \frac{1}{2}\left(|s\rangle + |p_x\rangle + |p_y\rangle + |p_z\rangle\right) \\
|h_2\rangle &= \frac{1}{2}\left(|s\rangle + |p_x\rangle - |p_y\rangle - |p_z\rangle\right) \\
|h_3\rangle &= \frac{1}{2}\left(|s\rangle - |p_x\rangle + |p_y\rangle - |p_z\rangle\right) \\
|h_4 &= \frac{1}{2}\left(|s\rangle - |p_x\rangle - |p_y\rangle + |p_z\rangle\right)
\end{aligned} \tag{9.74}$$

The first of these hybrids points in the (111) direction, while the other three point in the $\left(1\bar{1}\bar{1}\right)$, $\bar{1}1\bar{1}$, and $\bar{1}\bar{1}1$ directions, respectively (the bar over the top indicates a negative coefficient). The factor of $\frac{1}{2}$ is included to normalize the hybrids properly so that $\langle h_i | h_j \rangle = \delta_{ij}$. These hybrids are now directional and point in the proper directions for the tetrahedral bonding coordination of these semiconductors. Thus, the bonds are directed at the nearest neighbors in the lattice. For Ga in GaAs, the four hybrids point directly at the four As neighbors, which lie at the points of the tetrahedron. The locations of the various atoms for the tetrahedral bond are shown in Figure 9.3.

Each of the atomic levels possesses a distinct energy level that describes the atomic energy in the isolated atom, and these were shown in Table 9.1. Thus, the *s* levels possess an energy given by E_s and the *p* levels have the energy E_p. In general, these levels are properties of the atoms, so that the levels are different in the heteropolar compounds like GaAs. The levels will be marked with a superscript A or B, corresponding to the A–B compound that forms the basis of the lattice. In the following, the compound semiconductors will be treated, as they form a more general case, and the single component semiconductors, such as Si or Ge, are a special case that is easily obtained in a limiting process of setting A = B.

The *s* and *p* energy levels are separated by an energy that has been termed the *metallic* energy (Harrison, 1980). In general, this energy may be defined from the basic atomic energy levels through

$$\begin{aligned}
4V_1^A &= E_p^A - E_s^A, \\
4V_1^B &= E_p^B - E_s^B.
\end{aligned} \tag{9.75}$$

Here the A atom is the *cation* and the B atom is the *anion* in chemical terms. The hybrids themselves possess an energy that arises from the nature of the way in which they are formed. Thus, the *hybrid* energy is

$$E_h = \frac{\langle h_i | H | h_i \rangle}{\langle h_i | h_i \rangle} = \frac{1}{4}\left(E_s + 3E_p\right). \tag{9.76}$$

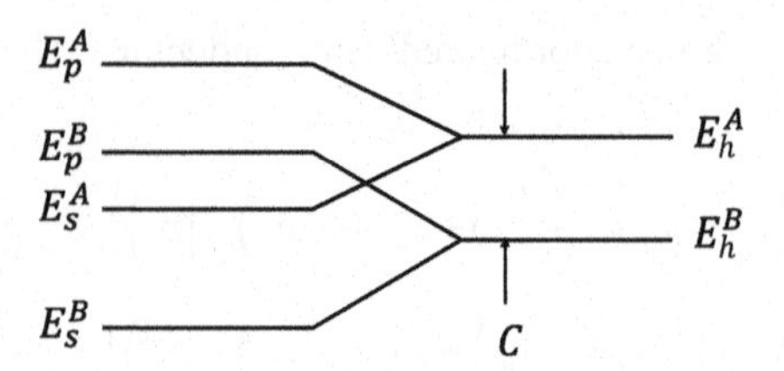

FIGURE 9.4 The atomic energies form hybrids, as indicated here, for each of the two atoms in the basis. These hybrids are separated by the *heteropolar* energy C.

This holds for all values of the index i, where H is the Hamiltonian operator for the Schrödinger equation representing the crystal but neglecting any interaction between the atoms (without these interaction terms, the cross terms that would arise in (9.76) will vanish by orthonormality). In Figure 9.4, the hybrid energy is shown pictorially in the manner in which it forms from the atomic energies in a compound semiconductor.

The two hybrid energies are separated by the *hybrid polar energy* or the *heteropolar energy*, depending on whose definitions one wants to use. The notation C has been used here for this energy. The heteropolar energy is a product of the ionic transfer of charge in the compound semiconductor (since Ga has only three electrons and As has five electrons, there is a charge transfer in order to get the average four electrons of the tetrahedral bond). Of course, this energy vanishes in a pure single compound such as Si or Ge, which are referred to as homopolar materials. That is, they are composed of a homogeneous set of atoms, while the general compound semiconductor is heterogeneous in that it contains two types of atom. The heteropolar energy C may be easily evaluated using (9.76) as follows:

$$C = E_h^A - E_h^B = \frac{1}{4}\left[\left(E_s^A - E_s^B\right) + 3\left(E_p^A - E_p^B\right)\right] \tag{9.77}$$

It is important to note that the hybrids are not eigenstates of either the isolated atom or of the crystal. Rather, they are constructed under the premise that they are the natural wave function for the tetrahedral bonds. But we have created them to be what seems a natural form (at least to us). In principle, they will be stable in the crystal once the interactions between the various atoms are included. In fact, they are not orthogonal under action of the Hamiltonian, since the matrix element between two hybrids is the metallic energy, as defined in (9.75). In this sense, the metallic energy describes the contribution to the energy of the itinerant nature of the electrons.

Now we have to turn to the interaction between hybrids localized on neighboring atoms. Only standing-wave interactions will be considered here ($k = 0$) and the propagating wave-function-dependent changes are calculated by the methods of Chapter 4. The interaction energy between two atoms on different sites can arise from, for example, one atom's hybrid pointed in the (111) direction and the nearest neighbor in that direction, displaced $\left(\frac{a}{4}, \frac{a}{4}, \frac{a}{4}\right)$, whose hybrid points in the opposite direction (there is a complete flip of the hybrid directions of the atoms as one moves along the body diagonal direction). The interaction energy between these nearest-neighbor hybrids is

$$-V_2 = \frac{1}{4}\left\langle s^A \left| H \right| s^B \right\rangle + \frac{\sqrt{3}}{4}\left[\left\langle p^A \left| H \right| s^B \right\rangle + \left\langle s^A \left| H \right| p^B \right\rangle\right] + \left\langle \frac{3}{4} p^A \left| H \right| p^B \right\rangle \quad (9.78)$$

Harrison (1980) has argued that these energies should depend only on the interatomic spacing $d\left(=\sqrt{3}a/4\right)$, where a is the edge of the face-centered cube, and that they should have the general form

$$V_2 \sim 4.37\frac{\hbar^2}{md^2} \quad (9.79)$$

Phillips (1973) also argues that V_2 should be a function of the interatomic spacing, but that it should also satisfy another scaling rule. Since the atomic radii are the same for each row of the periodic table, the value of V_2 should be the same in AlP as in Si, the same in GaAs and ZnTe as in Ge, and so on. In other words, the value for this quantity is set by the distance between the atoms, and this really does not change as one moves across a row of the table. Thus, this value is the same in heteropolar compounds as in homopolar compounds and Phillips has termed V_2 the homopolar energy, $E_{ho} = 2V_2$ (this should not be confused with the hybrid energy E_h, which differs for the two atoms). The *average energy gap* between bonding and antibonding orbitals is thus composed of a contribution from the homopolar energy E_{ho} and a contribution from the heteropolar energy C.

The bonding orbital will be composed of the two hybrids based on the atoms at each end of the bond, as suggested in Figure 9.4. These bonding orbitals can be written as a linear combination of the two hybrids at each atom, and then interact to give rise to an average energy in the bonding band and an average energy in the antibonding band, separated as

$$E_G = \frac{E_h^A + E_h^B}{2} \pm \sqrt{\left(2V_2\right)^2 + C^2} = \frac{E_h^A + E_h^B}{2} \pm \frac{\Delta}{2} \quad (9.80)$$

It is important to note that the energy gap in this equation is *not* the normal bandgap between valence and conduction bands, but the difference in the average energy of these two bands. The bonding (lower sign) and the antibonding (upper sign) energy levels are symmetrically spaced about the average hybrid energy of the two atoms. Of course, in a homopolar material, these two hybrid energies are the same and are equal to E_h. The significance of the average energy is unexpected in heteropolar materials but is important for getting the positions of the average bonding and antibonding energies correct. The actual bands spread around these two average energies according to the techniques of Chapter 4.

It is clear that once we know the atomic energy levels, we can immediately determine the hybrid energies, and the heteropolar and homopolar energies. Using V_2 as the constant value given in (9.79), the average energies are now known. However, Phillips (1973) treats the homopolar energy and the heteropolar energy C as adjustable parameters to get better scaling for the average energy gaps, building his results from a need to get the dielectric functions correct. His values are close to the Harrison (1980) values obtained by straightforward application of the atomic energies, but there are differences. In the heteropolar materials, the differences are even greater, and the average gap can differ by more than a volt between the two approaches. To be sure, it is not at all clear that these two methods are comparable or that the two authors are talking about exactly the same quantities, even though it appears to be so. That the numbers are close is perhaps remarkable and points out the basic correctness of the overall picture of the composition of the energy bands in semiconductors.

9.4 HYDROGENIC IMPURITIES IN SEMICONDUCTORS

One can achieve in semiconductors a situation in which an impurity atom is placed in the host lattice, with the special case that the impurity may have an additional electron or be short of one electron. When the impurity atom has an additional electron over those (four) required for tetrahedral bonding in the covalent lattice, this extra electron can be ionized rather easily. In this case, the impurity atom has a single positive charge, while the ionized electron has a single negative charge. In many ways this interaction is quite like that in the hydrogen atom. On the other hand, the impurity atom that is short of one electron allows for electrons to move from other atoms to this one in order to complete the tetrahedral bonding requirements. Here, we say that a 'hole' moves from the impurity to the other atoms. This hole (the absence of an electron in the valence band) can be ionized from the impurity, leaving a negatively charged impurity atom and a positively charged particle.

The energy required to ionize either the electron or the hole is generally much smaller than that required to lift an electron out of the valence band into the conduction band—the band gap energy. Thus, these impurities introduce defect levels within the bandgap region of the semiconductor. Because of the coulombic nature of the interaction between the ionized particle (electron or hole) and the central cell of the ionized impurity, the perturbing potential is simply a screened Coulomb potential:

$$V(r) = -\frac{e^2}{4\pi\varepsilon r} \tag{9.81}$$

Here, ε is the dielectric constant (times the free-space permittivity) of the semiconductor host crystal. It may generally be assumed that the carriers in the semiconductor are characterized by a scalar *effective mass* m^*, which accounts for the band structure nature of the electrons or the holes, as the case may be. Thus, we are concerned either with electrons near the minimum of the conduction band or holes near the maximum of the valence band. With this potential, and the effective mass, Schrödinger's equation for this reduced system becomes

$$-\frac{\hbar^2}{2m^*}\nabla^2\psi - \frac{e^2}{4\pi\epsilon r}\psi = E\psi \tag{9.82}$$

This is the same equation as for the simple hydrogen atom (with the appropriate changes in the dielectric constant and the mass). This means that the allowed energy levels for the bound electrons are simply given by the hydrogen energy levels (9.52) suitably adjusted.

The first ionization energy of the hydrogen atom is one Rydberg, or 13.6 eV. Here, this value is reduced by the square of the dielectric constant, which is of the order of 10, and by the effective mass, which is of the order of 0.1. Thus, the ionization energy of the impurity is of the order of 0.0136 eV. This value, of course, varies according to the specific dielectric constant and the effective mass. What is not accounted for in this hydrogenic model is variation according to just which atomic species is providing the impurity atom. In the simple model, the results should be the same for all impurities, but this is not the case. However, the hydrogenic model is a good approximation. Actual ionization energies vary from a few meV to a significant part of an eV in deep levels. The wave function for the first ionization state varies just like that for the hydrogen atom, but with an adjusted radius. In fact, we may write the wave function as

$$\psi(r) \sim e^{-r/a} \tag{9.83}$$

where

$$a = \frac{4\pi\epsilon\hbar^2}{e^2 m^*} = a_0 \frac{\epsilon}{\epsilon_0}\frac{m}{m^*} \tag{9.84}$$

and a_0 is the Bohr radius. For our hypothetical semiconductor, with a dielectric constant of 10 and an effective mass of 0.1, we find an effective radius of 5.3×10^{-7} cm, or 5.3 nm. This is about ten lattice constants, so the size of the orbit of the electron, when it is captured by the impurity, actually samples a great many unit cells of the crystal. This means that the effective mass approximation is a fairly good approximation for the hydrogenic impurity. Nevertheless, the potential probably deviates in reality from a coulombic one, and this difference is likely to account for the variation in ionization energy seen for different impurities in the same semiconductor.

REFERENCES

Cohen-Tannoudji, C., Diu, B., and Laloë, F. *Quantum Mechanics*, vol. 1, Wiley: New York, 1977, p.727

Darwin, C. G. 1931. The diamagnetism of the free electron. *Proc. Cambr. Phil. Soc.* **27** 86–90

Fock, V. 1928. Bemerkung zur Quantelung des harmonischen Oszillators im Magnetfeld. *Z. Phys.* **47** 446–448

Harrison, W. A. 1980. *Electronic Structure and the Properties of Solids.* Freeman: San Francisco, CA.

Herman, F. and Skillman, S. 1963. *Atomic Structure Calculations*. Prentice-Hall: Englewood Cliffs, NJ.

McEuen, P. L., Foxman, E. B., Meirav, U., Kastner, M. A., Meir, Y., Wingreen, N. S., and Wind, S. J. 1991. Transport spectroscopy of a Coulomb island in the quantum Hall regime. *Phys. Rev. Lett.* **66** 1926–1929

Phillips, J. C. 1973. *Bonds and Bands in Semiconductors*. Academic: New York, NY.

Rössler, U. 1991. Electronic Structure of Small Systems, in *Quantum Coherence in Mesoscopic Systems*, Ed. Kramer, B. Plenum Press: New York. Pp. 45–62

Sikorski, Ch. and Merkt, U. 1989. Spectroscopy of electronic states in InSb quantum dots. *Phys. Rev. Lett.* **62** 2164–2166

Tarucha, S., Austing, D. G., Honda, T., van der Haage, R. J., and Kouwenhoven, L. P. 1996. Shell filling and spin effects in a few electron quantum dot. *Phys. Rev. Lett.* **77**, 3613–3616

PROBLEMS

1 Find the wave functions for the $n = 2$, $m = 0$ level of the two-dimensional harmonic oscillator. Use the angular momentum operators and show that this level is cylindrically symmetric.

2 Suppose that the x-axis harmonic oscillator is characterized by a frequency ω_1, which is slightly larger than that of the y-axis $(\omega_1 \simeq \omega + \delta\omega)$. Compute the exact energy levels of the two-dimensional harmonic oscillator. Then, using the energy levels and wave functions for the symmetric two-dimensional harmonic oscillator, calculate the perturbation shift of these energy levels with the perturbation $m(\delta\omega)^2 x^2 / 2$.

3 Let us consider the application of an electric field $-Ea_x$ to the two-dimensional harmonic oscillator in the presence of a magnetic field. Determine the change in the energy levels introduced by this electric field.

4 What are the energy values of the lowest three energy levels in the hydrogen atom? (That of the lowest energy level, the ionization energy of the hydrogen atom, is termed one Rydberg.)

5 Verify that (9.42) is the correct equation for the angular variation of the hydrogen atom.

6 Develop the full wave functions (the form, neglect normalization) for the 3p states utilizing the angular momentum operators.

7 Show that L^2 in (9.66) commutes with each component of the angular momentum.

8 (a) Using the known lattice constants and atomic energy levels, compute the values of the parameters C, V_2, and E_G for Si, Ge, GaAs, AlAs, InAs, InSb, and InP.

9 From the known lattice constant and density of Si, compute the number of atoms per cubic meter. Then, use this value of N, and the known dielectric constant of Si, to compute the valence plasma frequency from $\omega_P^2 = Ne^2 / m_0\varepsilon_s$. What is the average energy gap required to satisfy the Penn dielectric function

$$\varepsilon_s = \varepsilon_0\left[1+\left(\frac{\hbar\omega_P}{E_G}\right)^2\right]?$$

10 Spin Angular Momentum

We have commented several times, in the previous chapters, that each of our quantum states found in the various systems to which we have guided our attention, that each electron quantum state will actually hold two electrons. But these two electrons must have opposite spin. This is constituted by the particular constraints that are introduced by the fact that the electrons are *indistinguishable and identical* particles. While we have only paid lip service to the fact that each electron also possesses a *spin angular momentum* about its own axis, this became important at several points in the book, such as when we discuss superconductivity and Cooper pairs as well as at the end of the last chapter. When we deal with single isolated electrons, these approximations do not get us into very much trouble. However, most atoms and solids are densely populated with electrons, and the neglected properties can introduce new effects that need to be considered. Even though the effects may be quite small, an understanding of them is necessary, if for no other reason than to be able to ascertain when they may properly be ignored.

In classical mechanics, it is quite easy to follow the individual trajectories of each and every electron. However, the Heisenberg uncertainty principle prevents this from being possible in quantum mechanics. If we completely understand the position of a given particle, we can say little about its momentum and most other dynamical variables. This principle of the indistinguishability of identical particles leads to other more complex forms of the wave functions. These more complicated forms can lead to effects in quantum mechanics that have no analog in classical mechanics. We use the word *identical* to indicate particles that can be freely interchanged with one another with no change in the physical system. While these particles may be distinguished from one another in situations in which their individual wave functions do not overlap, the more usual case is where they have overlapping wave functions because of the high particle density in the system. In this latter case, it is necessary to invoke a *many-particle wave function*. To describe properly the properties of the many-particle wave function, it is necessary to understand the properties of the interactions that occur among and between these single particles. If the one-electron problem were our only interest, this more powerful development that is used for the interacting system would be worthless to us, for its complexity is not worth the extra effort. However, when we move to multielectron problems, the power of the approach becomes apparent. In this chapter, we want to examine just the spin properties of these general wave functions for the case of electrons, which also possess their own spin angular momentum, and to introduce the interaction among the electrons.

We begin by considering just two electrons. Because these are indistinguishable particles, the physical state that is obtained by merely interchanging the positions of these two particles must be completely equivalent to the original one. This puts certain constraints upon the wave function. Let us consider the two-electron wave function $\psi(x_1, x_2)$. Here, the parameter x_1 refers to the position of the first particle and the spin orientation $\boldsymbol{\sigma}_l$ of that electron (which will be further explained below). Thus, we can refer to this as a four-vector in space—three components for the position and the fourth for the spin orientation. Similar notation is used for the second particle. Now, under exchange of the positions and spins of the two particles, we must have

$$\psi(x_1, x_2) = e^{i\vartheta}\psi(x_2, x_1) \tag{10.1}$$

The phase ϑ is some real constant. By repeating the interchange, we arrive at

$$\psi(x_1, x_2) = e^{i\vartheta}\psi(x_2, x_1) = e^{2i\vartheta}\psi(x_1, x_2) \tag{10.2}$$

But this requires $e^{2i\vartheta} = 1$, or $e^{i\vartheta} = \pm 1$. Thus, the possible forms for the wave function are

$$\psi(x_1, x_2) = \pm\psi(x_2, x_1) \tag{10.3}$$

The wave function that is found upon interchange of the two particles is going to be either symmetrical or antisymmetrical. By symmetrical, we mean that it is unchanged, while antisymmetrical implies a change of sign. Which of these two are we to choose?

The choice of either a symmetric or an antisymmetric wave function lies in the imposition of the Pauli exclusion principle (Pauli, 1925). We have constantly held that each eigenstate determined by solving the Schrödinger equation could hold two electrons, provided that they had opposite spin. In each situation, the positional wave functions for the two electrons are the same, but the spin wave functions must yield a spin eigenvalue σ that is oppositely *directed*. Now, this spin eigenvalue is one of the vectors included above in the description of the two-particle wave function. Thus, the electrons, which obey the Pauli exclusion principle, must have an *antisymmetric wave function*. We can summarize this by saying that particles that do not obey the exclusion principle (phonons, photons, etc.) are *bosons*, and are found to obey Bose–Einstein statistics. These bosons have symmetric wave functions under interchange of the particles. On the other hand, particles that obey the exclusion principle (electrons and some others, with which we will not be concerned) are *fermions* and are found to obey Fermi–Dirac statistics. Fermions must have antisymmetric wave functions under the interchange of particles. We will see below that the use of the antisymmetric wave function actually ensures that the Pauli exclusion principle is obeyed.

Now, what do we really mean by bosons and fermions? In the previous paragraph, we stated that bosons do not obey the Pauli exclusion principle but do obey the Bose–Einstein distribution. On the other hand, fermions obey the Pauli exclusion principle—no more than two fermions, and these are of opposite spin—can be accommodated in any quantum state. Bosons are particles with integer spin, such as phonons (zero spin) and photons (integer spin given by ±1, corresponding to right- and left-circularly polarized plane waves, for example). Fermions are particles with half-integer spin, such as electrons. These two distributions are given by

$$f_{BE}\left(E\right)=\frac{1}{e^{E/k_BT}-1} \tag{10.4}$$

and

$$f_{FD}\left(E\right)=\frac{1}{1+e^{(E-E_F)/k_BT}} \tag{10.5}$$

where E_F is the Fermi energy level, at which energy the distribution function has the value 0.5. There is a significant difference in the behavior of these two distributions at low energy. The Bose–Einstein distribution diverges at $E \to 0$, as there is no limit to the number of bosons that can occupy the lowest energy state. On the other hand, the Fermi–Dirac distribution approaches unity in this limit, as the state is certainly occupied if it lies well below the Fermi energy E_F. As pointed out, one way of achieving the Pauli exclusion principle is to ensure that the wave function for a fermion is antisymmetric, that is it satisfies (10.3).

If we insert our many-particle wave function into the Schrödinger equation, it is found that under the circumstances discussed above, it is possible to write this equation as

$$\left[H\left(x_1,x_2\right)-E\right]\psi\left(x_1,x_2\right)=0 \tag{10.6}$$

Here, we may assert that *the Hamiltonian itself is invariant under the exchange of the particles*, and the equivalence of the two physical states implies that the energy has the same invariance. Hence, the imposition of the exclusion principle appears only in the wave function (unless some special spin-dependent interaction, such as the spin-orbit interaction—discussed below—is introduced into the Hamiltonian to distinguish one spin state from another). Whether the wave function is symmetric or antisymmetric has no impact upon (10.6). The importance of this latter result is that, for simple Hamiltonians with no interaction among the electrons, it is usually possible to separate (10.6) into two equations, one for each particle, with the total energy being the sum of the single-particle energies. This separation is carried out in exactly the same manner as that in which separation of coordinates is done in a many-dimensional partial differential equation. As a result, it is possible to write the two-particle wave function as a product of the one-particle wave functions $\psi_1\left(x_1\right)$ and

$\psi_2(x_2)$, where we consider the situation in which the two functions are different. The subscripts on the wave functions themselves (as opposed to those on the variables) refer to the particle 'number'. Carrying this out produces

$$\psi(x_1, x_2) \sim \psi_1(x_1)\psi_2(x_2) \tag{10.7}$$

However, this wave function does not possess the proper symmetry for electrons; for example, interchanging the positions of particles '1' and '2' does not produce the necessary antisymmetry. However, we can achieve this with a somewhat cleverer summation; that is, we use

$$\psi(x_1, x_2) = \frac{1}{\sqrt{2}}\left[\psi_1(x_1)\psi_2(x_2) - \psi_2(x_1)\psi_1(x_2)\right] \tag{10.8}$$

This wave function has the desired antisymmetry under the interchange of the two electrons.

This method of forming a properly antisymmetric many-electron wave function from the single-electron wave functions has been extended to an arbitrarily large number of electrons by Slater (1929). The resulting wave function for N electrons is given by the *Slater determinant*

$$\psi(x_1, x_2, \ldots, x_N) = \frac{1}{\sqrt{N}}\begin{vmatrix} \psi_1(x_1) & \psi_1(x_2) & \ldots & \psi_1(x_N) \\ \psi_2(x_1) & \psi_2(x_2) & \ldots & \psi_2(x_N) \\ \ldots & \ldots & \ldots & \ldots \\ \psi_N(x_1) & \psi_N(x_2) & \ldots & \psi_N(x_N) \end{vmatrix} \tag{10.9}$$

Equation (10.9) has the added usefulness that it has the property that it vanishes if two of the ψ_i are the same. This implies that no two electrons can have the same state (recall that we have included spin angular momentum explicitly in the variables), and thus the Pauli exclusion principle is automatically satisfied. Thus, when we can separate the Hamiltonian into single-particle parts, and separate the wave function accordingly, the antisymmetrized product of these wave functions in a Slater determinant ensures that the Pauli exclusion principle is satisfied.

10.1 SPIN ANGULAR MOMENTUM

As we have pointed out previously, electrons can have two spin orientations. Typically, these are called spin-up and spin-down. But this is for an arbitrary orientation of the electron's own magnetic moment. Recognition of the two spin states arises from the introduction of the Pauli exclusion principle whenever we consider a fully quantized state. These two spin states must have opposite polarization, hence the designations as up and down. In fact, the spin can be oriented into any desired direction by the application of external forces, such as a magnetic field. The most common orientation, arising in the Zeeman (1897) effect, is along the z-axis which is achieved via an external magnetic field oriented in this direction. But, other directions are useful in a variety of applications such as the use of spin orientation in a qubit (quantum bit) for quantum computation, which we discuss in Chapter 11. Here, we want to discuss the values and orientations of the spin and its representation via the Pauli spin matrices.

General angular momentum for electrons around an atom was introduced in the previous chapter. For sure, the angular momentum of an electron orbiting in a centrally symmetric potential, such as the Coulomb potential around an atom, possesses a quantized value for the angular momentum. In this atomic case, both the total angular momentum L^2 and the z-directed angular momentum L_z can be made to commute. The fact that they could both be made to commute with the Hamiltonian tells us that they can both be diagonalized along with the Hamiltonian and therefore could both be measured at the same time as the total energy. In the present case, the only angular momentum to be considered is the spin angular momentum, which we may take to be oriented in the z-direction. We expect that, just as in the atomic case, the total spin S^2 will continue to commute with the z-component of spin S_z, just as in the atomic case.

Let us begin by recalling that, in any quantum confinement problem such as the atom or a quantum well, the total wave function is a sum over a set of eigenfunctions $\varphi_i(x)$. For each of these functions, which are defined by a set of quantized values due to the confinement, one has not considered the spin. If we now want to also include the spin, then we need an additional part of each eigenstate wave function. Typically, this is a multiplicative term describing the spin state. Traditionally, this is a two-component wave function called a *spinor*. From the Zeeman effect, we know that one typically denotes the extra energy for the spin-up state by the value 1/2, and the value of the spin-down state by the value −1/2. We use this to denote the two possible states and their spinors as

$$\varphi\left(\frac{1}{2}\right)=\begin{bmatrix}1\\0\end{bmatrix},\quad \varphi\left(-\frac{1}{2}\right)=\begin{bmatrix}0\\1\end{bmatrix} \tag{10.10}$$

where the first row refers to the *up* state and the second row refers to the *down* state. Thus, the eigenvalues correspond to those adopted in the Zeeman effect, as mentioned above.

Because the spin angular momentum has been taken to be oriented along the z-axis, we expect the spin matrix for the z-component of angular momentum must be diagonal for two reasons. First, it must commute with the total spin and with the Hamiltonian and, second, it must produce the eigenvalues found from the Zeeman effect. Thus, we simply state that

$$S_z=\frac{\hbar}{2}\begin{bmatrix}1 & 0\\0 & -1\end{bmatrix}, \tag{10.11}$$

and this gives

$$\begin{aligned} S_z\varphi\left(\frac{1}{2}\right) &= \frac{\hbar}{2}\begin{bmatrix}1 & 0\\ 0 & -1\end{bmatrix}\begin{bmatrix}1\\0\end{bmatrix} = \frac{\hbar}{2}\varphi\left(\frac{1}{2}\right) \\ S_z\varphi\left(-\frac{1}{2}\right) &= \frac{\hbar}{2}\begin{bmatrix}1 & 0\\ 0 & -1\end{bmatrix}\begin{bmatrix}0\\1\end{bmatrix} = -\frac{\hbar}{2}\varphi\left(-\frac{1}{2}\right) \end{aligned}. \tag{10.12}$$

We know from the study of normal angular momentum that the value of L^2 is given as $\ell(\ell+1)\hbar^2$. Thus, we expect a similar result for the spin angular momentum, and using $s = 1/2$, we get $S^2 = \ = 3\hbar^2/4$. More strictly, this value is for the square of the magnitude of the total spin angular momentum. Again, the matrix representation of this total spin angular momentum must be diagonal, and this is given by

$$|S|^2 = \frac{3\hbar^2}{4}\begin{bmatrix}1 & 0\\ 0 & 1\end{bmatrix}. \tag{10.13}$$

To find the other components of the spin angular momentum, and the other spin matrices, we introduce the rotating coordinates as we did in the previous chapter, and

$$S_+ = S_x + iS_y, \quad S_- = S_x - iS_y. \tag{10.14}$$

Then, we can write the squared magnitude of the total spin angular momentum as

$$S^2 = S_x^2 + S_y^2 + S_z^2 = S_+S_- + S_z^2 - i\left[S_x, S_y\right]. \tag{10.15}$$

Here, we have introduced the various commutator relations for the individual spin components, which are, in general

$$\left[S_x, S_y\right] = i\hbar S_z, \quad \left[S_y, S_z\right] = i\hbar S_x, \quad \left[S_z, S_x\right] = i\hbar S_y. \tag{10.16}$$

Using these results, we may rewrite (10.15) as

$$S^2 = S_+S_- + S_z^2 + \hbar S_z. \tag{10.17}$$

Similarly, if we reverse the two rotating terms, we get

$$S^2 = S_-S_+ + S_z^2 - \hbar S_z. \tag{10.18}$$

We can combine these last two equations, and use the results of Equations (10.11) and (10.13) to yield

$$\frac{1}{2}\left(S_+S_- + S_-S_+\right) = \frac{\hbar^2}{2}. \tag{10.19}$$

The operators S_+ and S_- act as creation and annihilation operators for the spin angular momentum. That is, operating on a spinor with the first of these operators will raise the angular momentum, which can only occur if it acts on the spin-down state and produces the spin-up state, or

$$S_+\varphi\left(-\frac{1}{2}\right) = \hbar\varphi\left(\frac{1}{2}\right) \rightarrow S_+ = \hbar\begin{bmatrix} 0 & 1 \\ 0 & 0 \end{bmatrix}. \tag{10.20}$$

Similarly, acting with the operator S_- removes a quantum of angular momentum and lowers the spin angular. This can occur only if the operator acts upon the spin-up state and produces the spin-down state, or

$$S_-\varphi\left(\frac{1}{2}\right) = \hbar\varphi\left(-\frac{1}{2}\right) \rightarrow S_- = \hbar\begin{bmatrix} 0 & 0 \\ 1 & 0 \end{bmatrix}. \tag{10.21}$$

We can now invert Equation (10.14) to give

$$\begin{aligned} S_x &= \frac{1}{2}\left(S_+ + S_-\right) = \frac{\hbar}{2}\begin{bmatrix} 0 & 1 \\ 1 & 0 \end{bmatrix} \\ S_y &= \frac{1}{2i}\left(S_+ - S_-\right) = \frac{\hbar}{2}\begin{bmatrix} 0 & -i \\ i & 0 \end{bmatrix} \end{aligned}. \tag{10.22}$$

The Pauli spin matrices are just the square matrices that appear in the definitions of the components of the spin angular momentum above. These are sometimes denoted as

$$\begin{aligned} \sigma_x &= \begin{bmatrix} 0 & 1 \\ 1 & 0 \end{bmatrix}, \\ \sigma_y &= \begin{bmatrix} 0 & -i \\ i & 0 \end{bmatrix}, \\ \sigma_z &= \begin{bmatrix} 1 & 0 \\ 0 & -1 \end{bmatrix}. \end{aligned} \tag{10.23}$$

Such spinors are used in many cases beyond spin to represent the behavior of two-level systems that have analogous behavior to spin systems. We deal with the two-level system in the next section.

10.2 TWO-LEVEL SYSTEMS

As we discussed in the previous section, the spin of the electron is characterized by its eigenstate, for which the spin wave function is a two-component spinor. It makes natural sense then to spend a little more time with the idea of two-level systems (Merzbacher, 1970). To simplify the notation slightly, let us use the generalized spinors (10.10)

$$\alpha = \varphi(\uparrow) = \begin{bmatrix} 1 \\ 0 \end{bmatrix}, \quad \beta = \varphi(\downarrow) = \begin{bmatrix} 0 \\ 1 \end{bmatrix}. \tag{10.24}$$

The equation of motion for these spinors is, as usual, given by the Schrödinger equation in which the Hamiltonian is a 2 × 2 matrix. If there are no external forces acting upon our simple two-level system, then the Hamiltonian is independent of time and the wave functions evolve as

$$\varphi(t) = e^{-iHt/\hbar}\varphi(0), \tag{10.25}$$

where φ corresponds to either of the spinors α or β. In general, we diagonalize the Hamiltonian to provide the eigenvalues, which we take to be E_α and E_β. The total wave function can be written as a sum of the two spinors as

$$\begin{gathered} \psi = c_1\alpha + c_2\beta, \\ |c_1|^2 + |c_2|^2 = 1. \end{gathered} \tag{10.26}$$

We can use the above properties of the Hamiltonian and the spinors to determine a number of intuitive properties. For example, if we want to determine how β evolves into α, we need only examine the total Hamiltonian, as

$$\begin{aligned} \beta^\dagger e^{-\frac{iHt}{\hbar}}\alpha &= \begin{bmatrix} 0 & 1 \end{bmatrix} e^{-\frac{iHt}{\hbar}} \begin{bmatrix} 1 \\ 0 \end{bmatrix} \\ &= \frac{H_{21}}{E_\beta - E_\alpha}\left(e^{-iE_\alpha t/\hbar} - e^{-iE_\beta t/\hbar}\right), \end{aligned} \tag{10.27}$$

where

$$H_{21} = \langle\beta|H|\alpha\rangle = \begin{bmatrix} 0 & 1 \end{bmatrix} H \begin{bmatrix} 1 \\ 0 \end{bmatrix}. \tag{10.28}$$

If we examine the magnitude squared value of this term, we discover that it oscillates, and that the occupation oscillates from one state to the other with a frequency given by the difference in the eigen-energies of these two states.

The oscillation that occurs above is suggestive, as it points out that we can define the total state Equation (10.26) itself as a single spinor whose components are the complex coefficients given in this equation. What this means is that the two spinors α and β define a two-dimensional space for our wave function. Where the quantum well was an infinite-dimensional space of eigenfunctions, the problem here is limited to just these two states. Thus, a spinor whose coefficients are the coefficients in Equation (10.26) is a vector in this two-component space, as

$$\psi = \begin{bmatrix} c_1 \\ c_2 \end{bmatrix} = \begin{bmatrix} |c_1| e^{i\vartheta_1} \\ |c_2| e^{i\vartheta_2} \end{bmatrix}. \tag{10.29}$$

But this description is not unique since the squares of the magnitude must sum to unity. Thus, the two phases can be rather arbitrary, and we can only specify in detail the relative phase $\vartheta_1 - \vartheta_2$. So, if we write $\vartheta_2 = \vartheta_1 + \phi$, we cannot then tell the difference between the states ψ and $e^{i\vartheta_1}\psi$, which means we have to somehow believe that these are the same state. In a sense, this all arises from the periodicities of the angles and that is a property we will use below.

A useful quantity to use in dealing with two-level systems is the corresponding density matrix, defined via

$$\rho = \psi\psi^\dagger = \begin{bmatrix} c_1 \\ c_2 \end{bmatrix} \begin{bmatrix} c_1^* & c_2^* \end{bmatrix} = \begin{bmatrix} |c_1|^2 & c_1 c_2^* \\ c_2 c_1^* & |c_2|^2 \end{bmatrix}. \tag{10.30}$$

An important property of this matrix is that its trace is unity, as required by Equation (10.26). Now, we come to an important point. A fundamental property of any 2×2 matrix, whose trace is unity, is that it can be written in terms of the Pauli spin matrices and a quantity known as the *polarization* as

$$\rho = \frac{1}{2}(I + \mathbf{P} \,.\, \boldsymbol{\sigma}), \tag{10.31}$$

where the individual Pauli spin matrices are given in (10.23). Comparing this last result with Equation (10.30), we can identify the various component connections through

$$\begin{aligned} |c_1|^2 &= \frac{1}{2}(1 + P_z) \\ |c_2|^2 &= \frac{1}{2}(1 - P_z). \\ c_1 c_2^* &= (c_2 c_1^*)^* = \frac{1}{2}(P_x - iP_y). \end{aligned} \tag{10.32}$$

We can now invert these equations to define the components of the polarization as

$$P_x = 2Re\{c_1c_2^*\}$$
$$P_y = -2Im\{c_1c_2^*\}. \quad (10.33)$$
$$P_z = |c_1|^2 - |c_2|^2$$

In addition to the trace of the density matrix being unity, it is also required to be the sum of the eigenvalues of the two wave function components. Since these are 0 and 1, we also satisfy this requirement. This now implies that one of the eigenvectors of this density matrix is ψ itself. The other eigenvector must be orthogonal to ψ so that their inner product yields 0, but this other eigenvector is still arbitrary at this point.

An important property arises from the expectation values of the various spinors. We examine this with the x-spinor as

$$\langle\sigma_x\rangle = Tr\{\rho\sigma_x\} = \frac{1}{2}\left(Tr\{I\sigma_x\} + \mathbf{P}\cdot Tr\{\boldsymbol{\sigma}\sigma_x\}\right)$$
$$= \frac{1}{2}P_x Tr\{\sigma_x^2\} = P_x. \quad (10.34)$$

We can repeat for each of the other components, which yields the important result that

$$\mathbf{P} = \boldsymbol{\sigma} = Tr\{\rho\boldsymbol{\sigma}\}. \quad (10.35)$$

That is, the polarization of our two-level system is defined by the expectation value of the spin vector. The direction of the spin is uniquely connected to the polarization of the system. Another important point is that since the wave function ψ is an eigenfunction of the density matrix, it is also an eigenfunction of the last term in the density matrix (10.31):

$$\mathbf{P}\cdot\boldsymbol{\sigma}\psi = \psi. \quad (10.36)$$

Note that the eigenvalue is unity in keeping with (10.26).

To examine the nature of the polarization itself, let us make some angular definitions of the various components of the wave function Equation (10.29), as

$$c_1 = e^{i\gamma}cos(\vartheta/2),$$
$$c_2 = e^{i(\gamma+\phi)}sin(\vartheta/2), \quad (10.37)$$

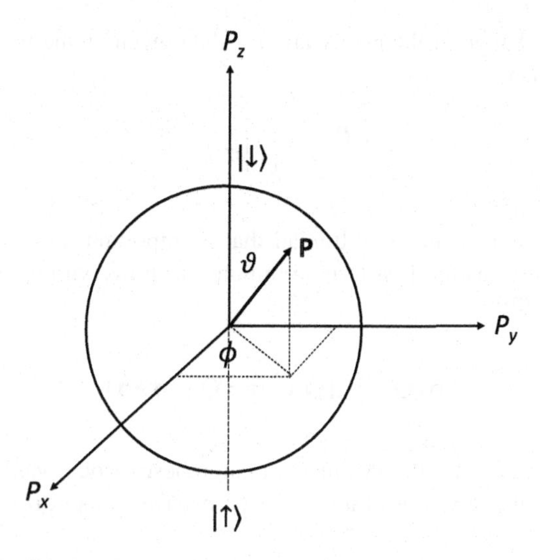

FIGURE 10.1 The Bloch sphere. All spin rotations are constrained to lie on the surface of this sphere, since the wave function is normalized to unity.

so that

$$\begin{aligned} P_x &= cos(\phi)sin(\vartheta) \\ P_y &= sin(\phi)sin(\vartheta) \\ P_z &= cos(\vartheta). \end{aligned} \tag{10.38}$$

We recognize these angles as the angles in a spherical coordinate system which relate to the normal rectangular coordinates. The angle ϑ is the polar angle and the angle ϕ is the azimuthal angle. Hence, moving around within the state ψ means that we move around on the surface of a sphere by varying the two angles. This sphere is called the *Bloch sphere*, and is pictured in Figure 10.1. From the definition of our spinors and the above coefficients, we recognize that our states may be defined as

$$\begin{aligned} \beta &= \left|\downarrow\right\rangle, \quad \vartheta = 0, \quad \mathbf{P} = \mathbf{a}_z \\ \alpha &= \left|\uparrow\right\rangle, \quad \vartheta = \pi, \quad \mathbf{P} = -\mathbf{a}_z \end{aligned} \tag{10.39}$$

One of the properties of the spin that we are familiar with is the fact that it precesses under a variety of different forces. We can investigate that by introducing a simple form for the Hamiltonian. This form is suggested by the density matrix itself, and we will denote it as

$$H = \frac{1}{2}\left(Q_0 I + \mathbf{Q}\cdot\boldsymbol{\sigma}\right). \tag{10.40}$$

The time rate of change of the polarization is then given by the well-known relation in quantum mechanics

$$\frac{d\boldsymbol{P}}{dt} = -\frac{i}{\hbar}[\boldsymbol{\sigma}, H] \tag{10.41}$$

The only component of the Hamiltonian that is important now is the second term in Equation (10.40), as the first term commutes with everything. Then, we need to evaluate the quantity

$$\boldsymbol{\sigma}(\mathbf{Q}\cdot\boldsymbol{\sigma}) - (\mathbf{Q}\cdot\boldsymbol{\sigma})\boldsymbol{\sigma} = \mathbf{Q}\times(\boldsymbol{\sigma}\times\boldsymbol{\sigma}). \tag{10.42}$$

If $\boldsymbol{\sigma}$ were a simple vector, the last term in parentheses would vanish, but this is not a simple vector—it is a vector of tensors, so it behaves rather differently. In fact, we find that

$$(\boldsymbol{\sigma}\times\boldsymbol{\sigma}) = \begin{vmatrix} \mathbf{a}_x & \mathbf{a}_y & \mathbf{a}_z \\ \sigma_x & \sigma_y & \sigma_z \\ \sigma_x & \sigma_y & \sigma_z \end{vmatrix} = 2i\boldsymbol{\sigma}, \tag{10.43}$$

where we have used the commutator relations (10.16) for the Pauli spinors. This now leads us to the result that

$$\frac{d\mathbf{P}}{dt} = \frac{1}{\hbar}\mathbf{Q}\times\mathbf{P}. \tag{10.44}$$

So, the precession of the spin polarization arises from any vector term in the Hamiltonian that is not parallel to the polarization itself. It is easy to show that this motion does not change the amplitude of the polarization, which is of course required for the spin itself. If the vector **Q** is a constant amplitude vector, then it defines a precession energy corresponding to a precession frequency $\omega_Q = Q/\hbar$.

10.3 SYSTEMS OF IDENTICAL PARTICLES

In the beginning of this chapter, the multiparticle wave function was introduced. In particular, this wave function was connected with products of single-particle wave functions through the Slater determinant to ensure the antisymmetry of the wave function. Here, we want to expand this topic, and to explore some of the properties of the many-electron wave functions in further detail.

Quite generally, we may introduce the *N*-particle wave function $\psi_N(x_1, x_2, \ldots, x_N)$, which represents the probability (when the magnitude squared is computed) of finding particles at positions and spins $x_1, x_2, \ldots, x_N$. This wave function must satisfy the condition

$$\langle \psi_N | \psi_N \rangle = \int dx_1 dx_2 \ldots dx_N \left| \psi_N (x_1, x_2, \ldots, x_N) \right|^2 = 1. \tag{10.45}$$

As we have defined it to this point, the Hilbert space for the N-particle system is simply the Nth tensor product of the single-particle Hilbert spaces and the corresponding spin spaces. The wave function of the N fermions is properly antisymmetric under the exchange of any two particles, and therefore for a large number of permutations must satisfy

$$\psi_N (x_1', x_2', \ldots, x_N') = (-1)^P \psi_N (x_1, x_2, \ldots, x_N). \tag{10.46}$$

Here, P is the total number of permutations required to reach the configuration of the wave function on the left from that on the right of (10.46). Thus, if we have for example, $\psi_4 (x_1, x_2, x_3, x_4)$, then to get to $\psi_4 (x_4, x_3, x_2, x_1)$ requires a total of six permutations (three to get x_4 moved to the beginning, then two to bring x_3 adjacent to it, and finally one more to interchange x_2 and x_4). Hence, this new example yields a symmetric product after the six interchanges. The general permuted wave function (10.46) then has the inner product with the original of

$$\langle \psi_N (x_1', x_2', \ldots, x_N') | \psi_N (x_1, x_2, \ldots, x_N) \rangle = (-1)^P. \tag{10.47}$$

It is important to note that the normalization of the many-electron wave function arises from the normalization of the single-electron wave functions that go into the Slater determinant (10.9). The factor of $(-1)^P$ arises from the need to get these into the 'right' product order for taking the associated inner products of the single-electron wave functions. The normalization factor in (10.9) then goes to cancel the multiplicity of terms that arise from the determinantal form. Let us consider just the two-electron case of (10.8) as an example (remember that the adjoint operator on the left of the inner product reverses the order of terms):

$$\begin{aligned}
\langle \psi_2 | \psi_2 \rangle &= \frac{1}{2} \langle [\psi_1(x_1)\psi_2(x_2) - \psi_2(x_1)\psi_1(x_2)] | [\psi_1(x_1)\psi_2(x_2) - \psi_2(x_1)\psi_1(x_2)] \rangle \\
&= \frac{1}{2} \Big[\langle \psi_1(x_1)\psi_2(x_2) | \psi_1(x_1)\psi_2(x_2) \rangle + \langle \psi_2(x_1)\psi_1(x_2) | \psi_2(x_1)\psi_1(x_2) \rangle \\
&\quad - \langle \psi_1(x_1)\psi_2(x_2) | \psi_2(x_1)\psi_1(x_2) - \psi_2(x_1)\psi_1(x_2) | \psi_1(x_1)\psi_2(x_2) \rangle \Big] \\
&= \frac{1}{2} \left[(-1)^0 + (-1)^0 - (-1)^1 0 - (-1)^1 0 \right] = 1
\end{aligned} \tag{10.48}$$

The zeros arise because the wave functions at the same variables are different and orthogonal to one another. Thus, one needs to be careful that the one-electron wave functions are orthonormal and that the permutations are properly computed.

Finally, we note that it is usually the case that the coordinates are the same for all of the single-electron wave functions; only the spin coordinates differ. Thus, we can think of the differences between the single-particle states as the particular member of the Hilbert space and the particular spin variable that is excited for each electron. No two combinations can be occupied by more than one electron. With this realization, it is possible to think of a simple set of parameters to characterize each single-electron wave function—its index in the Hilbert space and its spin angular momentum. Thus, we may write the single-particle wave functions as $\psi_{i\sigma}(x)$. Here, the index i signifies the particular member of the basis set, and σ denotes the spin state. These may be combined for simplicity into the index $\lambda = (i, \sigma)$. We will do this in the following discussion. However, we are making a major change of paradigm with this notational change. In the case treated above, it was assumed that the coordinates (including spin) of a particle defined a particular localized wave function in the position representation, and the index of the wave function described the type of wave function at that position. Thus, at the point $\xi = (x,\sigma)$, there may be many types of wave function available, and the $\psi_{i\sigma}$ correspond to these types. In the latter description, however, it is assumed that there is only a single type of wave function at each site, or that there is no localization and that there are a number of wave function types. An example of the first of these options is a Gaussian wave packet localized at x. This is characteristic of, for example, localized delta function wave packets. The latter formulation is characteristic of, for example, the wave functions of an electron in a quantum well, where the 'space' is the region that lies within the well, and the various wave functions are those that arise from the different energy levels. A second example of the latter, and one that will be heavily used, is that of momentum eigenfunctions, which are plane waves extending over all space. The indices are then the momentum values and the spin indices. It is important in evaluating the antisymmetrized wave functions to understand fully just which interpretation is being placed on the individual single-electron wave functions.

In the treatment of the harmonic oscillator, it was found to be useful to introduce a set of noncommuting operators which described the creation or annihilation of one unit of energy in the harmonic oscillator. This also changed the wave functions accordingly. Can we use a similar description to enhance our understanding of the many-electron picture? The answer is obviously yes, but we must carefully examine how the rules will be changed by the requirement of antisymmetry and the corresponding Pauli exclusion principle. New views that arise will be based upon the fact that the energy of the harmonic oscillator could be continuously raised by pumping bosons into the system; the introduction of n bosons could be achieved by using the operator $\left(a^\dagger\right)^n$. Here, however, operation with a single creation operator (we use the notation c for fermions) $c_\lambda^\dagger$ creates one fermion in state λ. This state may be a momentum eigenstate, a state in a quantum well, or any other state in a system in which the wave functions span the entire allowed variable space. If we try to create a second fermion in this state, the result must be forced to vanish because of the exclusion principle; for example

$$\left(c_\lambda^\dagger\right)^2|0\rangle = 0. \tag{10.49}$$

Here, $|0\rangle$ is the so-called vacuum state in which no fermions exist. In fact, however, Equation (10.49) must hold for any wave function in which the state λ is empty (or even filled). Similarly

$$\left(c_\lambda\right)^2|\ldots\rangle = 0. \tag{10.50}$$

Here, any state of the system will yield the zero value. If the state is full, with an electron in it, it cannot be emptied twice (remember, these are spin states, so only a single electron is allowed in the state). If it is empty, it cannot be emptied again.

This suggests that a different combination of products must be used for these fermion operators. Consider, for example, the product

$$c_\lambda^\dagger c_\lambda + c_\lambda c_\lambda^\dagger. \tag{10.51}$$

If the state $|\lambda\rangle$ is empty, the first term immediately gives zero. The second term creates a fermion in the state, then destroys it, so the result is $(0 + 1) = 1$. Similarly, if the state $|\lambda\rangle$ is occupied, the second term gives zero (another fermion cannot be created), while the first term is the number operator and yields 1. Thus, the result in either case is the *anticommutator* product

$$\left\{c_\lambda^\dagger, c_\lambda\right\} = c_\lambda^\dagger c_\lambda + c_\lambda c_\lambda^\dagger = 1. \tag{10.52}$$

Here, we use the curly brackets to indicate that the positive sign is used in the anticommutator, as opposed to the negative sign used in the commutator relation. This may be extended to operators on other states as

$$\left\{c_\lambda^\dagger, c_\mu\right\} = \delta_{\lambda\mu}, \quad \left\{c_\lambda^\dagger, c_\mu^\dagger\right\} = \left\{c_\lambda, c_\mu\right\} = 0. \tag{10.53}$$

We also note that the number operator is idempotent; that is,

$$n_\lambda = n_\lambda^2 = \left(c_\lambda^\dagger c_\lambda\right)^2 = c_\lambda^\dagger c_\lambda\left(1 - c_\lambda c_\lambda^\dagger\right) = c_\lambda^\dagger c_\lambda.$$

Note that the operators $c_\lambda^\dagger$ and $c\lambda$ used here are fermion operators. The relationship (10.49) ensures that no more than a single fermion can exist in the given state. The attempt to put a second particle in this state must yield zero, as given by (10.49). Similarly, only a single fermion can be removed from an occupied state, as indicated by (10.50). These statistics differ markedly from the operators used in previous chapters, which were for bosons. While the wave function differed for each state, we referred to a particular harmonic oscillator mode as being described by its occupation. In increasing the occupation, we raised the energy of the mode (given by the number of bosons in that particular harmonic oscillator mode). Thus, there could be a great many bosons occupying the mode, even though the distinct mode wave function would change. In the case of fermions, however, we limit the range of the occupation factor n_i to be only 0 or 1. This is imposed by using the same generation properties of the creation and annihilation operators, but with the limitations for fermions that are given by (10.49) and (10.50).

The general many-electron wave function is created by operating on the empty state, or vacuum state, with the operators for positioning the electrons where desired. For example, a three-electron state may be created as

$$|\mu\nu\lambda\rangle = c_\lambda^\dagger c_\nu^\dagger c_\mu^\dagger |0\rangle. \tag{10.54}$$

It should be noted that the order of creation of the particles is important, since changing the order results in a permutation of the indices, with its accompanying minus sign(s). Notice that the right-most creation operator creates the first electron state, while the second operator adds the second electron, and so on. It is possible to recognize that the creation and annihilation operators do not operate in a simple Hilbert space. In general, the creation operator operates in a space of n electrons and moves to a space with $n + 1$ electrons. Similarly, the annihilation operator moves to a space with $n - 1$ electrons. The general space may then be a product space of Hilbert spaces, and elements may be combined with a variety of partially occupied wave functions. This complicated structure is called a *Fock space*, but the details of this structure are beyond the simple treatment that we desire here.

10.4 SPIN EFFECTS IN SEMICONDUCTORS

The spin of an electron can be manipulated in a large number of ways, but in order to take advantage of current semiconductor processing technology, it would be preferable to find a purely electrical means of achieving this. For this reason, a great deal of attention has centered on the spin Hall effect in semiconductors. The idea was apparently first suggested by Dyakonov and Perel (1971) and later, and independently by Hirsch (1999). The basic idea was that, in the presence of scattering centers, it was possible for spins of one orientation to be scattered in a different direction than spins of the opposite orientation. This would lead different spins to accumulate on opposite sides of the sample, a result of the presence of anisotropic scattering in the presence of the spin-orbit interaction. Thus, a transverse spin current arises in response to a longitudinal charge current, without the need for magnetic materials or externally applied magnetic fields. The spin-orbit interaction is a common part of the energy bands of a semiconductor, where it splits the otherwise triply degenerate top of the valence band at the Γ point. But there are other forms of the spin-orbit interaction that are of interest in situations in which symmetries are broken in the semiconductor device. In the spin Hall effect, we achieve edge states, as in the quantum Hall effect (Section 5.4.3), but here these edge states are spin polarized. The impurity driven spin separation is known as the extrinsic spin Hall effect, and there can be an intrinsic spin Hall effect directly from the spin-orbit interaction when asymmetries exist in the device.

10.4.1 The Spin-Orbit Interaction

The quantum structure of atoms can lead to the angular momentum of the electrons mixing with the spin angular momentum of these particles. Since the energy bands are composed of both the *s*- and *p*-orbitals of the individual atoms in many of the semiconductors (Section 9.3), it has been found that the spin-orbit interaction affects band structure calculations. The spin-orbit interaction is a relativistic effect in which the angular motion of the electron interacts with the gradient of the confining potential of the atom to produce an effective magnetic field. This field couples to the spin in a manner similar to the Zeeman effect. Early papers, which used first-principles calculations of the band structure, clearly demonstrated that the spin-orbit interaction was important for the detailed properties of the bands. Not the least of these effects is the splitting of the threefold bands at the top of the valence band, producing the so-called split-off valence band. This latter band lies from a few meV to a significant fraction of an eV below the top of the valence band in various semiconductors.

We can illustrate how the spin-orbit interaction modifies the transport by considering a simple Hamiltonian for the electrons in a periodic potential $V(\mathbf{r})$ as

$$H(\mathbf{p},\mathbf{r}) = \frac{p^2}{2m} + V(\mathbf{r}) + \frac{\hbar}{4m^2c^2}(\sigma \times \nabla V)\cdot\mathbf{p}, \tag{10.55}$$

where $\mathbf{p}$ is the normal momentum operator and the last term represents the spin-orbit interaction. The quantity σ is a vector whose components are the normal Pauli spin matrices (10.23). It is clear from these matrices that the wave function is now more complicated with a spin component, as indicated in (10.10). The direct effect of the spin-orbit interaction may be qualitatively understood as an extra energy cost for the alignment of the intrinsic magnetic moment of the electron with the magnetic field that arises from its own orbital motion. As a result, this term leads to a modification of the momentum of the carriers, through a gauge transformation of the wave function much like the Peierls' modification from the presence of the magnetic field.

10.4.2 Bulk Inversion Asymmetry

Bulk inversion asymmetry arises in crystals which lack an inversion symmetry, such as the zinc-blende materials. In these crystals, the basis pair at each lattice site is composed of two dissimilar atoms, such as In and P, or Ga and Sb. Because of this, the crystal has lower symmetry than, for example, the diamond lattice, where the basis pair is two Si atoms. Without this inversion symmetry, one still can have symmetry of the energy bands $E(\mathbf{k}) = E(-\mathbf{k})$, but the periodic part of the Bloch functions no longer satisfies $u_k(\mathbf{r}) = u_k(-\mathbf{r})$. As a result of this, the normal twofold spin degeneracy is no longer required throughout the Brillouin zone (Dresselhaus, 1955). The importance of this interaction is already recognized by people who study the electronic band structure. The inclusion of this term via the spin-orbit interaction leads to the warped surface of the valence bands (Ferry, 2020). For the conduction band, the perturbing Hamiltonian can be written as

$$H_{BIA} = \eta\left[\left\{k_x, k_y^2 - k_z^2\right\}\sigma_x + \left\{k_y, k_z^2 - k_x^2\right\}\sigma_y + \left\{k_z, k_x^2 - k_y^2\right\}\sigma_z\right], \qquad (10.56)$$

where k_x, k_y, and k_z are aligned along the [100], [010], and [001] axes, respectively. The terms in curly brackets are anticommutation relations given by (10.52). The parameter η comes from the spin-orbit interaction and is given by Sakarai (1967)

$$\eta = \frac{4i}{3} PP'Q\left[\frac{1}{\left(E_G + \Delta\right)\left(\Gamma_0 - \Delta_c\right)} - \frac{1}{E_G \Gamma_0}\right], \qquad (10.57)$$

where E_G and Δ are the primary energy gap and the spin-orbit splitting of the valence band in a material in which the minimum of the conduction band and the maximum of the valence band both occur at the Γ point in the Brillouin zone. The quantities P, P' and Q are matrix elements in the spin-orbit interaction along the line of (Ferry, 2020)

$$P = \frac{i\hbar}{m_0}\left\langle S\left|p_z\right|Z\right\rangle, \qquad (10.58)$$

where S and Z are s-symmetry and p-symmetry wave functions (see Section 9.3) and p_z is the momentum operator. Finally, in Equation (10.57), Γ_0 is the splitting of the two lowest conduction bands at the zone center and Δ_c is the spin-orbit splitting of the lowest conduction band at the zone center. This interaction is stronger in materials with small band gaps, as may be inferred from Equation (10.57). Note that Equation (10.56) is cubic in the magnitude of the wave vector and is often referred to as the k^3 term.

While the above expressions apply to bulk semiconductors, much of the interest in recent years has been directed at quasi-two-dimensional systems in which the carriers are confined in a quantum well such as exists at the interface between AlGaAs and GaAs (see Section 2.4). Often, this structure is then patterned to create a quantum wire. For example, a common configuration is with growth along the [001] axis, so that there is no net momentum in the z-direction, and $\langle k_z\rangle = 0$, while $\langle k_z^2\rangle \neq 0$ is a representation of the quantization energy in the z-direction. Then, Equation (10.56) can be written as

$$H_{BIA} = \eta\left[\left\langle k_z^2\right\rangle\left(k_y\sigma_y - k_x\sigma_x\right) + k_x k_y\left(k_y\sigma_x - k_x\sigma_y\right)\right]. \qquad (10.59)$$

This is an important result. The prefactor of the first term in the square brackets is constant and depends upon the material and the details of the quantum well. This average over the z-momentum corresponds to the different subbands in the quantum well, so that only a single value will result when the carriers are only in the lowest subband. But, this structure has now split Equation (10.56) into a k-linear term and a k^3 term.

To explore Equation (10.59) a little closer, let us chose a set of spinors to represent the spin-up and -down states as follows:

$$|+\rangle = |\uparrow\rangle = \begin{bmatrix} 1 \\ 0 \end{bmatrix}, \quad |-\rangle = |\downarrow\rangle = \begin{bmatrix} 0 \\ 1 \end{bmatrix}, \tag{10.60}$$

as in (10.10). Then, the linear first term in (10.59) gives rise to an energy splitting according to

$$\Delta E_1 \sim -\langle \eta k_z^2 \rangle (k_x \pm i k_y) \tag{10.61}$$

in the rotating coordinates discussed in Section 10.1. Now, the spin 'up' state rotates around the z-axis in a right-hand sense (with the thumb in the z-direction) with the spin polarization tangential to the constant energy circle in two dimensions. On the other hand, the spin 'down' state rotates in the opposite direction, but with the spin polarization still tangential to the energy circle (we will illustrate this rotation below).

If we ignore the cubic terms in momentum (the last term in Equation (10.59)) for the moment and solve for the eigenvalues of the two spin states, we have

$$H = \begin{bmatrix} \dfrac{\hbar^2 k^2}{2m^*} & -\langle \eta k_z^2 \rangle (k_x + i k_y) \\ -\langle \eta k_z^2 \rangle (k_x - i k_y) & \dfrac{\hbar^2 k^2}{2m^*} \end{bmatrix}, \tag{10.62}$$

where we have assumed that the normal energy bands are parabolic for convenience. Then, we find that the energy levels of the two states are given as

$$E = \frac{\hbar^2 k^2}{2m^*} \pm \eta \langle k_z^2 \rangle k. \tag{10.63}$$

Thus, we find that not only is the energy splitting linear in k, but it is also isotropic in the two-dimensional momentum space. The resulting energy bands for the two states are composed of two interpenetrating paraboloids, and a constant energy surface is composed of two concentric circles (see Figure 10.2). The inner circle represents the

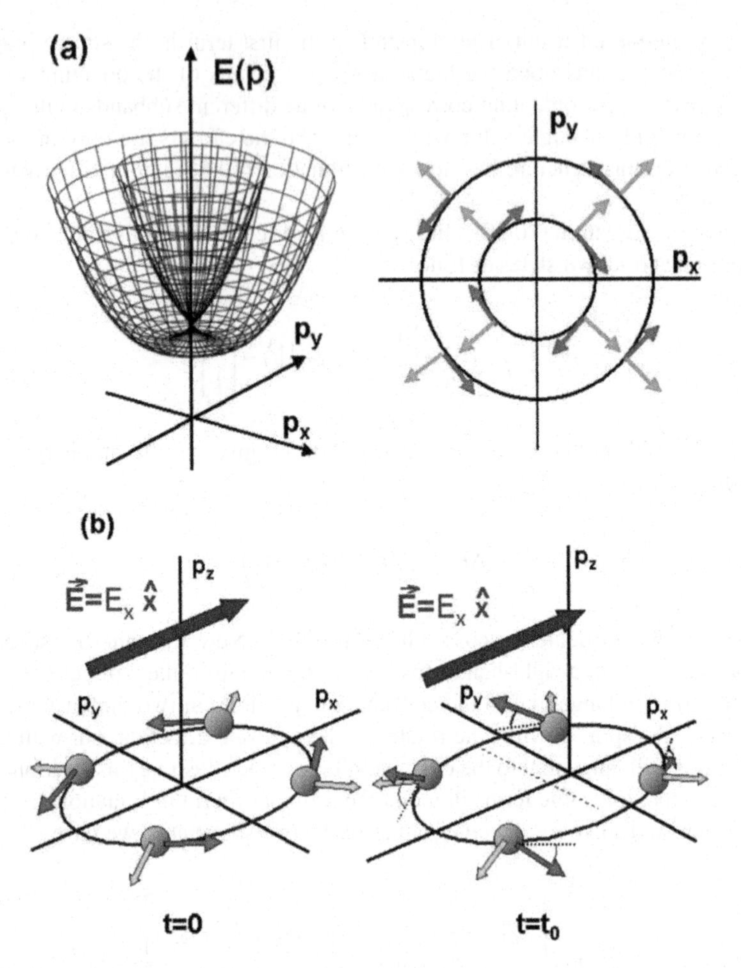

FIGURE 10.2 (a) Energy structure for the Rashba spin-orbit coupled system. The momentum is shown by light arrows and the spin polarization is shown by the dark arrows. (b) When an electric field is applied for some time, the spin is rotated out of the plane according to the direction of the transverse momentum. Reprinted with permission from Sinova et al. (2004). Copyright 2004, the American Physical Society.

positive sign in the above equation while the outer circle corresponds to the negative sign. The eigenfunctions are no long pure spin states, but are an admixture given by

$$\varphi_z^{(\pm)} = \frac{1}{\sqrt{2}} \begin{bmatrix} 1 \\ e^{\pm i\vartheta} \end{bmatrix}, \tag{10.64}$$

where ϑ is the angle that k makes with the [100] axis of the underlying crystal in the heterostructure quantum well. This angle is the angle defined by the polar coordinates in the two-dimensional momentum space. Hence, the root with the upper sign in Equation (10.63), which we think of as being mostly spin 'up', has the spin polarization tangential to the inner circle. Correspondingly, the root with the lower sign in Equation (10.63), which we think of as mostly spin 'down', has the spin polarization tangential to the outer circle.

If we now add in the cubic terms in Equation (10.59), the energy levels of Equation (10.63) are modified to be

$$= \frac{\hbar^2 k^2}{2m^*} \pm \eta\langle k_z^2 k\rangle \left[1 + \left(\frac{k^4}{\langle k_z^2\rangle^2} - 4\frac{k^2}{\langle k_z^2\rangle}\right) sin^2\vartheta cos^2\vartheta\right]^{1/2}. \tag{10.65}$$

This is a much more complicated momentum and angle dependence but suggests that the constant energy circles are now warped in the same manner as the valence band. The transport is no longer isotropic in the transport plane. Similarly, the phase on the down spin contribution to the eigenfunction Equation (10.64) is no longer simply the angle but acquires a 'wobble' as it rotates around the circle.

10.4.3 Structural Inversion Asymmetry

The spin Hall effect most commonly originates from the Rashba form of spin-orbit coupling (Bychov and Rashba, 1984), which is present in a two-dimensional electron gas formed in an asymmetric semiconductor quantum well. Such an asymmetric quantum well is the quasi-triangular well found at the interface between AlGaAs and GaAs or at the interface of a Si MOSFET shown in Section 2.4, although the spin-orbit interaction is small in Si (but still very important). This is known as structural inversion asymmetry. In either of the quantum wells mentioned above, the structure is asymmetric around the heterojunction (or oxide) interface. As a result, there is a relatively strong electric field in the quantum well, and motion normal to this can induce an effective magnetic field. This is the structural inversion asymmetry known as the Rashba effect. Just as in the bulk inversion asymmetry, the electric field in the quantum well can lead to spin splitting without any applied magnetic field, due to the spin-orbit interaction. If we take the z-axis as normal to the heterojunction interface, then the spin-orbit interaction Equation (10.55) can be written as

$$H_{SIA} = r\boldsymbol{\sigma}\cdot(\mathbf{k}\times\nabla V)\rightarrow\left[\boldsymbol{\sigma}\cdot(\boldsymbol{\sigma}\times\mathbf{k})\right]_z, \tag{10.66}$$

where

$$r = \frac{P^2}{3}\left[\frac{1}{E_G^2} - \frac{1}{(E_G+\Delta)^2}\right] + \frac{P'^2}{3}\left[\frac{1}{\Gamma_0^2} - \frac{1}{(\Gamma_0+\Delta_c)^2}\right] \tag{10.67}$$

arises from the spin-orbit interaction in the creation of the band structure, and the symbols have the same meanings as in Equation (10.57). The parameter is composed of the constants and the electric field, and for the configuration discussed, we have

$$H_{SIA} = \alpha_z\left(k_y\sigma_x - k_x\sigma_y\right), \tag{10.68}$$

where

$$\alpha_z = \frac{1}{4}\left(\frac{\hbar}{m_0 c}\right)^2 \frac{\partial V}{\partial z}. \tag{10.69}$$

If we continue to use the basis set defined in Equation (10.60), then the Rashba contribution to the energy is simply

$$E_R = \mp i\alpha_z\left(k_x \pm ik_y\right). \tag{10.70}$$

While the spin states are split in energy, this does not simply add to the bulk inversion asymmetry. First, the two spin states are orthogonal to each other, and then they are phase shifted (with opposite phase shift) relative to the previous results. It is easier to understand the effect of this Rashba term if we diagonalize the Hamiltonian for the two spin states with the Rashba contribution. We can write the Hamiltonian, for parabolic bands using Equation (10.70) as

$$H = \begin{bmatrix} \frac{\hbar^2 k^2}{2m^*} & \alpha_z\left(k_x + ik_y\right) \\ \alpha_z\left(k_x - ik_y\right) & \frac{\hbar^2 k^2}{2m^*} \end{bmatrix}. \tag{10.71}$$

When this Hamiltonian is compared to Equation (10.62), the two off-diagonal terms are shifted not only by the minus sign in front, but also by a phase factor of $-\pi/2$ within the term in parentheses in the upper right term and $\pi/2$ in the lower left term. Nevertheless, we can still diagonalize the Hamiltonian to give the new energies

$$E = \frac{\hbar^2 k^2}{2m^*} \pm \alpha_z k. \tag{10.72}$$

As in the case of the linear term in the bulk inversion asymmetry, the energy splitting is linear in k, and also isotropic with respect to the direction of $\mathbf{k}$. Thus, the energy bands also are composed of two interpenetrating paraboloids, and a constant energy surface is composed of two concentric circles (see Figure 10.2). The inner circle represents the positive sign in Equation (10.72), while the outer circle corresponds to the negative sign. The two eigenfunctions are given by

$$\varphi_z^{(\pm)} = \frac{1}{\sqrt{2}}\begin{bmatrix} 1 \\ e^{i(\vartheta \mp \pi/2)} \end{bmatrix}, \tag{10.73}$$

where ϑ is the angle that k makes with the [100] axis of the underlying crystal, within the heterostructure quantum well described in the previous section. The form of Equation (10.73) clearly shows the phase shift relative to the bulk inversion asymmetry wave function. The spin direction remains tangential to the two circles but pointed in the negative angular direction for the inner circle and in the positive angular direction for the outer circle. When both spin processes are present, the spin behavior becomes quite anisotropic in the transport plane. However, the Dresselhaus bulk inversion asymmetry is generally believed to be much weaker than the Rashba terms discussed here. The strength of the Rashba effect can be modified by an electrostatic gate applied to the heterostructure, as it modifies the potential gradient term in Equation (10.69).

REFERENCES

Bychov, Yu. A., and Rashba, E. I. 1984. Oscillatory effects and the magnetic susceptibility of carriers in inversion layers. *J. Phys. C: Sol. State Phys.* **17** 6039–6046

Dresselhaus, G. 1955. Spin-orbit coupling effects in zinc-blende semiconductors. *Phys. Rev.* **149** 580–586

Dyakonov, M. I. and Perel, V. I. 1971. Possibility of orienting electron spins with current. *JETP Lett.* **13** 467–469

Ferry, D. K. 2020. *Semiconductors: Bonds and Bands*. 2nd edn. IOPP: Bristol, UK. Sec. 2.5.2

Hirsch, J. E. 1999. Spin Hall effect. *Phys. Rev. Lett.* **83** 1834–1837

Merzbacher, E. 1970. *Quantum Mechanics*, 2nd ed. Wiley: New York, Chapter 13.

Pauli, W. 1925. Über der Zussamenhang des Abschlusses der Eleckronengruppen in Atom mit der Komplexstructur der Spektren. *Z. Phys.* **31** 765–783

Sakarai, J. J. 1967. *Advanced Quantum Mechanics*. Addison-Wesley: Reading, MA. Pp 85–87.

Sinova, J., Culcer, D., Niu, Q., Sinitsyn, N. A., Jungwirth, T., and MacDonald, A. H. 2004. Universal intrinsic spin Hall effect. *Phys. Rev. Lett.* **92** 126603

Slater, J. C .1929. The theory of complex spectra. *Phys. Rev.* **34** 1293–1322

Zeeman, P. 1897. On the influence of magnetism on the nature of light emitted by a substance. *Phil. Mag., Ser.* 5 **43** 226–239

PROBLEMS

1 Consider two electrons in a state in which the radius (around some orbit center) is set normalized to unity, and the angular wave function is $\varphi(\theta) = f\,[\cos(\theta)]$. If these two electrons have opposite spins and are located at $f\,[\cos(\theta)]$ and $f\,[\cos(\theta + \pi)]$, discuss the antisymmetric properties of these electrons. Can you infer the nature of the function describing the angular variations?

2 Show that the number operator

$$\langle n \rangle = \sum_i \langle c_i^\dagger c_i \rangle$$

commutes with the Hamiltonian.

3 Consider a perturbation that scatters an electron from state k to state k' while creating or annihilating one unit of lattice vibration. In Section 5.3, it was shown how the vibrating lattice could be represented in Fourier space as a summation of a set of harmonic oscillators. Devise a perturbing potential that accounts for the scattering of the above electron by the lattice (it will take three operators, two for the electrons and one for the boson). Carry out lowest-order perturbation and sketch the diagrams and Green's functions that will result.

11 An Introduction to Quantum Computing

The idea of using quantum mechanics to provide more power than classical computers seems to have begun around 1981–2 (Benioff, 1982), but Deutsch (1985) put the concept on a sounder footing a short time later. Deutsch reformulated the Turing (1937) thesis with the assertion that every finite realizable physical system can be perfectly simulated by a universal model computing machine operating by finite means. However, he asserted that the output of a quantum machine, although fully determined by the input state, is not an observable itself. Hence, the user must provide some cleverness if this state is to be determined. Benioff pointed out the need for the quantum state to experience interference, some indeterminism, and was what we would today call an entangled state. He also realized that nondissipative evolution was unphysical, so immediately it became clear that a great deal of effort would be required to minimize the decoherence of the physical states. Deutsch (1989) then provided some input as to what type of states or systems might be usable in this process. Jozsa (1991) mathematically showed that one could associate a quantum state to each function that one desired to evaluate, and that this should lead to a rapid solution of problems via the quantum computer (Deutch and Jozsa, 1992).

Progress beyond mathematical approaches came later when Lloyd (1993) suggested that arrays of weakly coupled quantum systems might be able to serve as the quantum computer if they were subjected to a sequence of controlled electromagnetic pulses, which could place the quantum bits into the desired superpositions. This shifted the problem to arranging for the superpositions, and the need for suitable *algorithms*. Perhaps the first such algorithm came from Peter Shor (1994), who introduced the idea of the quantum Fourier transform, a key element in breaking many encryption techniques that utilized large numbers generated by the product of prime numbers. Another algorithm, for fast data base searches, was developed by Grover (1996). Now, it turns out that, some three decades later, progress in creating and using quantum computers has been made, and these usually operate on the principles outlined by Lloyd (1993).

Our purpose here is not to cover the world of quantum computers. There are many other books that do this just fine (Nielsen and Chuang, 2000; Marinescu and Marinescu, 2004). However, the quantum mechanics of these processes is a natural evolution of the quantum mechanics discussed in the previous ten chapters. What we want to deal with is the quantum mechanical ideas of what creates the heart of the quantum computer, and how it functions with our normal approach discussed above. We will begin with the idea of the quantum mechanical replacement for the computer bit—the *qubit*, short for quantum bit. Then we will discuss some current (as of 2020) implementations of circuits and materials that create one or more qubits.

11.1 QUBITS AND ENTANGLEMENT

The quest for quantum computers begins with the qubits. Here, we will describe qubits and their important differences from classical bits in the next section, and the following sections will describe the interactions and entanglement in qubits. While it may not appear to be the case, due to the similarity of name between bit and qubit, these two quantities are quite different. To begin with, the normal bit, the basis of our binary computation in everyday classical computers, has only two values, 0 and 1, which form the binary base. While the qubit also has a 0 and a 1, it is analog in that we may define arbitrary complex values in any manner in which the magnitude of the qubit remains 1. In the following, we want to go through these ideas in some detail to understand how we get to qubits with the properties required to make quantum computers.

11.1.1 BITS AND QUBITS

It was John von Neumann (1945), working with the computer pioneers at the University of Pennsylvania, who suggested that the binary system should be used for simplicity in building computers. He also suggested the stored program approach, which along with binary arithmetic, formed the idea of the von Neumann architecture. Essentially, all computers from this time forward have used the binary system, as it is eminently suitable to two-state bits. In Figure 11.1, the idea of the bit/qubit is presented as two orthogonal axes in a configuration space (for reasons that will become apparent later). In this space, we define the two unit values for the bit as the states 0 and 1. Then, using quantum notation, we can define this space, and the bit, as the function

$$\psi = a|0\rangle + b|1\rangle, \quad a,b \sqsubset (0,1) \ni a+b=1\,. \tag{11.1}$$

The terms on the right just tell us that a and b must be either a 0 or a 1 and the sum must be unity. Hence, the bit value is merely 0 or 1, and not anything in between. Typically, in a computer, these levels are defined by either the system ground ($V = 0$), or a proper bias voltage usually denoted as V_{DD}. The secret of modern microchips is that the core ingredient, a transistor, is quite good at resetting these voltage values at each bit operation, minimizing any errors or noise (Mead and Conway, 1979).

In the computer itself, a particular state is defined by the values of the bits used to describe the state. Then, one can define a state transition diagram, in which each possible state is a node, and transitions between these nodes indicate the action taken when a control signal is applied. Various outputs arise from the set of transitions and

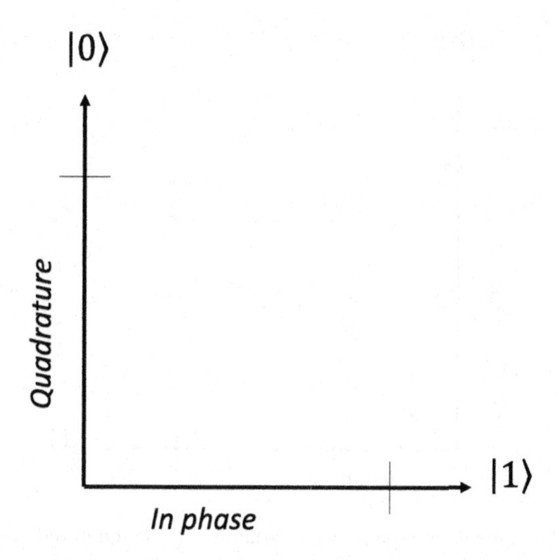

FIGURE 11.1 The coordinate set for an arbitrary bit, showing how the axes correspond to the values 0 and 1.

the state of the machine after the transition. Hence, an algorithm is a set of instructions that guide the system through a set of transitions, changing the various bits under the guidance of the control signals, to a desired result. Good algorithms lead to the machine halting at the end of the algorithm and yield a desired output (and with bad algorithms, the machine sometimes never stops). In fact, in the Turing (1937) model, a valid computer must have a "stop" state and must not go into interminable "loops." This same operational model will exist also in the quantum computer. That is, the qubits will define a state, while the state transitions will depend upon control signals.

When we move from the bit to the qubit, the change is rather subtle, although exceedingly important. We use the same diagram of Figure 11.1 for the qubit, but give it new properties, as shown in Figure 11.2. We still have the two coordinate axes representing the 0 and the 1. But now the wave function is analog and can take any value on the unit circle. Hence, (11.1) is still valid, but now the coefficients are related by

$$\psi = a|0\rangle + b|1\rangle,\ |a|^2 + |b|^2 = 1. \tag{11.2}$$

Thus, the two coefficients may be complex numbers, and (11.2) tells us that the net wave function must have a magnitude of 1, as required by any quantum wave function. The problem that arises is that this formulation is not completely descriptive for the actual wave function for the quantum qubit, as it says nothing about the phases of the two coefficients.

The two states $|0\rangle$ and $|1\rangle$ can be thought of as representing real energy levels of a two-level atom (this will connect us with the two-level system discussed in Section 10.2). Equally well, they can be thought of as the two possible spin states of

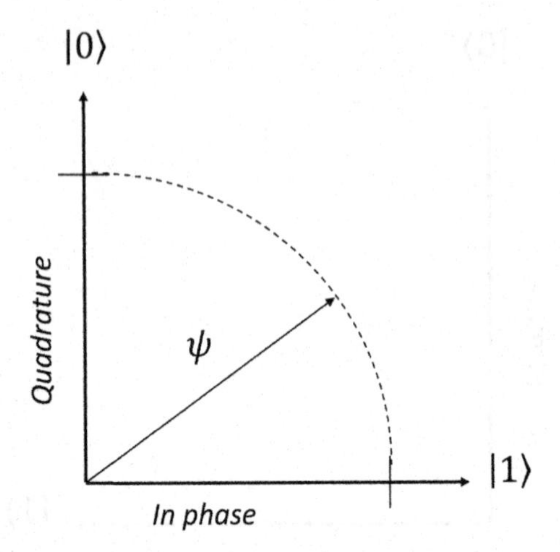

FIGURE 11.2 The value of the qubit now is analog and the magnitude of the wave function is now unity, but the phases are arbitrary, but controllable.

a fermion—spin up and spin down. Normally, these are aligned along the *z*-axis in spherical coordinates. From the latter concept, we then tend to write the Hamiltonian representing this two-level system as

$$H = \frac{1}{2}\Delta\sigma_z, \tag{11.3}$$

where Δ is the energy spacing between the two levels and σ_z is a Dirac spinor, given by (10.23). Hence, we have mapped the (0,1) states into the states (-1,1), in terms of $\Delta/2$. The upper level is taken to be the 1 state and the lower level the -1 state, in keeping with the normal spin wave functions. In terms of these two states, we can then rewrite the wave function as

$$\psi = \begin{bmatrix} a \\ b \end{bmatrix} = \begin{bmatrix} |a| e^{i\phi_1} \\ |b| e^{i\phi_2} \end{bmatrix}. \tag{11.4}$$

Now, each of the two coefficients has its own unique phase.

It is convenient to write the wave function in a slightly different form which characterizes the system but makes the two phases somewhat less arbitrary. Hence, we write the two-level system in terms of the density matrix

$$\rho = \begin{bmatrix} |a|^2 & ab^* \\ a^*b & |b|^2 \end{bmatrix}, \tag{11.5}$$

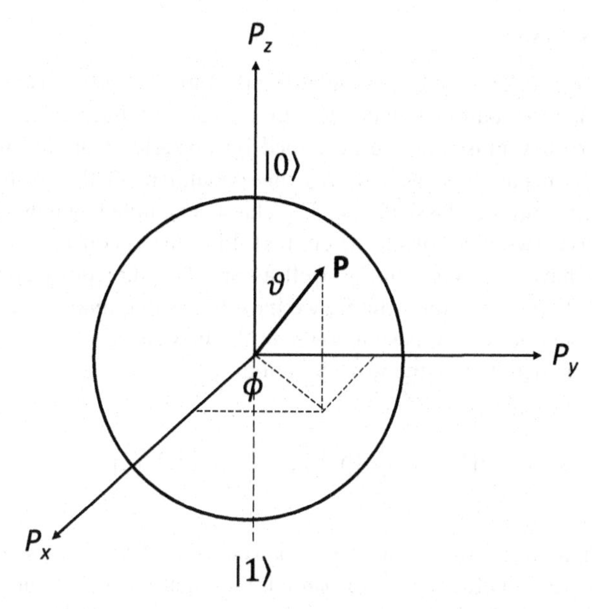

FIGURE 11.3 The Bloch sphere allows the polarization to rotate in three directions. Then, many qubit operations are expressible as rotations around one of the three axes. This is the same Bloch sphere from Figure 10.1, but the spin directions are replaced with the qubit values.

which is just the density matrix for the two-level system of Section 10.2, and given as (10.30), with a slightly different notation. Hence, we can use all the properties developed there to describe the qubit here, particularly the polarization representing the vector direction of the qubit on the unit surface of the Bloch sphere shown in Figure 11.3. As previously, the two angles are defined as the polar angle and the azimuthal angle, where the former is the angle away from the $+z$-axis, and the latter lies in the (x,y) plane and is measured from the x-axis, as is normal for spherical coordinates. This many qubit operations are easily expressible as rotations around the three axes shown in the figure.

We talked about the qubit as a two-level system, specifically a two-level atom. This could be two specific levels of a real atom, or in an artificial atom such as a quantum dot. The important point is that the phrase "two-level atom" is often a "code word" for the qubit system. This is because the energy levels in an atom are nonlinearly spaced; that is, they vary as $1/n^2$. Hence, the two desired levels are separated by an energy that is not the same as the separation of any other two levels. If the levels were linearly spaced, as in a normal harmonic oscillator, one has a problem. For example, if the lower two levels are those of interest, they are separated by exactly the same energy as any other two adjacent levels. This means that excitation of the lower two levels can also be absorbed by any other two levels, and this constitutes a sizable source of decoherence in the qubit. Hence, one desires the energy levels to be nonlinearly spaced, so that the two levels of the qubit do not couple to other unused energy levels. Since, the atom has this property, the phrase "two-level atom" is often applied to any qubit system to indicate it has the desired properties.

11.1.2 Entanglement

Erwin Schrödinger (1935) called entanglement the most important aspect of quantum mechanics. When two particles interacted, they were no longer independent of each another. Rather, they must now be described by a single entangled wave function incorporating both particles. And it was this entanglement that differentiated the quantum system from the classical one. The classic example is when a photon is absorbed and causes two electrons to be emitted. In order to conserve spin, these two electrons must have opposite spin, as well as opposite directions and momentum. Since we don't know what this spin is, we have to assume that either the first electron is spin up and the second is spin down, or the first electron is spin down and the second is spin up, denoted by the wave function

$$\psi(1,2) = \frac{1}{\sqrt{2}}\left(|1,\uparrow\rangle|2,\uparrow\rangle - |1,\uparrow\rangle|2,\uparrow\rangle\right). \tag{11.6}$$

This wave function describes our lack of knowledge of the spin states, but if the spin of one of the two electrons is measured, then that measurement proscribes the opposite spin for the other particle.

Needless to say, it is entanglement that has become the crucial ingredient in quantum computing that gives rise to the possibility of major speedup in the computation. For example, if we take the Fourier transform, we compute this transform for each possible frequency component. If there are N frequencies, then we repeat the transform N times. On the other hand, if we use qubits for the quantum Fourier transform, then in principle we can entangle all N frequencies into the wave function of each of the qubits and do the transform once. This, of course, may be the wrong answer since we are dealing with a probability distribution with the wave function, so we need to redo the transform many times to arrive at the most likely answer. But the theory goes that this will be much less than N times, and so using the quantum Fourier transform will speed up the computation dramatically, especially if N is a large number.

In the quantum computer, we want qubits to interact with each other under controlled circumstances. When two qubits interact, they become entangled, just as the two particles discussed above. They remain entangled until some mechanism gives rise to decoherence, which breaks up the entanglement. As a result, one needs long decoherence times in the qubits. In general, the quantum computer works on controlled gates, as these gates provide the interaction between the qubits. As an example, we take the so-called CNOT gate, or controlled NOT. In the classical system, any transistor (or CMOS gate) is naturally a NOT gate. This is because the transistor inverts its input. If we take a high voltage as a logical 1 and a low voltage as a logical 0, then a 1 input to the gate causes the transistor to turn on, which lowers the output voltage, and this gives a 0 output. Similarly, when the input voltage is low, the transistor is in the off state and the output voltage is high. Thus, the input is inverted, which means that the logical output is NOT(input). The CNOT differs from this slightly, in that the action of the transistor gate is controlled in a manner that makes the classical XOR

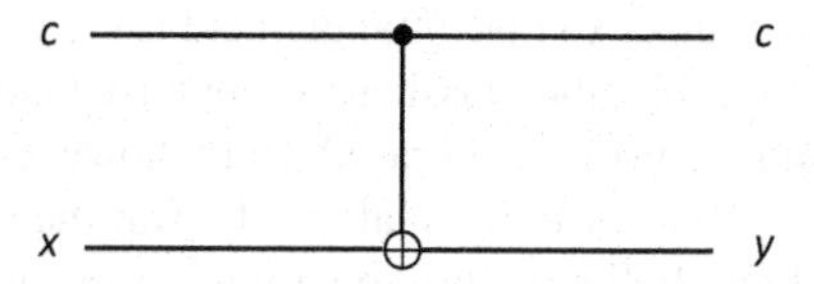

FIGURE 11.4 The CNOT gate, where c is the control input and x is the target input. The two are added modulo 2 to produce the output y.

gate. Now, the XOR gate has two inputs, which are called a and b. If either a or b is high (a 1), then the output is also high. But, if both inputs are either high or low, the output of the gate is low. In other words, the output of the XOR gate is 1 if either a or b is high, but not when both are high.

In the quantum CNOT gate, one of the two inputs is taken to be the control signal, which then operates upon the other input, as shown in Figure 11.4. Here, the control input (our a, for example) is labeled as c. The input wave function (our b, for example), from a previous qubit, is denoted by x. The symbol on the bottom line indicates binary arithmetic modulo 2. So, if the control bit is high, the current state of the signal is inverted (the carry is ignored). On the other hand, if the control is 0, the current state of the signal x is passed unchanged to the output. This then gives us the XOR function. We can illustrate this further with a "truth table" as

$$\begin{array}{ccc} x & c & y \\ 0 & 0 & 0 \\ 1 & 0 & 1. \\ 0 & 1 & 1 \\ 1 & 1 & 0 \end{array} \tag{11.7}$$

Here, we can see that the output y is high when either of the two inputs is high, but not when both are the same. We can write the output as a wave function, in which the state of the control bit is the first signal and the state of x is the second variable exactly with (11.6) (the sign between the terms can actually either positive or negative in the qubit case).

If we take the control bit as a two-state Hilbert space, and the input bit as a two-state Hilbert space, then the entangled output lies in a tensor product Hilbert space, denoted as $H^2 \otimes H^2$. But note that the wave (11.6) is not a simple product, and it cannot be separated into parts that lie in only one of the individual Hilbert spaces. This is what defines the entanglement. Hence, the wave function in (11.6) is an entangled wave function. The plus/minus sign gives two possibilities, which both give the same value of 1 for the square magnitude of the wave function (the factor of ½ provides proper normalization).

The introduction of the control signal is really accomplished by the manner in which the interaction between qubits is accomplished and managed. To carry out an algorithm, one has to manipulate the control signals, which actually change the interactions in a specific manner so that entanglement is created, manipulated, or destroyed. This is no different than the classical computer where bit strings are pushed around, manipulated, and then stored or erased. Here, however, we have the entanglement, and the increased power of individual qubits. But the qubit is the heart of the quantum computer, and we shall now turn to various forms of creating qubits.

11.2 THE JAYNES-CUMMINGS MODEL

In the previous section, we described the qubit in terms of a "two-level atom." Moreover, in the quantum computer, the qubit is adjusted, from the algorithm, often by an electromagnetic signal. The coupling between a two-level atom and a resonant (quantized) electromagnetic cavity is actually a very common situation in chemistry, and is referred to the Jaynes-Cummings (1963) model. More recently, the process of coupling a coherent optical beam to such an atom is of great interest in quantum optics. In this section, we want to discuss the model in some detail, because it is a simple model for coupling electromagnetic radiation to a qubit. A schematic diagram of the Jaynes-Cumming model is shown in Figure 11.5.

The principle of the model is analogous to the problem of two coupled classical pendulums, that are connected to one another in order to produce an interaction. This interaction leads to a coupling of the two oscillating modes in a way in which all of the energy oscillates between the two pendulums. At a given instant of time, pendulum 1 may be static while all the energy is in the oscillation of pendulum 2, and at a later time the process is in the opposite state. In the quantum case, one pendulum is the harmonic oscillator mode of the electromagnetic cavity, while the second is the atom,

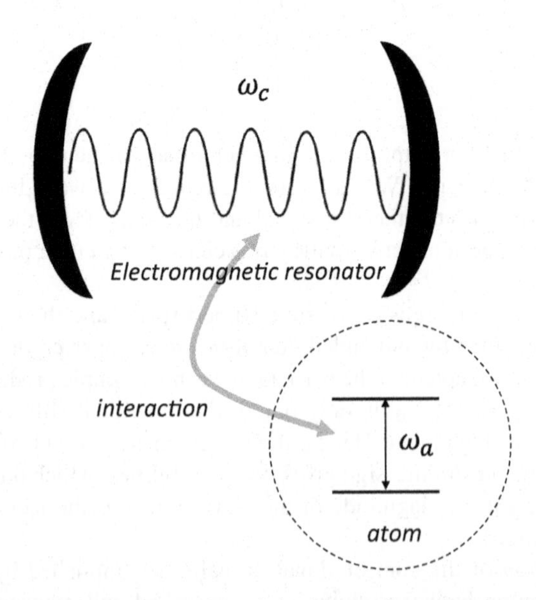

FIGURE 11.5 Schematic representation of the Jaynes-Cummings model in which a two-level atom (lower right) interacts with a mode of a resonant cavity (top). Details are in the text.

which is described as an oscillator by its Rabi frequency for oscillation of probability between the two states (energy levels). Hence, the field can interact with the atomic state and lead to Rabi oscillations of the atomic state population. As with the pendulums, the Rabi oscillation is not static in amplitude. Something different happens, in that the Rabi oscillations collapse and then reappears periodically, just as the energy moves from one pendulum to the other. The temporal time over which this collapse and reappearance occurs has been termed the *revival time* (Norozhny et al., 1981). When the field in the cavity is in a coherent state, and the atom is prepared in an excited state, it is found that the atomic and field states become rapidly entangled, and then subsequently disentangle at one-half the revival time (Gea-Banacloche, 1990). That is, the maximum entanglement of the atomic states occurs when the Rabi oscillations are strongest, and the minimum entanglement occurs when the Rabi oscillations are minimal. Thus, the model shows the full capability of providing entanglement, and this under the control of an electromagnetic field.

The Hamiltonian can be described relatively simply in terms of the various components, as discussed above, and shown in Figure 11.5. First, the electromagnetic field, or the laser field, is quantized as usual in terms of a series of harmonic oscillators that describe each of the modes that can exist in the cavity or wave, with each harmonic oscillator described as in Chapter 5. Since we are interested in a single mode, we write the wave energy as

$$H_{EM} = \hbar\omega_c a^\dagger a, \tag{11.8}$$

where the operators are the normal creation and annihilation operators (5.10) for the field harmonic oscillator representation of the single mode of the cavity, and ω_c is the frequency of this mode. The atomic levels are just the two levels of the atom, which we describe with a pseudo-spin index with the spin up state as the upper level and the spin down state as the lower level, and we can write the Hamiltonian as

$$H_a = \frac{1}{2}\hbar\omega_a\sigma_z, \tag{11.9}$$

Where ω_a is the frequency determined from the difference between the two atomic energy levels and σ_z is the Pauli spinor from (10.23).

Now, we recall that the total electric flux density arises from the applied electric field and the polarization of the atom constituents of the dielectric medium, which here is the atom. This polarization is given by the two states of the atom. We arbitrarily align this polarization in the x-direction and use (10.22) to write this as

$$P = P_x = \sigma_x = \frac{1}{2}(\sigma_+ + \sigma_-). \tag{11.10}$$

The electric field is arbitrarily oriented in the z-direction, so that the energy is $eE_z z$. We absorb the charge into a coupling constant, and write the interaction energy as

$$H_{int} = \hbar\Omega \cdot E_z z \cdot P = \frac{1}{2}\hbar\Omega E_z \left(a^\dagger + a\right)\left(\sigma_+ + \sigma_-\right). \tag{11.11}$$

The parameter Ω incorporates all the various constants arising from the connection between the operators (from the harmonic oscillators) and their corresponding position operators. The creation and annihilation operators each have their time variation determined by their respective frequencies, so the four cross terms will have slow variations due to the difference of the two frequencies and fast variations due to the sum of the two frequencies.

Generally, the fast components are ignored and only the slow components are retained in what is called the *rotating wave approximation*. The Hamiltonian is generally separated into two commuting parts, which are described by

$$\begin{aligned} H_1 &= \hbar\omega_c\left(a^\dagger a + \frac{\sigma_z}{2}\right), \\ H_2 &= \hbar\delta\frac{\sigma_z}{2} + \frac{1}{2}\hbar\Omega\left(a\sigma_+ + a^\dagger\sigma_-\right). \end{aligned} \tag{11.12}$$

Here

$$\delta = \omega_a - \omega_c \tag{11.13}$$

is the so-called detuning of the frequency between the field and the atom. This creates a proper 2×2 Hamiltonian, whose eigenvalues are given in terms of the number operator for the field as

$$E_\pm = \hbar\omega_c\left(n + \frac{1}{2}\right) \pm \frac{1}{2}\hbar\Omega_n, \tag{11.14}$$

where

$$\Omega_n = \sqrt{\delta^2 + \Omega^2(n+1)}. \tag{11.15}$$

This last expression defines the Rabi frequency for oscillation of density between the two atomic levels, in response to the interaction with the electromagnetic field. It should be noted that, as the mode occupation builds up, the Rabi frequency increases, and the amplitude of these oscillations increases. When the mode occupation is small, the Rabi frequency is basically just the detuning frequency of the system.

In this system, the electromagnetic wave, and its interaction with the two-level atom, provides the control over the state of the atom. By modulating, or changing, the interaction process, one changes the nature of the qubit itself. Here, the Jaynes-Cummings model provides a single example of how this process is accommodated. In the following sections, we will examine actual experimental techniques for the production of various types of qubits, hopefully explaining each approach sufficiently well for the reader to understand the process.

11.3 QUANTUM DOTS FOR QUBITS

The spin-orbit interaction is relatively small in Si, but it is large enough to consider making spin-based qubits in this material. The idea for a qubit based upon Si effectively dates from the suggestion of Kane (1998) to use a ^{31}P dopant in isotopically pure ^{28}Si. In this approach, the half-integer spin of an electron bound to the ^{31}P dopant couples to the spin-1/2 nuclear spin of the dopant atom via the hyperfine interaction, with a characteristic frequency of 98 MHz, when a magnetic field is used to split the various spin states. This gives rise to a four-level system, which will be described further in the following paragraphs.

The idea of using spin to create a qubit with quantum dots was apparently first proposed by Loss and DiVincenzo (1998). They developed a detailed scheme to achieve quantum computation with a pair of single-electron quantum dots (as discussed in Section 9.1). The qubit is realized with the spin of an excess electron in one of the quantum dots. Two-qubit quantum-gate operation is achieved by merely adjusting the barrier existing between the two dots. If the potential barrier is high, the two qubits, one in each dot, do not interact. If the potential is lowered, then the two qubits are allowed to interact, and the spins are affected by a coupling due to the spin–spin coupling energy. Many approaches burst upon the scene after this seminal paper.

11.3.1 ^{31}P Donors

As mentioned above, it was suggested to use the nuclear spin of the P atom. Such a qubit is interesting because it has the promise of a long spin-relaxation time, and is scalable (O'Brien et al., 2001). The process of getting a P atom to a precise location turns out not to be so difficult. One method is to hydrogenate the dangling bonds on the Si surface, then to remove particular hydrogen atoms using a scanning tunneling microscope (in ultra-high vacuum) (Lyding et al., 1994). Removing the hydrogen allows a phosphorus molecule to be chemically attach to the dangling Si bond at that particular site. In the case mentioned above, the molecules from phosphine gas are used to deposit the P atom at the desired position, then the difficult task is to deposit isotopically pure Si while maintaining the P atom at the desired position.

The operation of a single donor qubit under control of surface gates and measured with a single-electron transistor (described in Section 3.4.4) was evaluated by Dehollain et al. (2016). In Figure 11.6, we show the donor qubit formed on the P atom and describe some of the operation. Here, they use gate set tomography (GST), which is a tool for characterizing logic operations in the qubit. The qubit itself is formed by the spin states of an electron bound to a ^{31}P atom, implanted into the isotopically pure silicon substrate. The spin states are split by an applied magnetic field, and switching is achieved via coupling of the electron spin to the nuclear spin, resulting

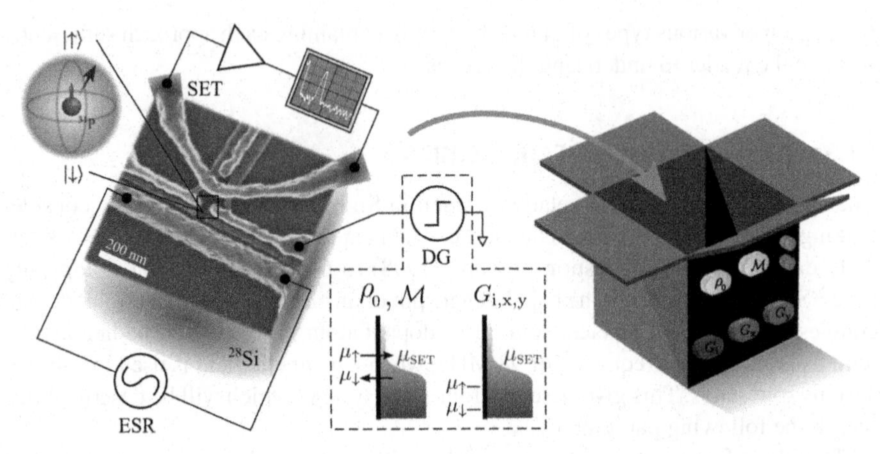

FIGURE 11.6 Diagram of qubit device and the gate set tomography (GST) model of a qubit. SEM image of the on-chip gate structure is shown to the left. The aluminum gates have been false colored for clarity. Depicted in darkest grey are the source-drain n^+ regions, which the single-electron transistor (SET) connected to the current measurement electronics. For initialization and measurement, the donor gates are pulsed such that $\mu > \mu_{SET} > \mu_{\downarrow}$, inducing spin-dependent tunneling between the donor and the SET. When applying a gate sequence, the DG are pulsed to higher voltage to prevent the donor electron from tunneling to the SET. The inset diagram—(upper left) zoomed from the approximate donor location—represents the Bloch sphere of the qubit, consisting of the spin of an electron confined by an implanted ^{31}P donor, with its nuclear spin frozen in an eigenstate. The GST model treats the qubit as a black box with buttons (right), which allow one to initialize (ρ_0), apply each gate in the gate set ($G_{i,x,y}$) and measure (M) in the observable basis (Reproduced under the Creative Commons Attribution 3.0 licence from J. P. Dehollain et al., "Optimization of a solid-state electron spin qubit using gate set tomography," New J. Phys. **18**, 103018 (2016) DOI:10.1088/1367–2630/18/10/103018.)

in a two-spin, four-level system. Qubit preparation is handled by tunneling to/from a single-electron transistor, which is also used for sensing the state. The aluminum gates are placed on top of a SiO_2 layer. Manipulation of the qubit state is done with an electromagnetic signal tuned to the electron spin resonance frequency that itself is set by the magnetic field.

As mentioned above, a magnetic field of 1.55 T is applied to the system, and this splits the various spin states, with the electron spin coupled to the nuclear spin by the hyperfine interaction. The relaxation rate for the nuclear spin is orders of magnitude slower than that for the electron spin, so that the latter spin levels can be used as the two levels for the qubit. The qubit states $|1\rangle$ and $|0\rangle$ are the electron spin levels $|\uparrow\rangle$ and $|\downarrow\rangle$, respectively. Preparation of the initial state is done via single-electron tunneling to/from the donor atom, a process controlled by the aluminum gates. Then, during the state manipulation with the spin resonance signal, the pair of gates, just above the electromagnetic line connected to the electromagnetic source, are raised to a higher voltage to disconnect the donor spin from the SET. Finally, the SET can be used to sense the qubit state.

11.3.2 Double Quantum Dots

It was pointed out above that two coupled quantum dots could be used for a spin qubit (Loss and DiVincenzo, 1998). We can illustrate the double-quantum-dot approach with some relatively recent work (Brunner et al., 2011). Here, the structure is fabricated on a GaAs/AlGaAs heterostructure with Schottky barrier metallic gates used to define the active region, which may be glimpsed below the magnetic material in Figure 11.7(a). Along the bottom of this panel, one can identify the various gates (dashed lines connecting to the gray areas of the open region) as left barrier (L), plunger (PL), control (C), and right barrier (R). As above, this is a gate-defined double-quantum-dot (see Section 3.4.4), to which has been added a split Co micromagnet (the light regions in this figure), in order to provide a magnetic field. To the outside of the active double-dot, a quantum point contact (QPC) has been added (indicated in the figure by the current path arrow for I_{QPC}). When the dot on the right has an electron (or more) in it, the Coulomb interaction causes the opening for the QPC to narrow, and the resulting decrease in current can signal the charge state. This

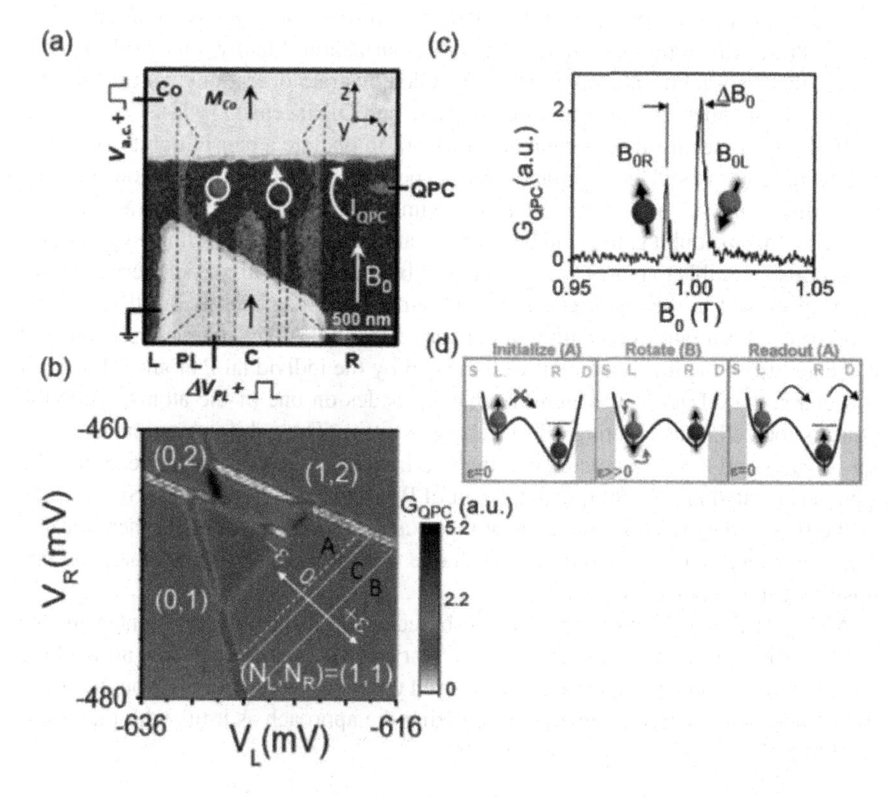

FIGURE 11.7 (a) A image of the device showing the two quantum dots and the electron spin. (b) The stability diagram around the area of interest. (c) Spin resonance signals for the left and right dots. (d) Measurement cycle for controlled spin rotations. (Reprinted with permission from Brunner et al., 2011). Copyright 2011, the American Physical Society.

sensor can be used to map out the stability regime for the double-dot system, shown in Figure 11.7(b). Essentially, this is a plot of the conductance through the QPC as the potentials on the left (L) and right (R) barriers are varied. Here, N_L and N_R denote the charge in the left and right dots, and are indicated by the numbers in parentheses. The single-spin rotations and qubit interactions are carried out in the (1,1) state (lower right of the stability diagram). To rotate each spin, electrically driven spin resonance is used, although the Co nanomagnets provide a spatially varying magnetic field due to the shape of these magnets. The spin resonance signal, for the two dots, is seen in panel (c) of this figure. Panel (d) shows the shape of the double-well potential, which is controlled by the various gate electrodes. During operation, a static magnetic field of 2 T in the left dot and 1.985 T in the right dot is applied, and the dot is excited with a microwave signal of 11.1 GHz. As indicated in the figure, various gate voltage pulses are also applied to sequence the interaction. For stage A in the figure, the two dots are set in the spin blockade condition, where the spins cannot interact, and both spins are either up or down. In stage B, the dots are isolated from one another and one of the spins is rotated by applying a pulsed microwave signal. Finally, at the second stage A, to the right, the spin can be read out with the QPC sensor in which the signal is proportional to the probability that the two spins are oppositely polarized. This demonstrates the ability to control the spin polarization. Finally, this work demonstrates that two-qubit gates may be used to demonstrate operations via the interdot spin exchange operation, as suggested by Loss and DiVincenzo (1998).

If all of this seems to be a great deal of work to operate a pair of qubits, it is. Yet, this technology is still in its infancy, and experiments such as this demonstrate the feasibility of the scientific basis. In the experiments, the authors demonstrated that an all electrical two-qubit gate could be realized, and it could perform simple operations.

It is also possible to marry the previous double-dot qubit with the single-atom transistor discussed in the last section. A double-dot qubit has been realized using two single phosphorus atoms recently (Hollenberg et al., 2004). The double-well potential is created from the local potential well created by the individual P atoms. This qubit is a charge-based qubit in which the charge resides on one of the atoms, while the other is ionized, thus creating a P–P^+ charge system. The qubit is manipulated by a set of gates and a SET, operated at a high ac frequency, is used for the readout. The device is created in a Si–SiO_2 with the pair of P atoms implanted into the Si substrate. In the figure, the structure, the control gates, and the SET are shown schematically. The control gates are used to move the charge from one P atom to the other, and this position can be sensed by the SET.

Much progress has been made in the Si-based quantum dot qubit in recent years, but most of this work remains in the academic world. That is, there is to date no working quantum computing based circuit that is based upon this technology, although several industries have asserted an interest in utilizing the approach as it fits with their classical Si technology.

11.3.3 NV Centers

The nitrogen-vacancy (NV) complex centers have been considered for possible qubits for the past few years (Jelezko and Wrachtrup, 2006). In some sense, this is like the donor qubit, as the complex produces an energy level in the band gap of wide band gap materials. The most common material in which this has been created is diamond, a wide band gap semiconductor (5.47 eV). These NV centers are one kind of defect often referred to as a "color center." There are a variety of actual N and V configurations within the crystal structure, each of which produces a different range of defect energy levels. While the center also has been found in GaAs and SiC, diamond remains the most common carrier. The various levels for these centers can be in the infrared or visible for use in these qubit applications. And, the nature of the qubit structure varies, as some approaches use the coupling of the electronic and nuclear spins, while others couple the electronic spin to a strain in the crystal. In the latter case, the divacancy has also been used.

Normally, the configuration used is with a substitutional nitrogen atom on the diamond lattice with an adjacent vacancy. This NV center is usually discussed in terms of a six-electron model. In this model for the negatively charged center, there is a ground state of an ^{3}A spin triplet (usually denoted as a ^{3}A level), with a zero-field splitting of 11.88 μV between the three angular momentum levels: $m_S = 0$ and the $m_S = \pm 1$. The degeneracy of the $m_S = \pm 1$ can be lifted with an external magnetic field, so that either one of these levels, together with the $m_S = 0$ level, form a well-defined two-level system. This two-level system is the physical realization of the qubit. The spin states can be polarized by excitation of an electron to the upper of another energy level, denoted as the ^{3}E triplet, with a green laser, and then readout is performed by using the florescence of this excited level. The quantum state of this spin qubit can be manipulated with the application of a microwave signal, especially to affect the interaction between this two-level system and the nuclear spins of the with neighboring ^{13}C atoms via the hyperfine interaction. This role of the microwaves can be reversed in that the qubit can be used to sense the presence of the microwaves.

11.4 JOSEPHSON JUNCTIONS

We have previously mentioned Josephson junctions and superconductivity in Chapters 3 and 4. The Josephson junction (Section 3.4.5) is a tunnel junction in which the materials on either side of the insulator are superconductors (Section 4.4). In the Josephson junction, these superconducting leads produce zero loss in dc operation. The formation of the Cooper pairs lowers the energy of the entire electron gas below the Fermi energy, and this leads to a gap opening at the Fermi energy. Normally, with Josephson junctions, the tunneling is carried out by these Cooper pairs, but both the Cooper pair tunneling and single-electron tunneling (Section 3.4.4) can occur under the appropriate conditions. The Josephson junction itself operates so that, generally no tunneling occurs until $eV_a > 2\Delta$ at low temperature. As the temperature increases, both unpaired single electrons and Cooper pairs exist below the gap, and the normal (unpaired) can tunnel giving a very small current. As with normal tunneling, if the barrier is relatively thin, say 1–2 nm, then the superconducting wave functions can extend through the barrier, so that they interact with those wave functions on the other side. An unusual effect in the Josephson junction is that this coherent mixing of the wave functions on either side of the junction can produce a current at zero bias! This is termed the dc Josephson current.

In recent years, qubits have been made from superconducting circuits that include Josephson junctions. The step from a Josephson junction toward a qubit based upon this device was taken in 1999, with macroscopic qubits based on quantum levels representing collective degrees of freedom such as the persistent current in a superconducting ring (Mooij et al., 1999) or excess charge on a superconducting island (Nakamura et al., 1999). When superconducting components are used in electrical circuits, then these circuits become quantum objects, which can then be used in either type of qubit. The pursuit of superconducting qubits can be a Cooper pair box charge qubit, the persistent current flux qubit, or a hybrid charge-flux qubit.

11.4.1 The SQUID

The Superconducting QUantum Interference Device (SQUID) is basically a combination of a ring resonator and a pair of embedded Josephson junctions (Jaklevic et al., 1964). When there are two junctions, then one can see interference between the two oscillatory behaviors under current bias. With the ring, electromagnetic waves can be coupled into the circuit, and the dc bias leads to a dc SQUID, Figure 11.8. First, we remind ourselves that the Josephson current can be expressed as

$$I_J(B) = I_{J0}cos\left(\frac{eBA}{h}\right), \quad (11.16)$$

where $BA = \Phi$ is the flux flowing through the area of the junction. As before, h/e is the quantum unit of flux, so that the flux can be quantized in this quantum system. Shown in Figure 11.8 is a bias current I_a that flows in from the left and out to the right. This current splits between the two arms of the ring interferometer, so that $I_a/2$ flows through each Josephson junction, denoted as JJ_1 and JJ_2. An important point is that all of the leads are made from superconducting material. Initially, there is no flux through the superconducting loop that contains the two tunnel junctions. If we induce a flux in the loop that is less than $\Phi_0/2$, where $\Phi_0 = h/e$ is the quantum unit of flux, this creates a circulating current, which we designate as I_c, that flows in the loop. The flux can be created by coupling the loop to an external loop that provides

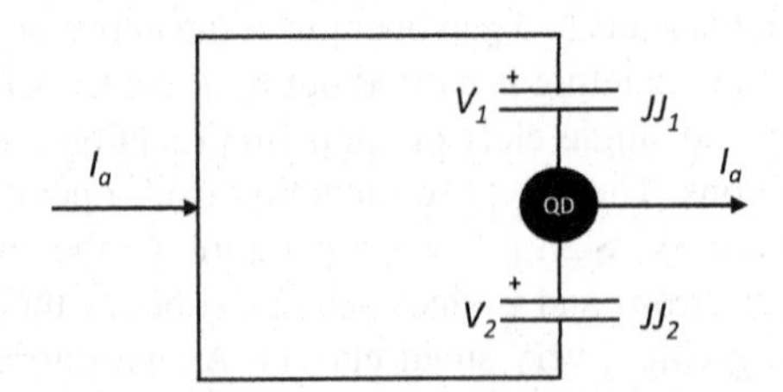

FIGURE 11.8 Schematic diagram of a d.c. SQUID made of a superconducting ring (large square) and two Josephson junctions, above and below the large black dot, designated as QD. Various forms of operation of this SQUID qubit depend upon the size of QD. See text for operation.

the flux with an electromagnetic field. Thus, for example, the currents through the two capacitors become

$$I_1 = \frac{1}{2} I_a + I_c$$
$$I_2 = \frac{1}{2} I_a - I_c \tag{11.17}$$

When the flux is increased to > $\Phi_0/2$, the induced circulating current has to change sign according to (11.16). And, when the flux reaches a full flux quantum unit, the cycle repeats. As a result, the SQUID is very sensitive to magnetic fields and can measure fields as small as a few times 10^{-18} T, so are used in a great many applications (Drung et al., 2007). As the currents oscillate with the flux, the induced voltage, that arises from the junction resistance and the loop inductance, measured between the QD and the input current-splitting point, becomes

$$\Delta V = \frac{R}{L} \Delta\Phi. \tag{11.18}$$

Most applications of the SQUID for quantum computing applications use the two (or more) junction version of the dc SQUID. In the drive to make the qubits small, one actually introduces the quantum dot, QD, shown in Figure 8.23. That is, there is a long part of the ring and a short part of the ring, the latter of which forms the quantum dot. Hence, there is a competition in various energies in the system. There is the single-electron (pair) energy (Section 3.4.4)

$$E_c = e / 2C, \tag{11.19}$$

If we assume the two junctions are equal capacitances, and the factor of 2 arises for Cooper pair tunneling. There is also the Josephson energy arising from the current (11.16), which becomes

$$E_J = \Phi_0 I_{J0} / 2\pi. \tag{11.20}$$

The operation of the qubit and the SQUID depends upon the relative size of the two energies in (11.19) and (11.20). To explain this further, one needs to distinguish two types of qubits: microscopic and macroscopic. The microscopic qubit is based upon an internal charge or spin, much like the Si qubits of the previous section, or upon two natural levels as the two-level atom in the Jaynes-Cummings model. Macroscopic qubits are based quantum levels representing collective degrees of freedom (often called *macroscopic quantum states*) like the persistent current in a superconducting ring or excess charge on a superconducting island, such as the quantum dot in Figure 11.8. These give different types of qubits, which we discuss below.

11.4.2 Charge Qubits

The charge qubit is composed primarily of the Cooper pair box, which itself is the region between the two junctions when this region is small such that there are a set of distinguishable energy levels. This box plays the role of our two-level atom. The box is accessed through a pair of small Josephson junctions, so that single-electron tunneling governs the excitation of the box. A single Cooper pair box is a unique artificial solid-state system, analogous to the single-electron box in Section 3.4.4. Although there are quite a few electrons on the island (box), they all form Cooper pairs in the superconducting state and then condense into a single macroscopic quantum state. This state is separated from the normal electrons by the superconducting gap Δ, discussed above. The only low-energy excitations arise from Cooper pair tunneling when the gap is larger than the charging energy $E_C = e^2/2C$. Notice that this differs by a factor of 2 from Section 3.4.4, due to two electrons in each Cooper pair. On the other hand, if the charging energy is larger than the gap energy, and the thermal fluctuations are suppressed, the system can be considered to be a two-level system in which the lowest two energy states differ by one Cooper pair. The separation of the two levels of interest can be controlled by an additional gate voltage.

When the charge qubit is shunted by a capacitor, as a method of reducing current noise, the device has been called a *transmon* (Koch et al., 2007). Transmon is an abbreviation of the *trans*mission line shunted plas*mon* oscillation qubit (note that this qubit does not employ a SQUID but uses striplines instead). Measurement and control of the box qubit is commonly done by means of microwave resonators. This design allows one to reverse the inequality above and make the Josephson energy much smaller than the single-electron tunneling energy, which is a characteristic of the transmon qubit.

In Figure 11.9, we illustrate a modern transmon qubit (Bergerbos et al., 2020). In general, a Cooper pair box, discussed above, has large charge dispersion (a wide variation of charge), while the transmon island charge dispersion is exponentially in the ratio of E_J/E_c (Koch et al., 2007). If the Josepson tunneling is mediated by a single, high transparency channel, the junction itself becomes characterized by a narrowly avoided level crossing at $\phi = \pi$, and an obstruction to tunnel trajectories reaching $\phi = 2\pi$. Then, the dispersion of charge vanishes completely when the transparency of the wire approaches unity. The qubit circuitry of Figure 11.9 nearly achieves this perfect situation. The qubit is characterized by the island's ground to first excited state transition, which is controlled and sensed by the coplanar waveguide for the electromagnetic signals.

11.4.3 Flux Qubits

One of the most interesting aspects of circuits is that a resonant circuit can be a quantum oscillator, much akin to the harmonic oscillator, discussed in Chapter 5 and in the Jaynes-Cummings model above. In the absence of dissipation, the Hamiltonian can be written as

$$H = \frac{q^2}{2C} + \frac{\Phi^2}{2L}, \tag{11.21}$$

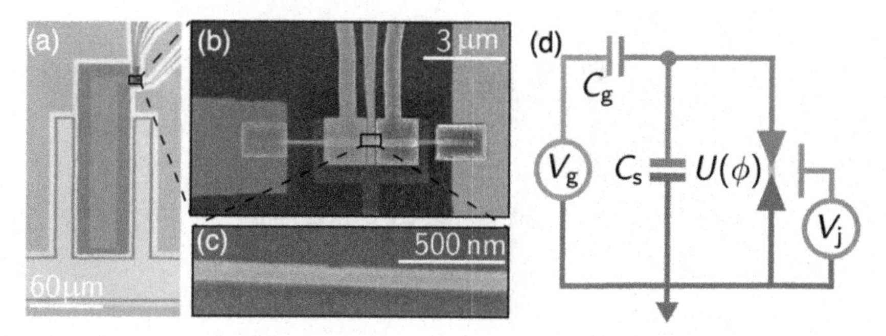

FIGURE 11.9 (a) A optical microscope image of the qubit. It consists of an island capacitively coupled to the ground plane (dark) and a coplanar waveguide resonator (two vertical light grey bars). (b) A scanning electron microscope (SEM) image, which is a blowup of the small black box in panel (a). A small superconducting nanowire connects the island (left) to the ground plane (right). Its weak link Josephson junction box is tuned by the gate voltage (center bar of the three light grey bars that are vertical), while the island gate tunes the number of electrons on the island. (c) SEM image of the nanowire prior to deposition of the top gates. The InAs core (light grey) and the Al shell (dark grey). (d) Effective circuit diagram of the qubit. The Josephson weak link potential $U(\phi)$ is shunted by the island capacitance C_s, while C_g tunes the number of electrons on the island (the small dot on the upper line). V_J tunes the transparency of the weak link. (Reprinted with permission from (Bargerbos et al., 2020). Copyright 2020, the American Physical Society.

where the charge q and flux Φ are the conjugate variables and satisfy a commutator relation with one another (see Section 5.2). Of course, to reach the quantum limit, we must work at low temperature, which naturally makes a connection with superconducting circuits. In the presence of dissipation, we need to have a good quality factor Q (= $\omega\tau >> 1$). Hence, when we couple a Josephson junction to a superconducting ring, we obtain the persistent current flux qubit, sometimes called an RF superconducting quantum interference device (RF-SQUID, discussed above). Here, we need to have $E_J >> E_C$, so the charging energy cannot be too large. The eigenstates of the ring represent two counterrotating persistent currents, corresponding to a fixed number of flux quanta (h/e) in the loop. The inductance of the ring gives rise to a parabolic potential, like the harmonic oscillator, and adding the Josephson oscillating potential provides the needed nonlinearity to separate off the lowest states from the linear chain, as discussed earlier. For the two levels of the qubit, the Hamiltonian can be written as

$$H = 4E_C n^2 + E_L\left(\varphi - \varphi_e + \varphi_{int}\sigma_z\right)^2 - E_J cos(\varphi), \tag{11.22}$$

where the interaction term arises from the effect of an external loop of electromagnetic flux that is coupled into the SQUID ring and affects the two persistent current states. When $\varphi_e = \Phi_0/2$, the lower part of the potential is a symmetric double-well, creating nearly degenerate $|L\rangle$ and $|R\rangle$ states which lead to bonding and antibonding

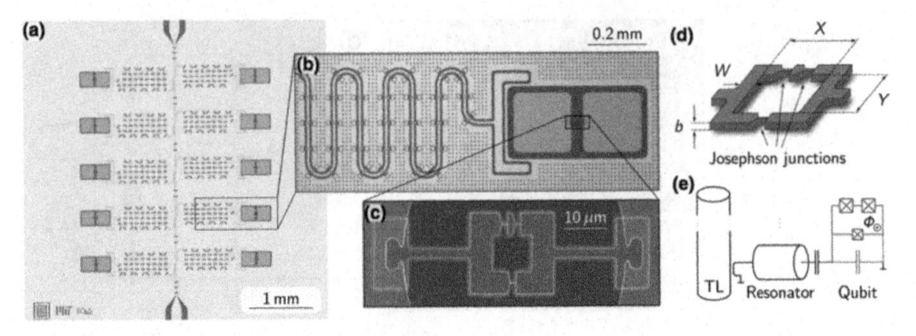

FIGURE 11.10 (a) Each chip holds ten uncoupled capacitively shunted qubits with individual readout resonator, which feature five different SQUID configurations at a twofold redundancy. (b) Optical micrograph of one of the qubits and part of its readout resonator (left-hand side). The capacitive shunt is the two medium grey areas region. (c) Electron microscope image of the SQUID qubit (3 narrow areas in central region). Three junctions are in each ring loop of the SQUIDs. (d) Schematic representation of the SQUID, showing the parameters that are varied in the different qubits. The SQUID dimensions X and Y are measured along the inner edge of the SQUID. W is the width of superconducting leads and b is the thickness of the film. (e) Effective schematic circuit of the qubit-resonator pair coupled to the common transmission line. (Reprinted with permission from Braumüller et al., 2020). Copyright 2020, the American Physical Society.

hybrids of these two states. These latter two states are created by the macroscopic tunneling through the Josephson junction which couples the two persistent current states, and these hybrid states provide the two levels of the qubit. The Wigner function in number-phase representation shows that the state of the system evolves into a quantum superposition of two coherent states which clearly demonstrate interference and negative values for the Wigner function description of the system. Yet, when the system is represented in the number-phase coordinates, the Wigner function evolves in a classical manner (Ferry and Nedjalkov, 2018). Dissipation is incorporated by coupling the system to a reservoir, and this expanded system is then projected back onto the reduced density matrix for the ring-junction system. Hence, the two coherent states survive even in the presence of dissipation, at least for weak dissipation.

The flux qubit can also be prepared with a three-junction ring, much like the hybrid qubit of the next section. This qubit can then be coupled to a transmission line resonator to produce cavity QED interactions. The capacitance between the qubit and the resonator can be controlled by varying the width of the capacitance line, so that the coupling depends upon the number of qubits placed in the overall circuit. We note that this can also be done with single junction qubits.

As usual, the qubit is sensitive to the noise, as the latter affects the dephasing time, which itself is a measure of the dissipation. Recently, the noise has been measured in a range of resonator coupled qubits (Braumüller et al., 2020). The 10-qubit chip and the devices are shown in Figure 11.10. The effective periphery of the SQUID is given as

$$P = 2X + 2Y + 4W, \tag{11.23}$$

where the individual X, Y, and W are shown in Figure 11.10(d). The values for these parameters were varied around commonly used values of $X = 9$ μm, $Y = 8$ μm, $W = 1$ μm. The qubit dephasing rate is proportional to the noise amplitude $\sqrt{A_\Phi}$. The noise power was measured as the various parameters were varied, and it was found that the noise amplitude increased linearly with the periphery $\sqrt{P}$, when W was fixed, and decreased as $1/\sqrt{W}$ when X and Y were fixed. Surprisingly, this suggests that lower noise, and hence less decoherence, is favored by smaller SQUID edge dimensions and wider superconducting lines in the SQUID. However, this is self-limiting, as pursuing both of these leads to no opening in the SQUID, which obviously doesn't lead to a proper device. Nevertheless, the guidelines are clear from these experiments, which examined some 50 different devices.

11.4.4 The Hybrid Charge-Flux Qubit

With a proper SQUID (at least two junctions), we can have both the charge box between the two junctions and flux coupled through the ring. If the Coulomb energy E_C dominates the Josephson energy, we have the charge qubit limit. If the Josephson energy dominates the Coulomb energy, we reach the flux qubit limit, but if we have $E_J \geq E_C$, we are in the charge-phase, or hybrid charge-flux qubit regime (Wendin, 2003). In many situations, the system acts as a two-level atom, which is called a *quantronium*.

A more extensive circuit is shown in Figure 11.11, in which a Cooper-pair transistor is coupled to a dc SQUID (Fay et al., 2011a). When the current-biased SQUID is noninductive and the Cooper pair transistor is symmetrical, one recovers the quantronium. When the SQUID is operated in the nonlinear regime, the circuit returns to being a charge qubit. The authors have conducted a thorough theoretical analysis of this system: a two-level quantum system (the Cooper pair transistor) coupled to two anharmonic oscillators (the d.c. SQUID). The dynamics of this system is basically a qubit coupled to a single anharmonic oscillator, a system that is well represented in terms of the Jaynes-Cummings model discussed above. In this circuit, one has two basic quantum devices that are building blocks of basic superconducting quantum devices. In this circuit, the two are controllably coupled. The first element is a d.c. SQUID (Section 11.4.1), in which each of the two junctions has its own critical current I_0 and capacitance C_0, and the total inductance of the loop consists of the three inductors shown in the figure (L_{1-3}). The second device is the asymmetric Cooper pair transistor, which is placed in parallel with the SQUID, as can be seen in Figure 11.11. The asymmetry arises from the unavoidable difference in the effective area of each of the two junctions that are in series to make the transistor. These two junctions have their own critical currents and capacitances, denoted by the superscript "T." The asymmetry parameters are

$$\mu = \frac{I_2^T - I_1^T}{I_2^T + I_1^T},$$
$$\lambda = \frac{C_2^T - C_1^T}{C_2^T + C_1^T}. \tag{11.24}$$

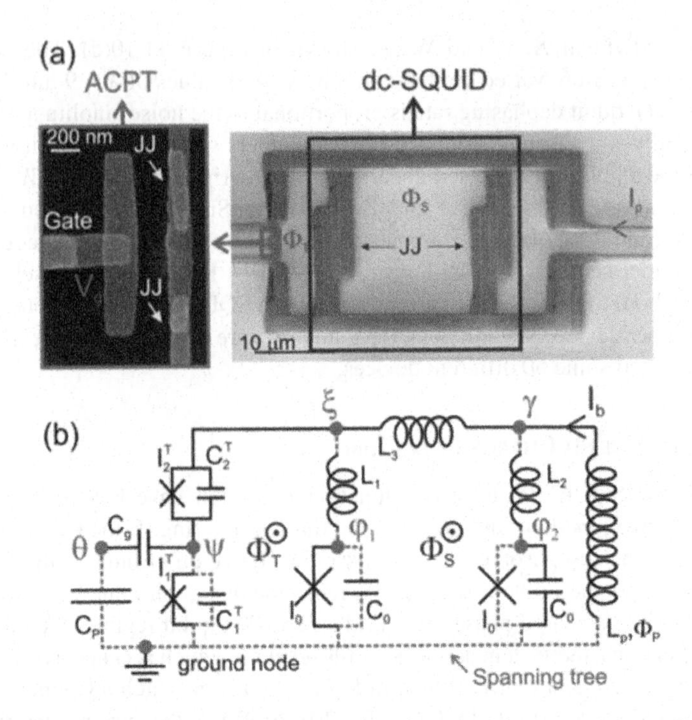

FIGURE 11.11 (a) On the right-hand side, a SEM image shows the two parts of the circuit. The dc SQUID is in the grey box on left of the right-hand figure box. An asymmetric Cooper pair transistor (zoom of small square box) is shown in detail on the left. The asymmetry arises from the different areas of the two series junctions in the transistor. The properties of the circuit can be modified by the bias current, I_p, in (a) (I_b in (b)), the flux Φ_s in the SQUID, Φ_T in the outer loop, and the voltage bias V_g (shown in the left-hand panel). (b) The schematic of the overall circuit. Each junction is said to have its pure properties as well as a shunt capacitance, and those in the SQUID have a series inductance arising from the ring. The voltage bias of the transistor gate and the current bias of the circuit are schematically indicated by an infinite capacitance C_P and an infinite inductance L_P, respectively. (Reprinted with permission from Fay et al., 2011). Copyright 2020, the American Physical Society.

The asymmetric Cooper pair transistor is coupled to an infinite capacitance C_p with a charge $Q_g \rightarrow \infty$, so that $\frac{Q_g}{C_g} = V_g$. The circuit is also current-biased with I_b modeled by the ratio of an infinite inductance with an infinite flux coupled through it.

The coupling between the two quantum devices results from both a capacitive and a Josephson coupling between the two devices, and this coupling can be tuned, by the gate voltage and bias current, from the very strong to the zero-coupling regime. That is, it spans from the quantronium regime to the adiabatic quantum readout regime (Fay et al., 2011b).

11.5 OPTICAL QUBITS

It is not surprising that optical qubits have been developed as well, and these are usually closely based upon the Jaynes-Cummings model. The optical gates are not as far advanced as those for silicon and superconducting technologies. But the optical approach has the advantage that it can be foreseen to operate efficiently at room temperature, an advantage that should not be overlooked. In addition, optical qubits are natural for use in optical communications systems. Clearly, one can extend the Jaynes-Cummings model to the case of multiple (qubit) atoms and a resonator that yields the extension of (11.8–11.11) to

$$H(t)=\hbar\omega_C\hat{a}^\dagger\hat{a}+\frac{\hbar\omega_a}{2}\sum_{i=1}^{N}\sigma_z^i+\frac{\hbar\Omega}{2}E_z\sum_{i=1}^{N}(\hat{a}+\hat{a}^\dagger)\hat{\sigma}_x^i, \tag{11.25}$$

where it is assumed that there are N atoms in this extended system. The prefactor of the second term is now interpreted as the qubit/dimer splitting in each atom. This has been used to show that a pulsed source can be used to generate many-body correlations in light-matter systems and to determine the time evolution of an N qubit system coupled to the global resonator (Gómez-Ruiz et al., 2018). This is just one advance, and most of the qubits described above have their analogs in optical systems.

For example, coupling an optical waveguide to a ring resonator can lead to entangled quantum pulses. The ring resonator is an optical waveguide in which the required periodicity around the ring leads to resonances that give quite small line widths, with the lines equally spaced in wavelength. Hence, a signal in the waveguide can couple into the ring through a tunneling process, and then the third-order nonlinearity in the polarization gives rise to a spontaneous four-wave mixing process. In this process, two incident (tunneled) photons are taken from the wave to produce two new phonons, one at a higher frequency and one at a lower frequency (Vernon and Sipe, 2015). These two new photons are entangled and can exit back to the waveguide to be used as an entangled pair in a quantum optical communications system.

Another approach is to use so-called color centers, of which the NV center above is one example. For optical applications, one approach is to use the spin of the color center to create an optically addressable qubit. Recently, it has been suggested that SiC, with its wide bandgap, provides a semiconductor substrate in which the neutral divacancy (denoted VV^0 in the literature), the silicon vacancy (denoted V_{Si}), or a substitutional transition metal ion that produces a state in the gap (such as Cr^{4+} or V^{4+}) (Crook et al., 2020). In such a system, the electron spin on the defect acts as a qubit node, which is interconnected to other nodes by single photons acting as the carrier of optical communications. Entanglement is limited for the intrinsic emission into a zero-phonon state transition, which produces indistinguishable photons for interference between spatially separated spins. One can enhance a coupled defect's zero-phonon emission by the use of nano-cavities. The enhancement is called the Purcell factor, which actually is quantified by the enhancement of the excited state's lifetime reduction via (Crook et al., 2020)

$$F \equiv \frac{\Gamma_{cavity}}{\Gamma_{bulk}} = F_1 F_2 \frac{3Q}{4\pi^2 V}\left(\frac{\lambda_{cavity}}{n}\right)^3 + 1, \tag{11.26}$$

where Γ_{cavity} and Γ_{bulk} are enhanced cavity and unenhanced emission rates, so that $F = 1$ means no enhancement. For the photonic cavity, Q is the quality factor, V is the mode volume, λ_{cavity} is the mode volume, and n is the index of refraction.

An array of photonic cavities is shown in Figure 11.12 (Crook et al., 2020). In panel (a) is a simulation of the nanobeam photonic cavity created by the array of holes in the metal surface (small circles). The fabrication process steps are illustrated in panel (b). The upper left drawing in panel (b) shows the semiconductor structure, which is an NINPN structure, where N denotes n-type, P denotes p-type, and I denotes intrinsic, undoped material. Electron beam lithography defines a 25 nm

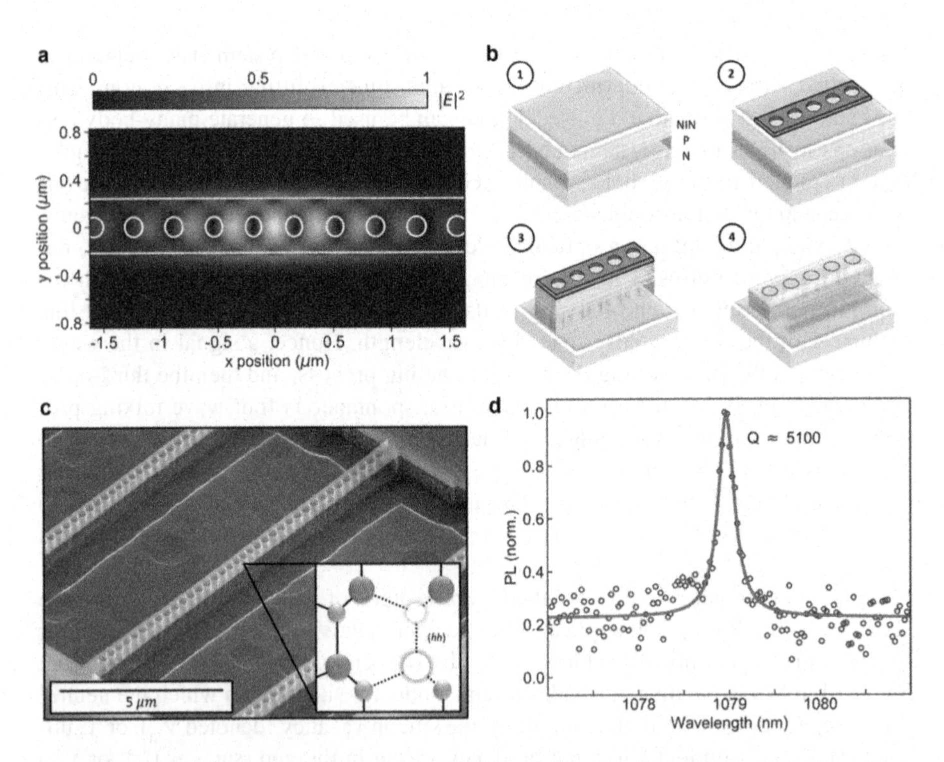

FIGURE 11.12 Nanophotonic cavities in 4H SiC. (a) Simulated nanobeam photonic cavities, with a simulated $Q \sim 3 \times 10^5$. (b) Outline of fabrication procedure, discussed in the text. (c) SEM image of fabricated nanobeam cavity structures. The inset shows a lattice representation of the (*hh*) VV^0 defect states. (d) Photoluminescence spectrum of a nanobeam cavity at room temperature. The linewidth indicates a Q ~ 5,100, as indicated by the full width of the line at one-half the maximum value. Reprinted with permission from Crook et al., *Nano Lett.* **20**, 3427 (2020). Copyright 2020, the American Chemical Society.

thick nickel mask, in which the holes are created. An inductively coupled plasma etch then removes the NIN and P layers everywhere except where the mask exists (the lower left drawing). A photoelectrochemical etch is then used to selectively remove part of the P layer in order to undercut the upper layers (the lower right drawing). Panel (c) shows an SEM image of the completed structure and a lattice drawing of the *hh* (hexagonal-hexagonal) structure of the VV^0 defect in the SiC. Finally, panel (d) shows a photoluminescence signal from the photonic cavity. Using a Lorenzian line fit to the observed luminescence allows one to use the full width at one-half the maximum value of the peak to estimate a quanlity factor of ~ 5,100 for the cavity. The defects mentioned above are created in the SiC by irradiating the material with 2 Mev electrons. There are several different types of defect structure that are possible, and the inset to Figure 11.12(c) shows the one in which each of the two sites in the VV^0 structure has a hexagonal character (as opposed to an quasi-cubic nature), as they both are part of the hexagonal array of atoms. Optical evaluation of the structures indicates that the Purcell enhancement may be as much as 53, which is determined by irradiating the cavity both on resonance and away from the resonance.

REFERENCES

Bargerbos, A., Uilhoorn, W., Yang, C.-K., Krogstrup, K., Kouwenhoven, L. P., de Lange, G., van Heck, B., and Kou, A. 2020. Observation of vanishing charge dispersion of a nearly open superconducting island. *Phys. Rev. Lett.* **124** 246802

Benioff, P. A. 1982. Quantum mechanical Hamiltonian models of discrete processes that erase their own histories: Application to Turing Machines. *Int. J. Theor. Phys.* **21** 177–201

Braumüller, J., Ding, L., Vepsäläinen, A. P., Sung, Y., Kjaergaard, M., Menke, T., Winik, R., Kim, D., Niedzielski, B. M., Melville, A., Yoder, J. L., Hirjibehedin, C. F., Orlando, T. P., Gustavsson, S., and Oliver, W. D. 2020. Characterizing and optimizing qubit coherence based on SQUID geometry. *Phys. Rev. Appl.* **13** 054079

Brunner, R., Shin, Y.-S., Obata, T., Pioro-Ladrière, M., Kubo, T., Yoshida, K., Taniyama, T., Tokura, Y., and Tarucha, S. 2011. Two-qubit gate of combined single-spin rotation and interdot spin exchange in a double quantum dot. *Phys. Rev. Lett.* **107** 146801

Crook, A. L., Anderson, C. P., Miao, K. C., Bourassa, A., Lee, H., Bayliss, S. L., Bracher, D. O., Zhang, X., Abe, H., Ohshima, T., Hu, E. L., and Awschalom, D. D. 2020. Purcell enhancement of single silicon carbide color center with coherent spin control. *Nano Lett.* **20** 3427–3434

Dehollain, J. P., Muhonen, J. T., Blume-Kohut, R., Rudinger, K. M., Gamble, J. K., Nielsen, E., Laucht, A., Simmons, S., Kalra, R., Dzurak, A. S., and Morello A. 2016. Optimization of a solid-state electron spin qubit using gate tomography. *New J. Phys.* **18** 103018

Deutsch, D. 1985. Quantum theory, the Church-Turing principle, and the universal quantum computer. *Proc. Roy. Soc. London A* **400** 97–117

Deutsch, D. 1989. Quantum computational networks. *Proc. Roy. Soc. London A* **425** 73–90

Deutsch, D. and Jozsa, R. 1992. Rapid solution of problmes by quantum computation. *Proc. Roy. Soc. London A* **439** 553–8

Drung, D., Abmann, C., Beyer, J., Kirste, A., Peters, M., Ruede, F., and Schurig, Th. 2007. High sensitivity and easy-to-use SQUID sensors. *IEEE Trans. Appl. Supercond.* **17** 699–704

Fay, A., Guichard, W., Buisson, O., and Hekking, F. W. J. 2011a. Aymmetric Cooper-pair transistor in parallel to a dc SQUID: Two coupled quantum systems. *Phys. Rev. B* **83** 184510

Fay, A., Hoskinson, E., Lecocq, F., Lèvy, L. P. Hekking, F. W. J., Guichard, W., and Buisson, O. 2011b. Strong tunable coupling between a superconducting charge and phase qubit. *Phys. Rev. Lett.* **100** 187003

Ferry, D. K. and Nedjalkov, M. 2018. *The Wigner Function in Science and Technology*. IOP Publishing: Bristol.

Gea-Banacloche, J. 1990. Collapse and Revival of the state vector in the Jaynes-Cummings model: An example of state preparation by a a quantum apparatus. *Phys. Rev. Lett.* **65**, 3385–3388

Gómez-Ruiz, F. J., Acevedo, O. L., Rogriguez, F. J., Quiroga, L., and Johnson N. F. 2018. Pulsed generation of quantum coherences and non-classicality in light-matter systems. *Frontiers Phys.* **6** 92

Grover, L. K. 1996. A fast quantum mechanical algorithm for database search, in *Proc. Ann ACM Symp. Theory Comp.* ACM: New York. pp. 212–219

Hollenberg, L. C. L., Dzurak, A. S., Wellard, C., Hamilton, A. R., Reilly, D. J, Milburn, G. J., and Clark, R. G. 2004. Charge-based quantum computing using single donors in semiconductors. *Phys. Rev. B* 69 113301

Jaklevic, R. C., Lambe, J., Silver, A. H., and Mercereau, J. E. 1964. Quantum interference effects in Josephson tunneling. *Phys. Rev. Lett.* **12** 159–160.

Jaynes, E. T. and Cummings, F. W. 1963. Comparison of quantum and semiclassical radiation theories with applications to the beam maser. *Proc. IEEE* **51** 89–109

Jelezko F. and Wrachtrup D. 2006. Single defect centers in diamond: a review. *Phys. Stat. Sol. (A)* **203** 3207–3225

Jozsa, R. 1991. Characterizing classes of functions computable by quantum parallelism. *Proc. Roy. Soc. London A* **435** 563–574

Kane, B. E. 1998. A silicon-based nuclear spin quantum computer. *Nature* **393** 133–137

Koch, J., Yu, T. M., Gambetta, J., Houck, A. A., Schuster, D. I., Majer, J., Blais, A., Devoret, M. H., Girvin, S. M., and Schoelkopf, R. J. 2007. Charge-insensitive qubit design derived from the Cooper pair box. *Phys. Rev. A* 76 042319

Lloyd, S. 1993. A potentially realizable quantum computer. *Science* **261** 1569–1571

Loss, D. and DiVincenzo, D. A. 1998. Quantum computation with quantum dots. *Phys. Rev. A* **57** 120–126

Lyding, J. W., Albein, G. C., Shen, T.-C., Wang, C., and Tucker, J. R. 1994. Nanoscale patterning and oxidation of silicon surfaces with an ultrahigh vacuum scanning tunneling microscope. *J. Vac. Sci. Technol. B* **12** 3735–3740

Marinescu, D. C. and Marinescu, G. M. 2004. *Approaching Quantum Computing*. Pearson: London

Mead, C. A. and Conway, L. 1979. *Introduction to VLSI Systems*. Addison-Wesley, Reading, MA

Mooij, J. E., Orlando, T. P., Levitov, L., Tian, L., van der Wal, C. H., and Lloyd, S. 1999. Josephson persistent-current qubit. *Science* **285** 1036–1039

Nakamura, Y., Pashkin, Yu. A., and Tsai, J. S. 1999. Coherent control of macroscopic quantum states in a single-Cooper-pair box. *Nature* **398** 786–788

Nielsen, M. A.; Chuang, I. L. 2000 *Quantum Computing and Quantum Information*. Cambridge Univ. Press: New York

Norozhny, N. B., Sanchez-Mondragon, I. I., and Eberly J. H. 1981. Coherence versus incoherence: Collapse and revival in a simple quantum model. *Phys. Rev. A* **23** 236–247

O'Brien, J. L., Schofield, S. R., Simmons, M. Y., Clark, R. G., Dzurak, A. S., Curson, N. J., Kane, B. E., McAlpine, N. S., Hawley, M. E., and Brown, G. W. 2001. Towards the fabrication of phosphorus qubits for a silicon quantum computer. *Phys. Rev. B* **64** 161401

Schrödinger, E. Die gegenwärtige situation in der quantenmechanik. *Naturwiss.* **23** 807–812, 823–828, 844–849

Shor, P. W. 1994. Algorithms for quantum computation: Discrete logarithms and factoring, in *Proc. 35th Ann. Symp. Fndtns. Comp. Sci.* IEEE Press: New York. pp. 124–134

Turing, A. 1937. On computable numbers, with an application to the Entscheidungsproblem. *London Math. Soc., Series II*, **42** 230–265

Vernon, Z. and Sipe, J. E. 2015. Spontaneous four-wave mixing in lossy microring resonantors. *Phys. Rev. A* **91** 053802

Von Neumann, J. 1945. *First draft of a report on the EDVAC*. Moore School of Engineering: Philadelphia, PA

Wendin, G. 2003. Scalable solid state qubits: Challanging decoherence and read-out. *Phil. Trans. Roy. Soc. London A* **361** 1323–1338

Solutions to Selected Problems

CHAPTER 1

2. We use de Broglie's wave expression $\lambda = h/p = h/\sqrt{2mE}$. The frequency is always given by $E = hf$, so $f = 2.42 \times 10^{16}$ Hz. The wavelengths are

$$\lambda_e = 1.22 \times 10^{-10}\,\text{m} \quad \lambda_p = 2.85 \times 10^{-12}\,\text{m}.$$

The group velocity is $h/\lambda m$ ($= \hbar k/m$) and the phase velocity is half of this value. This leads to

$$\begin{array}{lll} & \upsilon_\phi & \upsilon_{gr} \\ \text{electron} & 2.97 \times 10^6 & 5.95 \times 10^6\,\text{ms}^{-1}. \\ \text{proton} & 6.93 \times 10^4 & 1.38 \times 10^5\,\text{ms}^{-1} \end{array}$$

3. We work in momentum space, with

$$\Psi(x) = \frac{1}{\sqrt{2\pi}} \int_{-\infty}^{\infty} \varphi(k) e^{ikx}\, dk.$$

The expectation value is then

$$\langle x \rangle = \int_{-\infty}^{\infty} \Psi^*(x) x \Psi(x) dx$$

$$= \frac{1}{2\pi} \int_{-\infty}^{\infty} \left\{ \int_{-\infty}^{\infty} \varphi^*(k') e^{-ik'x} dk' \right\} x \left\{ \int_{-\infty}^{\infty} \varphi(k) e^{ikx} dk \right\} dx$$

which then is manipulated as

$$\langle x\rangle = \frac{1}{2\pi}\int_{-\infty}^{\infty}\varphi^*(k')\mathrm{d}k'\int_{-\infty}^{\infty}\mathrm{d}x\mathrm{e}^{-ik'x}\int_{-\infty}^{\infty}\mathrm{d}k\varphi(k)\left(-i\frac{\partial}{\partial k}\mathrm{e}^{ikx}\right)$$

$$= \frac{1}{2\pi}\int_{-\infty}^{\infty}\varphi^*(k')\mathrm{d}k'\int_{-\infty}^{\infty}\mathrm{d}x\mathrm{e}^{-ik'x}\int_{-\infty}^{\infty}dke^{ikx}\int_{-\infty}^{\infty}\left(i\frac{\partial}{\partial k}\varphi(k)\right)$$

$$= \int_{-\infty}^{\infty}\varphi^*(k')\mathrm{d}k'\int_{-\infty}^{\infty}\mathrm{d}k\left(i\frac{\partial}{\partial k}\varphi(k)\right)\delta(k'-k)$$

$$= \int_{-\infty}^{\infty}\mathrm{d}k\varphi^*(k)\left(i\frac{\partial}{\partial k}\varphi(k)\right)$$

$$= \left\langle i\frac{\partial}{\partial k}\right\rangle = \left\langle i\hbar\frac{\partial}{\partial p}\right\rangle.$$

CHAPTER 2

1. We can summarize the picture, by defining the momentum wave function as

$$\phi(k) = 2a\left[1-\left|\frac{ka}{\pi}\right|\right].$$

This, however, is not normalized properly, and we must check the proper normalization:

$$\int_{-\pi/a}^{\pi/a}|\phi(k)|^2\,\mathrm{d}k = 8a^2\int_{-\pi/a}^{\pi/a}\left(1-\left|\frac{ka}{\pi}\right|\right)^2\mathrm{d}k = \frac{8\pi a}{3}.$$

The normalized wave function is then

$$\phi(k) = \sqrt{\frac{3a}{8\pi}}2a\left[1-\left|\frac{ka}{\pi}\right|\right] = \sqrt{\frac{3a^3}{2\pi}}\left[1-\left|\frac{ka}{\pi}\right|\right].$$

This is now Fourier transformed into real space to give the wave function as

$$\psi(x) = \frac{1}{\sqrt{2\pi}}\int_{-\infty}^{\infty}\phi(k)e^{ikx}dk = \frac{\sqrt{3a^2}}{2\pi}\int_{-\pi/a}^{\pi/a}\left[1-\left|\frac{ka}{\pi}\right|\right]e^{ikx}dk$$

$$= \sqrt{\frac{3}{a}}\left(\frac{a}{\pi x}\right)^2\left[1-\cos\left(\frac{\pi x}{a}\right)\right].$$

We note first that this wave function is *symmetrical* about $x = 0$, so $\langle x \rangle = 0$. Thus, we then find that $(\Delta x)^2 = \langle x^2 \rangle$, and

$$(\Delta x)^2 = \frac{3a^4}{\pi^4}\int_{-\infty}^{\infty}\frac{1}{x^2}\left[1-\cos\left(\frac{\pi x}{a}\right)\right]^2 dx = \frac{6a^2}{\pi^2}.$$

Using the momentum wave function, we similarly find

$$(\Delta k)^2 = \int_{-\pi/a}^{\pi/a} k^2 \frac{3a^3}{2\pi}\left[1-\left|\frac{ka}{\pi}\right|\right]^2 dk = \frac{\pi^2}{a^2 10}$$

so the uncertainty relation is found to be

$$\Delta x \Delta k = \sqrt{\frac{3}{10}}.$$

2. In general, one thinks that the kinetic energy is reduced once the wave passes over the barrier, and this leads to a smaller value of k', and hence a longer wavelength ($2\pi/k$). However, how does this appear in a wave packet? The important point is that *anything that narrows the momentum distribution will broaden the spatial distribution just due to the properties of the Fourier transform.* Hence, we consider a Gaussian wave packet, centred at k_1, as in (following (2.71))

$$\phi(k) = \left(\frac{2}{k}\right)^{1/4}\sqrt{\sigma}\exp\left[-\sigma^2(k-k_0)^2\right].$$

For this wave packet, $\langle k \rangle = k_0$, and $\Delta p = \hbar/2\sigma$ from (2.78). *We note that Δp is a function only of the width of the packet and not the centroid.* If we now pass this wave packet over the barrier $\left(k_0 > k_V = \sqrt{2mV_0/\hbar^2}\right)$, then not much happens until the barrier begins to eat away part of the Gaussian distribution. This narrows the distribution, which means that the real space wave function (obtained from the Fourier transform) must become broader.

6. For this problem, we note that $\langle T \rangle = \langle V \rangle$ (which is true for any oscillating system, where the energy oscillates between being purely potential energy and purely kinetic energy), and this leads to (since we assume that $\langle x \rangle = \langle p \rangle = 0$)

$$(\Delta x)^2 = \frac{1}{m^2\omega^2}(\Delta p)^2 = \frac{1}{m^2\omega^2}\left(\frac{\hbar}{2\Delta x}\right)^2$$

and

$$(\Delta x) = \sqrt{\frac{\hbar}{2m\omega}} \quad (\Delta p) = \sqrt{\frac{m\omega\hbar}{2}}$$

for which

$$E = H = \frac{1}{2m}(\Delta p)^2 + \frac{m\omega^2}{2}(\Delta x)^2 = \frac{\hbar\omega}{2}.$$

CHAPTER 3

1–3. For a linear axis, these may be plotted as

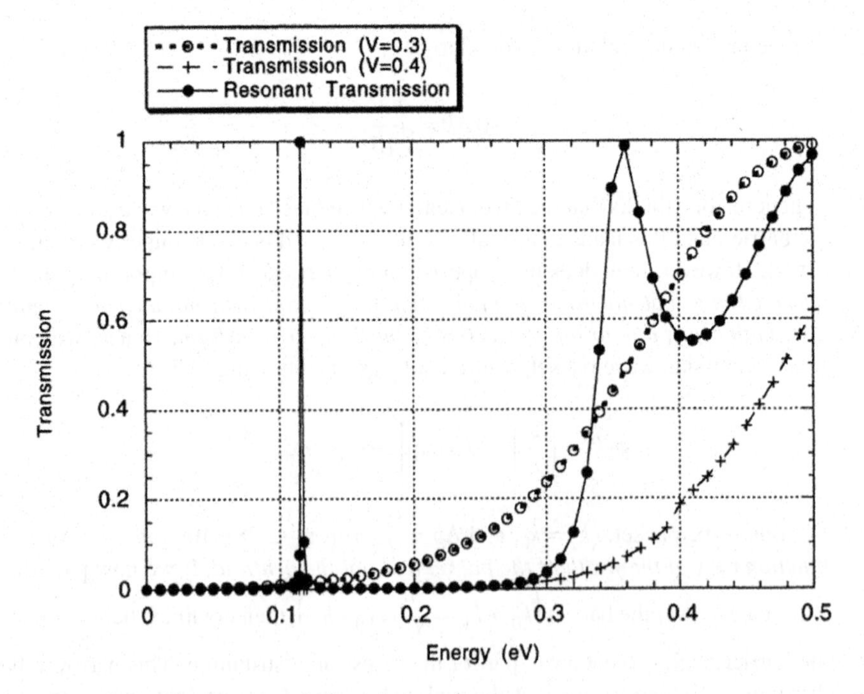

For a logarithmic transmission, the results are

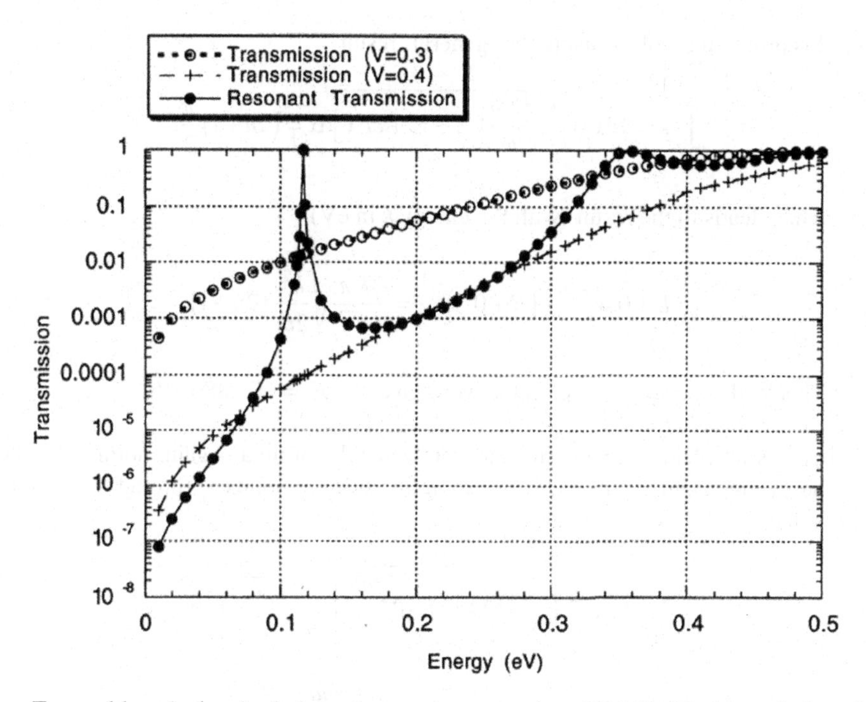

For problem 1, the single bound-state resonance is at 0.1166 eV, although there is a resonance at about 0.36 eV.

4. The resonant peak is now at 0.124 eV, and the transmission at this point is 0.029. The values from problem 1 and problem 2 at this energy are 0.0156 and 0.00011685, respectively, so $4T_{min}/T_{max}$ is 0.0299, which is about as close as one can get on such a problem.

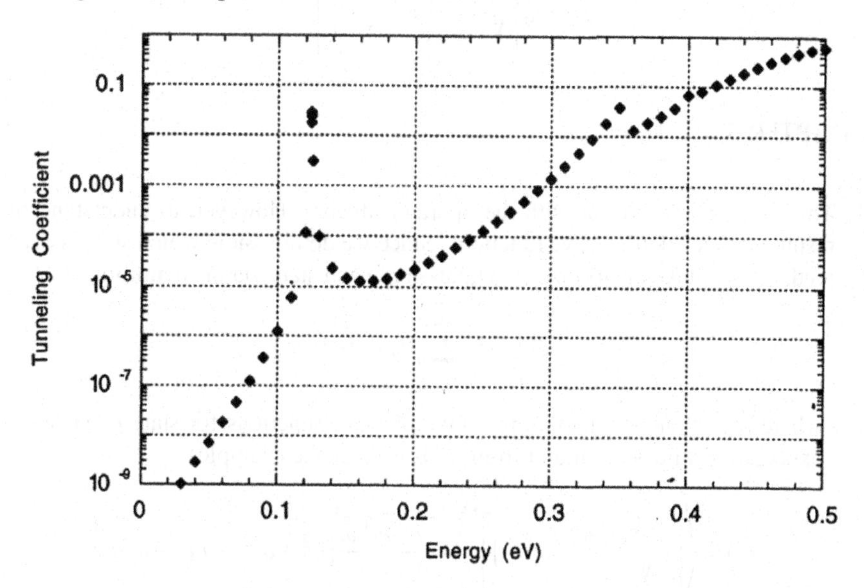

5. This problem is solved using the 'general' result:

$$\int_0^a k(x)\mathrm{d}x = \int_0^a \sqrt{\frac{2m}{\hbar^2}\left[E + V_1 + eEx\right]}\mathrm{d}x = (2n+1)\frac{\pi}{2}.$$

which leads to the result (with the energy ε in eV)

$$(E+0.4)^{3/2} - (E+0.3)^{3/2} = \frac{3E\pi}{4e^{1/2}}\sqrt{\frac{\hbar^2}{2m}}(2n+1)$$

This leads to a single level in the well, given by $E_0 = -0.2886$ eV.

8. For this problem, we can return to our 'general' formula. Again, noting that the energy eigenvalues will have corresponding turning points x_n through $E_n = a(x_n)^4$, we have

$$(2n+1)\frac{\pi}{2} = \int_{-x_n}^{x_n} k(x)\mathrm{d}x = \sqrt{\frac{2ma}{\hbar^2}}\int_{-x_n}^{x_n} \sqrt{x_n^4 - x^4}\mathrm{d}x$$

$$= 2x_n^3\sqrt{\frac{2ma}{\hbar^2}}$$

and the energy eigenvalues are given by

$$E_n = a\left[\sqrt{\frac{\hbar^2}{2ma}}\left(n+\frac{1}{2}\right)\frac{\pi}{2}\right]^{4/3}.$$

CHAPTER 5

1. This may be carried out with the operator algebra. However, to understand the result, we must define wave functions. Since we do not know which state is occupied, we will take an arbitrary occupation of each state via the definition

$$\Psi(x) = \sum_n a_n \psi_n(x)$$

where the ψ_n are the harmonic oscillator wave functions for state n. Then, the expectation values are found from (p^2 is used as the example)

$$\langle p^2 \rangle = \left\langle \left[-\sqrt{\frac{m\hbar\omega}{2}}\left(a - a^+\right)\right]^2 \right\rangle = \left\langle -\frac{m\hbar\omega}{2}\left(a^2 + a^{+2} - aa^+ - a^+a\right)\right\rangle$$

$$= -\frac{m\hbar\omega}{2}\left\{-\left\langle 2a^{+}a+1\right\rangle + \left\langle a^{2}\right\rangle + \left\langle a^{+2}\right\rangle\right\}$$

for which the *nk* matrix element is

$$\left\langle p^{2}\right\rangle_{nk} = \frac{m\hbar\omega}{2}\left\{(2n+1)\,\delta_{nk} - \sqrt{(n+1)(n+2)}\,\delta_{n+2,k}\right.$$
$$\left. - \sqrt{n(n-1)}\delta_{n-2,k}\right\}.$$

Similarly,

$$\left\langle x^{2}\right\rangle_{nk} = \frac{\hbar}{2m\omega}\left\{(2n+1)\,\delta_{nk} + \sqrt{(n+1)(n+2)}\,\delta_{n+2,k}\right.$$
$$\left. + \sqrt{n(n-1)}\delta_{n-2,k}\right\}.$$

Thus, in both cases, only states either equal in index or separated by 2 in index can contribute to the expectation values.

2. Using the results of problem **1**, we have

$$T_{nk} = \frac{1}{2m}\left\langle p^{2}\right\rangle nk$$
$$= \frac{\hbar\omega}{4}\left\{(2n+1)\,\delta_{nk} - \sqrt{(n+1)(n+2)}\delta_{n+2,k} - \sqrt{n(n-1)}\delta_{n-2,k}\right\}$$

$$V_{nk} = \frac{m\omega^{2}}{2}\left\langle x^{2}\right\rangle_{nk}$$
$$= \frac{\hbar\omega}{4}\left\{(2n+1)\,\delta_{nk} + \sqrt{(n+1)(n+2)}\delta_{n+2,k} + \sqrt{n(n-1)}\delta_{n-2,k}\right\}.$$

6. We take $B = B_0 a_z$, which we can get from the vector potential $A = (0, Bx, 0)$. This will give a harmonic oscillator in the *x*-direction, and for homogeneity in the *y*-direction, we take $\psi(x, y) = \varphi(x) \exp(iky)$. Then the Schrödinger equation becomes

$$\left[\frac{1}{2m}\left(\hbar k - eB_0 x\right)^{2} - \frac{\hbar^{2}}{2m}\frac{d^{2}}{dx^{2}} + \frac{m\omega_0^{2}}{2}x^{2}\right]\phi(x) = E\phi(x)$$

which with the change of variables

$$x_0 = \frac{\hbar k}{m}\frac{\omega_c}{\omega_0^2+\omega_c^2} \qquad \omega_c = \frac{eB_0}{m} \qquad \Omega^2 = \omega_0^2+\omega_c^2$$

gives

$$-\frac{\hbar^2}{2m}\frac{d^2\phi}{dx^2}+\frac{1}{2}m\Omega^2\left(x-x_0\right)^2\phi=\left[E-\frac{\hbar^2k^2}{2m}+\frac{1}{2}m\Omega^2x_0^2\right]\phi$$

which is a shifted harmonic oscillator and leads to the energy levels

$$E_n=\left(n+\frac{1}{2}\right)\hbar\Omega+\frac{\hbar^2k^2}{2m}-\frac{1}{2}m\Omega^2x_0^2.$$

CHAPTER 6

1. The Hamiltonian is given as $H = p^2 + Ax + Bpx$. From (6.10)

$$\frac{d\langle p\rangle}{dt}=\frac{i}{\hbar}\langle[H,p]\rangle=\frac{i}{\hbar}\left\langle\left[p^2,p\right]+A[x,p]+B[px,p]\right\rangle$$

$$=\frac{i}{\hbar}\left\langle 0+i\hbar A+B\left(pxp-p^2x\right)\right\rangle=\frac{i}{\hbar}\left\langle 0+i\hbar A+Bp\left(xp-px\right)\right\rangle$$

$$=-A-B\langle p\rangle$$

$$\frac{d\langle x\rangle}{dt}=\frac{i}{\hbar}\langle[H,x]\rangle=\frac{i}{\hbar}\left\langle\left[p^2,x\right]+A[x,x]+B[px,x]\right\rangle$$

$$=\frac{i}{\hbar}\left\langle P^2x-xp^2+0+B\left(px^2-xpx\right)\right\rangle$$

$$=\frac{i}{\hbar}\left\langle p\left(xp-i\hbar\right)-\left(px-i\hbar\right)p-i\hbar Bx\right\rangle$$

$$=2\langle p\rangle+B\langle x\rangle.$$

2. The eigenvalues are found from

$$\frac{1}{2}\begin{vmatrix} \left(\sqrt{2}-2E\right) & \sqrt{2} & 0 \\ -\sqrt{2} & \left(\sqrt{2}-2E\right) & 0 \\ 0 & 0 & \left(2-2E\right) \end{vmatrix} = 0$$

(the factor of 2 multiplying the E is due to the pre-factor) and this leads to $E_3 = 1$, with the other two roots being given by

$$\left(\sqrt{2}-2E\right)^2 + 2 = 0$$

$$E_{1,2} = \frac{1}{\sqrt{2}} \pm \sqrt{\frac{1}{2}-1} = \frac{1\pm i}{\sqrt{2}}$$

3. (i) $H = 2x^2p^2$

$$\frac{d\langle x\rangle}{dt} = \frac{i}{\hbar}\left\langle\left[2x^2p^2, x\right]\right\rangle = 4\left\langle x^2p\right\rangle$$

$$\frac{d\langle p\rangle}{dt} = \frac{i}{\hbar}\left\langle\left[2x^2p^2, p\right]\right\rangle = 4\left\langle xp^2\right\rangle$$

(ii) $H = x^2p^2 + p^2x^2$

$$\frac{d\langle x\rangle}{dt} = +2\left\langle x^2p\right\rangle + \frac{i}{\hbar}\left\langle p^2x^3 - xp^2x^2\right\rangle = -4\left\langle xpx\right\rangle$$

$$\frac{d\langle p\rangle}{dt} = -2\left\langle xp^2\right\rangle + \frac{i}{\hbar}\left\langle p^2x^2p - p^3x^2\right\rangle = -4\left\langle pxp\right\rangle$$

(iii) $H = 2(xp)^2$

$$\frac{d\langle x\rangle}{dt} = \frac{i}{\hbar}\left\langle[2xpxp, x]\right\rangle = 4\left\langle x^2p\right\rangle - 2i\hbar\left\langle x\right\rangle$$

$$\frac{d\langle p\rangle}{dt} = \frac{i}{\hbar}\left\langle[2xpxp, p]\right\rangle = -4\left\langle p^2x\right\rangle + 2i\hbar\left\langle p\right\rangle$$

4. $[A, B] = C$, in which C is a c-number. Thus,

$$A^3B^3 - \left(AB\right)^3 = A^3B^3 - ABABAB$$

$$= A^3B^3 - A(AB-C)(AB-C)B$$

$$= A^3B^3 - A^2BAB^2 + 2CA^2B^2 - C^2AB$$

$$= A^3B^3 - A^2(AB-C)B^2 + 2CA^2B^2 - C^2AB$$

$$= 3CA^2B^2 - C^2AB = 2C^2AB + 3C(AB)^2.$$

CHAPTER 7

1. We have $V(x) = 0$ for $0 < x < a$, and $\rightarrow \infty$ elsewhere. The perturbing potential is given by $V_1 = (x-a)^2/2$. The unperturbed wave functions and energies are then given by

$$\varphi_n(x) = \sqrt{\frac{2}{a}}\sin\left(\frac{n\pi x}{a}\right) \quad 0 \le x \le a$$

$$E_n = \frac{n^2\pi^2\hbar^2}{2ma^2}.$$

For the effect of the perturbation, we find the matrix elements

$$V_{nm} = \frac{1}{a}\int_0^a \sin\left(\frac{n\pi x}{a}\right)(x-a)^2 \sin\left(\frac{m\pi x}{a}\right)dx$$

$$= \begin{cases} \dfrac{a^2}{\pi^2}\left[\dfrac{1}{(n-m)^2} - \dfrac{1}{(n+m)^2}\right] & n \ne m \\ \dfrac{a^2}{6}\left[1 - \dfrac{3}{2n^2\pi^2}\right] & n = m. \end{cases}$$

The first term can be simplified, but the perturbed wave function can be written as

$$\varphi_n(x) = \sqrt{\frac{2}{a}}\sin\left(\frac{n\pi x}{a}\right)$$

$$+\sum_{l \ne n} \frac{a^2}{\pi^2}\frac{4nl}{(n^2-l^2)^2}\left\{\frac{\hbar^2\pi^2}{2ma^2}(l^2-n^2)\right\}^{-1}\sqrt{\frac{2}{a}}\sin\left(\frac{l\pi x}{a}\right)$$

$$= \sqrt{\frac{2}{a}} \sin\left(\frac{n\pi x}{a}\right) - \frac{2ma^2}{\hbar^2 \pi^4} \sqrt{\frac{2}{a}} \sum_{l \neq n} \frac{4nl}{\left(n^2 - l^2\right)^3} \sin\left(\frac{l\pi x}{a}\right).$$

The energy levels are now given by

$$E_n = E_n^{(0)} + E_n^{(1)} + E_n^{(2)}$$

$$= \frac{n^2 \pi^2 \hbar^2}{2ma^2} + \frac{a^2}{6}\left[1 - \frac{3}{2n^2\pi^2}\right] - \left(\frac{a^2}{\pi^2}\right) \frac{2ma^2}{\pi^2 \hbar^2} \sum_{l \neq n} \frac{16 n^2 l^2}{\left(n^2 - l^2\right)^5}.$$

4. The lowest energy level is for $n = m = 1$,

$$E_{11} = \frac{\pi^2 \hbar^2}{2ma^2}\left(1^2 + 1^2\right) = \frac{\pi^2 \hbar^2}{ma^2}$$

since the wave functions are

$$\varphi_{x,n}(x) = \sqrt{\frac{2}{a}} \sin\left(\frac{n\pi x}{2} + \frac{n\pi}{2}\right) \quad \varphi_{y,m}(y) = \sqrt{\frac{2}{a}} \sin\left(\frac{m\pi y}{2} + \frac{m\pi}{2}\right).$$

The lowest degenerate energy level occurs for the sets of $(n, m) = (2,1), (1, 2)$, which both give the energy level of

$$E_{21} = E_{12} = \frac{\pi^2 \hbar^2}{2ma^2}\left(2^2 + 1^2\right) = \frac{5\pi^2 \hbar^2}{2ma^2}.$$

If we call the (2,1) set a and the (1, 2) set b, then the matrix element, which couples these two wave functions is

$$V_{ab} = V_0 \int_{-a/4}^{a/4} dx \int_{-a/4}^{a/4} dy\, \varphi_{21}^*(x,y)\, \varphi_{12}(x,y)$$

$$= \frac{4V_0}{a^2} \int_{-a/4}^{a/4} dx \sin\left(\frac{2\pi x}{a}\right) \cos\left(\frac{\pi x}{a}\right) \int_{-a/4}^{a/4} dy \cos\left(\frac{\pi y}{a}\right) \sin\left(\frac{2\pi y}{a}\right)$$

$$= \frac{4V_0}{a^2} \left\{ \frac{1}{2} \int_{-a/4}^{a/4} dx \left[\sin\left(\frac{3\pi x}{a}\right) + \sin\left(\frac{\pi x}{a}\right) \right] \right\} = 0.$$

Hence this potential does not split the degeneracy. The reason for this is that the potential is still a *separable* potential (it can be split into x and y parts). Since the potential is an even function around the center of the well, it can only couple two even or two odd wave functions, hence it does not split the lowest level which is composed of even and odd functions.

5. We use the fact that $[a, a^+] = aa^+ - a^+a = 1$ to get

$$x = \left(\frac{\hbar}{2m\omega}\right)^{1/2}\left(a + a^+\right)$$

$$x^2 = \frac{\hbar}{2m\omega}\left(a + a^+\right)^2 = \frac{\hbar}{2m\omega}\left(a^2 + \left(a^+\right)^2 + 2n + 1\right)$$

$$x^4 = \left(\frac{\hbar}{2m\omega}\right)^2\left(a + a^+\right)^4$$

$$= \left(\frac{\hbar}{2m\omega}\right)^2\left[a^4 + \left(a^+\right)^4 + 2(2n+1)\left(a^2 + \left(a^+\right)^2\right) + 4n^2 + 4n + 1\right].$$

This leads to the matrix elements

$$\left\langle k\left|a^4\right|n\right\rangle = \delta_{k,n-4}\sqrt{n(n-1)(n-2)(n-3)}$$

$$\left\langle k\left|\left(a^+\right)^4\right|n\right\rangle = \delta_{k,n,+4}\sqrt{(n+1)(n+2)(n+3)(n+4)}$$

$$\left\langle k\left|a^2\right|n\right\rangle = \delta_{k,n-2}\sqrt{n(n-1)}$$

$$\left\langle k\left|\left(a^+\right)^2\right|n\right\rangle = \delta_{k,n+2}\sqrt{(n+1)(n+2)}.$$

The matrix has elements on the main diagonal $[4n^2 + 4n + 1]$, and elements given two and four units off the main diagonal, given by the above matrix elements (plus the pre-factor listed above).

6. By the same procedures as the previous problem, the perturbation can be written as

$$V_1 = \alpha x^3 = \alpha\left(\frac{\hbar}{2m\omega}\right)^{3/2}\left(a + a^+\right)^3$$

$$= \alpha\left(\frac{\hbar}{2m\omega}\right)^{3/2}\left[a^3 + \left(a^+\right)^3 + (3n+1)a + (3n+2)a^+\right].$$

We note that the nn-term is zero, so there is no first-order correction to the energy. Thus, we need only calculate the second-order correction using the matrix elements

$$(V_1)_{kn} = \alpha\left(\frac{\hbar}{2m\omega}\right)^{3/2}\left[\delta_{k,n-3}\sqrt{n(n-1)(n-2)} + \delta_{k,n-1}(3n+1)\sqrt{n}\right]$$

$$+\delta_{k,n+3}\sqrt{(n+1)(n+2)(n+3)} + \delta_{k,n+1}(3n+2)\sqrt{n+1}$$

which leads to

$$E_n^{(2)} = \sum_{k\neq n}\frac{(V_1)_{nk}(V_1)_{kn}}{E_n^{(0)} - E_k^{(0)}}$$

$$= \alpha^2\left(\frac{\hbar}{2m\omega}\right)^3\left\{\frac{n(n-1)(n-2)}{3\hbar\omega} - \frac{(n+1)(n+2)(n+3)}{3\hbar\omega}\right.$$

$$\left.+\frac{1}{\hbar\omega}\left[(3n+1)n^2 - (3n+2)^2(n+1)\right]\right\}$$

$$= \frac{\alpha^2\hbar^2}{8m^3\omega^4}\left\{-\left(\frac{9n^2+13n+6}{3}\right) - \left[15(n+1)n+4\right]\right\}.$$

CHAPTER 8

2. From (8.29), we may write the general propagator as

$$U(t,0) = \exp\left[-\frac{iEx\tau}{\hbar}\right]$$

for $t > \tau$. Thus, the total wave function is given by

$$\Psi(t) = U(t,0)\varphi_3(x) = \sqrt{\frac{2}{a}}\mathrm{e}^{-iEx\tau/\hbar}\sin\frac{3\pi}{a}\left(x+\frac{a}{2}\right)$$

and the connection to an arbitrary state $|k$ becomes

$$(\varphi_k, \Psi(t)) = \frac{2}{a}\int_{-a/2}^{a/2}\sin\left[\frac{k\pi}{a}\left(x+\frac{a}{2}\right)\right]\left[\cos\left(\frac{E\tau x}{\hbar}\right) - i\sin\left(\frac{E\tau x}{\hbar}\right)\right]$$

$$\times \sin\left[\frac{3\pi}{a}\left(x+\frac{a}{2}\right)\right].$$

These integrals may be computed with some care. It should be noted, however, that the field term does not have the periodicity of the quantum well but does have symmetry properties. Thus, for example, the initial state has even symmetry in the well. Thus, the cosine term in the field will only couple to states that have even symmetry. On the other hand, the sine term will only couple to states that have odd symmetry since it is odd. In general, all states are coupled to the $n = 3$ state by this perturbation.

CHAPTER 9

1. This zero-angular-momentum state requires $n_a = n_b = 1$. Thus, the state is excited as

$$|1,1\rangle = a^+ b^+ |0,0\rangle.$$

The various quantities are given by

$$a^+ = \frac{e^{i\varphi}}{2}\left[Br - \frac{1}{B}\frac{\partial}{\partial r} - \frac{i}{Br}\frac{\partial}{\partial \varphi}\right]$$

$$b^+ = \left(a^+\right)^* \quad |0,0\rangle = \frac{B}{\sqrt{\pi}}\exp\left(-\frac{B^2 r^2}{2}\right) \quad B = \sqrt{\frac{m\omega}{\hbar}}.$$

This leads to

$$|0,0\rangle = \frac{B}{\sqrt{\pi}}\left[B^2 r^2 - 1\right]\exp\left(-\frac{B^2 r^2}{2}\right).$$

2. The perturbation may be found by expanding the frequency in the form $V_1 = m(\omega 0$ $\delta\omega + \delta\omega^2)x^2/2$. The operator **x** may be expressed in terms of the various creation and annihilation operators as

$$x = \sqrt{\frac{\hbar}{2m\omega}}\left(a_x + a_x^+\right) = \frac{1}{2}\sqrt{\frac{\hbar}{m\omega}}\left(a + a^+ + b + b^+\right).$$

This leads to the crucial part, which is the x^2-operator:

$$x^2 = \frac{\hbar}{4m\omega}\Big[a^2 + b^2 + \left(a^+\right)^2 + \left(b^+\right)^2$$

$$+2\left(ab + ab^+ + a^+b + a^+b^+\right) + 2\left(n_a + n_b + 1\right)\Big].$$

We now define the matrix elements between the state $|k, l$ and the state $|n_a, n_b$ to be

$$(\bullet)_{n_a n_b}^{kl}$$

and the various terms in x^2 can be evaluated as

$$\left(a^2\right)_{n_a n_b}^{kl} = \frac{1}{\sqrt{2}}\delta_{k,n_a-2}\delta_{l,n_b} \quad \left(a^{+2}\right)_{n_a n_b}^{kl} = \frac{1}{\sqrt{2}}\delta_{k,n_a+2}\delta_{l,n_b}$$

$$\left(b^2\right)_{n_a n_b}^{kl} = \frac{1}{\sqrt{2}}\delta_{k,n_a}\delta_{l,n_b-2} \quad \left(b^2\right)_{n_a n_b}^{kl} = \frac{1}{\sqrt{2}}\delta_{k,n_a}\delta_{l,n_b+2}$$

$$\left(ab\right)_{n_a n_b}^{kl} = \delta_{k,n_a-1}\delta_{l,n_b-1} \quad \left(a^+b^+\right)_{n_a n_b}^{kl} = \delta_{k,n_a+1}\delta_{l,n_b+1}$$

$$\left(ab^+\right)_{n_a n_b}^{kl} = \delta_{k,n_a-1}\delta_{l,n_b+1} \quad \left(a^+b\right)_{n_a n_b}^{kl} = \delta_{k,n_a+1}\delta_{l,n_b-1}.$$

The term in the number operators is diagonal. Thus, there are a variety of shifting operations in the perturbation, which generally mixes a number of modes.

3. We write the Schrödinger equation with the potential $V = -eEx$ as

$$\left[-\frac{\hbar^2}{2m}\left(\frac{\partial^2}{\partial x^2} + \frac{\partial^2}{\partial y^2}\right) + \frac{m\omega_0^2}{2}\left(x^2 + y^2\right) - eEx\right]\psi(x,y) = E\psi(x,y).$$

Introducing the parameter

$$x_0 = \frac{eE}{m\omega_0^2}$$

this may be written as a shifted harmonic oscillator in the form

$$\left[-\frac{\hbar^2}{2m}\left(\frac{\partial^2}{\partial x^2}+\frac{\partial^2}{\partial y^2}\right)+\frac{m\omega_0^2}{2}\left[\left(x-x_0\right)^2+y^2\right]\right]\psi(x,y)$$

$$=\left[E+m\omega_0^2\frac{x_0^2}{2}\right]\psi(x,y)$$

so the new energies are given by

$$E_{n_xn_y}=\left(n_x+n_y+1\right)\hbar\omega_0-\frac{e^2E^2}{2m\omega_0^2}.$$

Index

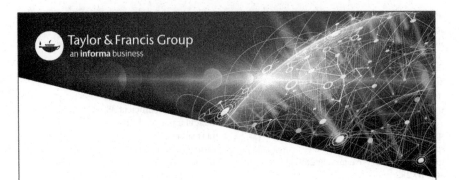

Printed in the United States
By Bookmasters